Water Science and Technology Library

Volume 125

The aim of the *Water Science and Technology Library* is to provide a forum for dissemination of the state-of-the-art of topics of current interest in the area of water science and technology. This is accomplished through publication of reference books and monographs, authored or edited. Occasionally also proceedings volumes are accepted for publication in the series. *Water Science and Technology Library* encompasses a wide range of topics dealing with science as well as socio-economic aspects of water, environment, and ecology. Both the water quantity and quality issues are relevant and are embraced by *Water Science and Technology Library*. The emphasis may be on either the scientific content, or techniques of solution, or both. There is increasing emphasis these days on processes and *Water Science and Technology Library* is committed to promoting this emphasis by publishing books emphasizing scientific discussions of physical, chemical, and/or biological aspects of water resources. Likewise, current or emerging solution techniques receive high priority. Interdisciplinary coverage is encouraged. Case studies contributing to our knowledge of water science and technology are also embraced by the series. Innovative ideas and novel techniques are of particular interest.

Comments or suggestions for future volumes are welcomed.

Vijay P. Singh, Department of Biological and Agricultural Engineering & Zachry Department of Civil and Environment Engineering, Texas A&M University, USA
Email: vsingh@tamu.edu

All contributions to an edited volume should undergo standard peer review to ensure high scientific quality, while monographs should also be reviewed by at least two experts in the field.

Manuscripts that have undergone successful review should then be prepared according to the Publisher's guidelines manuscripts: https://www.springer.com/gp/authors-editors/book-authors-editors/book-manuscript-guidelines

Akhilesh Kumar Yadav · Saba Shirin ·
Vijay P. Singh
Editors

Advanced Treatment Technologies for Fluoride Removal in Water

Water Purification

Springer

Editors
Akhilesh Kumar Yadav
Department of Environmental Engineering and Management
Chaoyang University of Technology
Taichung, Taiwan

Department of Mining Engineering
Indian Institute of Technology (Banaras Hindu University)
Varanasi, India

Environmental Science and Engineering Department
Indian Institute of Technology Bombay
Mumbai, India

Saba Shirin
Department of Environmental Science
School of Vocational Studies and Applied Sciences
Gautam Buddha University
Greater Noida, India

Department of Mining Engineering
Indian Institute of Technology (Banaras Hindu University)
Varanasi, India

Vijay P. Singh
Department of Biological and Agricultural Engineering
Texas A&M University
College Station, TX, USA

ISSN 0921-092X ISSN 1872-4663 (electronic)
Water Science and Technology Library
ISBN 978-3-031-38844-6 ISBN 978-3-031-38845-3 (eBook)
https://doi.org/10.1007/978-3-031-38845-3

This Springer imprint is published by the registered company Springer Nature Switzerland AG
The registered company address is: Gewerbestrasse 11, 6330 Cham, Switzerland

Dedicated
to
Environmental Engineers and Scientists
around the World

Foreword

It is well known that the excess fluoride in water is detrimental to aquatic life as well as to humans. *Advanced Treatment Technologies for Fluoride Removal in Water* systematically covers all the aspects of fluoride pollution. Appreciable efforts are made to provide a deeper understanding of fluoride pollution by discussing its status, remediation strategies, and removal treatments. The book has been structured into four parts, namely *Fluoride Status and Remediation Strategies*, *Fluoride Removal Techniques*, *Fluoride Effect on Human Health*, and *Future Framework and Advance Technologies*. Though there are various other books on fluoride removal methods, this book has a distinction in providing an updated knowledge on the subject. I congratulate the editors Akhilesh Kumar Yadav, Saba Shirin, and Vijay P. Singh for bringing out this thought-provoking work. I am sure this book will be useful to graduate students, faculties, and policy-makers in environmental impact and public health.

Dinesh K. Aswal, Ph.D.
Director
Health, Safety and Environment Group
Bhabha Atomic Research Center
Mumbai, India

Preface

The motivation of this volume, entitled *Advanced Treatment Technologies for Fluoride Removal in Water*, is mainly to present techniques to protect fluoride contamination by fluoride. Fluoride contamination affects the quality of water resources around the world due to human activities, such as mining and the use of pesticides. As a result of the high risk of fluoride exposure, specific water treatment processes are required to meet more stringent water quality standards. A better understanding of the currently available methods is necessary to develop economical, efficient, and effective methods for the removal of fluoride. It is the identification of causes of degradation, the impact of degradation on human health, the identification of suitable treatment or remediation technologies for mitigation, and strategies for the removal of fluoride in water.

The growing population, expanding urbanization, and industrialization pose a great threat to the most vital natural resources, water and air, for all living. Imminent climate change brings about a new level of complexity and challenges in the demand and supply chain. Modeling is a better approach to address these issues and remediate problems, but environmental modeling and simulation carry complexities with them, as there are various applications. These complexities may include the assimilation of parameters, dimension, amount of data, distribution, heterogeneity, and interdependence.

This book presents an overview of fluoride treatment techniques in water. The book content has been categorized into four parts. Part I is an introduction to the book and gives an overview on fluoride status and remediation strategies. Part II, consisting of nine chapters, deals with fluoride removal treatments, while Part III consists of six chapters under the title fluoride effect on human health; Part IV is devoted to the future framework and advance technologies of two chapters. This book aims to address water treatment by applying modern techniques and simulated solutions to a wider set of environmental problems that are becoming more relevant to environmental engineers and scientists. Hopefully, this book will open the dimensions for explicit programming dynamics to develop decision-making tools to help manage future environmental problems related to water.

First, we thank the Almighty for His blessings throughout this project achievement, which envisioned us successfully exploring this challenging research domain with a technique perspective. We are also sincerely thankful to Springer Nature for allowing us to publish this monograph on a contemporary research domain of modeling and simulations.

Special thanks to all who contributed to making this volume a source of knowledge and reporting the latest findings in their areas. Against the title of the proposed book, the total number of abstracts for the chapter was received 75, of which 53 were selected for a full chapter contribution. In this book, the project accommodated only 19 chapters according to the scope of the book, which were of better quality. The authors put a lot of effort during all phases of book production, starting from writing, revising based on reviewers' comments and Springer evaluation reports, and finally checking the proofs of the chapters. The project follows a review process in which the identities of the authors and reviewers are not disclosed to avoid bias decisions. We also thank and acknowledge the Springer Nature publication team for their quick responses and for providing a proper guideline on time. Finally, we acknowledge Jaydeep Kumar (BCAS, New Delhi, India) and Advocate Dinesh Chandra Gupta (High Court, Allahabad, India) and Uday Pratap Chaudhary for their continuous help and support during the execution of the project.

The editors would be happy to receive any comments to further improve future editions. Comments, feedback, suggestions for further improvement, or proposals for the new chapters for next editions are welcome and should be sent directly to the volume editors.

Taichung, Taiwan — Akhilesh Kumar Yadav, Ph.D., MIE
Greater Noida, India — Saba Shirin, Ph.D.
College Station, USA — Vijay P. Singh, Ph.D., D.Sc.

Contents

About the Editors

Akhilesh Kumar Yadav obtained his B.Tech. in Electronics and Communication Engineering from Chaudhary Charan Singh University, Meerut, India; M.Tech. in Environmental Engineering from Madan Mohan Malaviya Engineering College, Gorakhpur (Dr. APJ AKTU, Lucknow), India; and doctoral degree with research domain of air pollution from Indian Institute of Technology (BHU) Varanasi, India. He served several institutional bodies, i.e., IIT(BHU), India; IIT Bombay, Mumbai, India; BARC (research collaborator), Mumbai, India; and CYUT, Taichung, Taiwan, and his credit with more than 30 SCI/SCIE/Scopus research articles, three book chapters, eight proceeding articles, five edited/authored books, and one patent granted. He received a recognition award as the "*Young Engineer Award*" by the Institution of Engineers (India) and the "*Young Scientist Award*" by *VDGOOD Technology Factory*, Kolkata. His research interests are air, water, and soil pollution; climate change; vulnerability; human health risk assessments; and GIS applications in environmental pollution and management. He is an active member of reputed international professional bodies such as the IEI, Kolkata; ECI, New Delhi; AEACI, Mumbai; MSI, New Delhi; and IASTA, Mumbai.

Saba Shirin obtained her undergraduate degree in Chemistry, Zoology, and Botany from DDU Gorakhpur University, Gorakhpur, India, in 2005. She obtained a postgraduate degree in Environmental Engineering at MMM Engineering College Gorakhpur, India, in 2011. She completed her doctoral degree at the Indian Institute of Technology (BHU) Varanasi, India, in 2019. Presently, she is working as an Assistant Professor (OCFD), at Gautam Buddha University, Greater Noida, India. Her research interests are water and wastewater treatment, air quality, risk assessment, environmental impact assessment, environmental pollution, environmental monitoring, and solid waste management. She has published articles in peer-reviewed journals and was a reviewer in reputed journals. She has attended many international and national conferences and workshops around the world. She is Associate Member of the Association of Environmental Analytical Chemistry of India, Analytical Chemistry Division, Bhabha Atomic Research Centre, Mumbai, India.

Vijay P. Singh Ph.D., D.Sc., D.Eng. (Hon.), Ph.D. (Hon.), P.E., P.H., Hon.D.WRE, Dist.M.ASCE, Dist.Hon. M.IWRA, Dist.F.AGGS, Hon. Member AWRA, NAE, holds Caroline and William N. Lehrer Distinguished Chair in Water Engineering and is Distinguished Professor and Regents Professor, Department of Biological, and Agricultural Engineering, and Zachry Department of Civil and Environmental Engineering at Texas A&M University, USA. He has been recognized for four decades of leadership in research, teaching, and service to the hydrologic and water resources engineering profession. His contribution to the state of the art has been significant in many different specialty areas, including hydrologic science and engineering, hydraulic engineering, water resources engineering, environmental engineering, irrigation science, soil and water conservation engineering, entropy-based modeling, copula-based modeling, and mathematical modeling. His extensive publications in these areas include 32 textbooks, 1425 refereed journal articles, 115 book chapters, 330 conference proceedings papers, and 72 technical reports.

For his seminal contributions, he has been honored with more than 105 national/international awards from professional organizations. As a sample, he is a recipient of the Arid Land Hydraulic Engineering Award, Ven Te Chow Award, Torrens Award, Norma Medal, Lifetime Achievement Award, and OPAL Award, all given by ASCE. He was awarded the Ray K. Linsley Award for outstanding contributions to surface water hydrology and the Founders Award of AIH. He has been awarded three honorary doctorates. He is a fellow of ASCE, AWRA, IE, ISAE, IWRS, IASWC, and IAH; and a member of AGU, IAHR, IAHS, and WASER. He is a member/fellow of 12 engineering/science academies. He is licensed as Professional Engineer (PE), Professional Hydrologist (PH), and Honorary Diplomate, AAWRE.

Part I
Fluoride Status and Remediation Strategies

Chapter 1
Effect of Fluoride Contamination on Living Beings: Global Perspective with Prominence of India Scenario

Arya Johnny Shah, Oorv Sumant Devasthali, and Sachin Vijay Jadhav

Abstract Water is the most crucial component and the building block of life on Earth. The world is facing water scarcity, affecting millions of people. Most of the population uses groundwater for drinking purposes. Groundwater can be contaminated by natural anthropogenic and industrial sources such as heavy metals, manganese, radioactive materials, nitrates, sulfates, iron, fluorides, arsenic, etc. In the Earth's crust, fluorine ranks as the 13th most abundant element. Groundwater contains fluoride derived primarily from fluorspar, sedimentary rocks, cryolite, granite, fluorapatite, sellaite, dolomite, etc. The simultaneous presence of fluoride in groundwater is related to volcanic, geothermal, mining, and other activities. Although small amounts of fluorides are beneficial for dental and bone health in humans, fluoride in higher concentrations is detrimental to human health. Fluorides and other contaminants can affect health independently, synergistically, or antagonistically. The degree of leaching of crystalline minerals is directly related to the level of fluoride contamination. China, India, Mexico, and Pakistan are the countries most affected by groundwater fluoride contamination. Millions of people in China and India are at risk of fluoride. The following methods are currently used for fluoride removal: Chemical precipitation of coagulation, which also includes electrocoagulation floatation, Adsorption on special activated solids, ion exchange, and Membrane techniques, which include nanofiltration, reverse osmosis, electrodialysis methods, etc. Electrocoagulation and flotation techniques are relatively expensive due to high operational costs. Adsorption methods are the most widespread, but produce a significant amount of wastewater to recharge the adsorption beds. Based on the concentration of fluoride in water, these removal techniques can be applied sequentially to achieve the

---

A. J. Shah · O. S. Devasthali · S. V. Jadhav (✉)
Department of Chemical Engineering, Institute of Chemical Technology, Matunga, Mumbai, Maharashtra 400 019, India
e-mail: sv.jadhav@ictmumbai.edu.in

A. J. Shah
e-mail: 20cheaj.shah@ug.ictmumbai.edu.in

O. S. Devasthali
e-mail: 20cheos.devasthali@ug.ictmumbai.edu.in

A. K. Yadav et al. (eds.), *Advanced Treatment Technologies for Fluoride Removal in Water*, Water Science and Technology Library 125,
https://doi.org/10.1007/978-3-031-38845-3_1

desired fluoride concentration that is safe to use. This chapter discusses in depth the global scenario of fluoride contamination in depth and briefly discusses its effects and possible ways to reduce and treat fluoride contamination.

Keywords Groundwater · Fluoride · Contamination · Human health · Global perspective

1.1 Introduction

As a fundamental compound that is essential for life on planet Earth. Denoted by a chemical formula of H_2O, the water molecule consists of one oxygen atom and two hydrogen atoms. It can exist in all physical states of matter, namely liquid (water), solid (ice), and gas (steam).

Water on the Earth is found mainly as surface water or groundwater. Surface water refers to water found in sources such as ponds, lakes, rivers, oceans, etc. Water buried below the surface of the Earth and stored in porous rock formations is called groundwater. Groundwater is accessed through wells. From a historical perspective, groundwater is commonly exploited as a relatively clean water source for industrial, agricultural and human consumption. Groundwater is replenished by recharge processes that occur when rain/precipitation or surface water infiltrates the ground, refilling the aquifer. Overuse of groundwater leads to its depletion and can cause permanent damage to the aquifer.

Exposure of groundwater to hazardous substances or pollutants refers to groundwater pollution and contamination. Groundwater contamination has severe consequences as it is a significant source of water for human consumption and agricultural purposes. Nitrates and other nitrogen compounds, fluoride, arsenic, organic compounds, pathogens, etc., are significant groundwater contaminants. The sources can be natural, anthropogenic, or industrial. Arsenic and fluoride are the most serious inorganic contaminants recognized by the World Health Organization (WHO), affecting human health.

1.2 Fluoride and Its Health Effects

Fluoride is a mononegative ion, the simplest of the fluorine atoms. In terms of size, the charge is very similar to that of the hydrogen ion. Fluoride salts, usually white in solid form and colorless in solution, are odorless. In the Earth's shell, fluorine is the 13th most abundant element. Groundwater contains fluoride derived primarily from fluorspar, sedimentary rocks, cryolite, granite, fluorapatite, sellaite, and dolomite.

Fluoride is necessary for human health in small quantities. It is required for the development of teeth and bones. Concentrations above 1.5 mg/L are considered hazardous to human health and can trigger dental and skeletal fluorosis. Children are

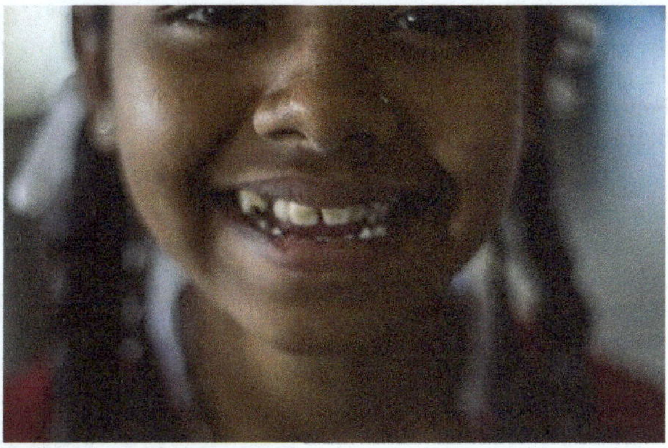

Fig. 1.1 Patient with dental fluorosis. Photo by Jal Seva Charitable Foundation, New Delhi, India. http://www.wateraidindia.in

more susceptible as fluoride is deposited in tissues in their early stages. Fluorosis is not medically treated and is not healable.

In dental fluorosis, as shown in Fig. 1.1, the development of the enamel that covers the tooth surface is disturbed due to overexposure to fluoride ions. Narrow white striations and a yellow-brownish smear on the teeth identify dental fluorosis. Enamel development occurs in the early stages of life, where mineralization increases and the protein matrix is reduced. Excess exposure to fluoride disrupts amelogenesis and dentinogenesis, which deforms the crystalline structure (Yadav et al., 2021).

Skeletal fluorosis, as shown in Figs. 1.2 and 1.3, occurs only after prolonged exposure to high concentrations when bone structure is modeled and remodeled, concentrations higher than 4 mg/L, increasing bone density and bone mass (osteosclerosis). Skeletal fluorosis occurs in three stages:

1. The symptoms of initial skeletal fluorosis include mild joint pain and stiffness, weak muscles accompanied by regular pain, and recurring fatigue.
2. The symptoms of the intermediate step include the Poker back, where the spine becomes stiff. This is because the bones and ligaments are calcified due to excess fluoride deposition.
3. The terminal step of skeletal fluorosis epitomizes restricted joint movement and permanently deformed bones due to severe hardening of bones and ligaments. Some patients even show some neurological defects (Yadav et al., 2021). In the final stage of skeletal fluorosis, the bones of the patient are extremely deformed and crippled.

Skeletal and dental fluorosis is one of the main health defects that occur due to excessive fluoride consumption. Other symptoms may include red blood cells, low haemoglobin levels, reduced immunity, and disruption of the patient, such as symptoms such as Alzheimer's and tingling toes and fingers. Other medical conditions that can arise from excessive fluoride consumption are headaches, gastrointestinal problems, nausea, depression, and abdominal pain.

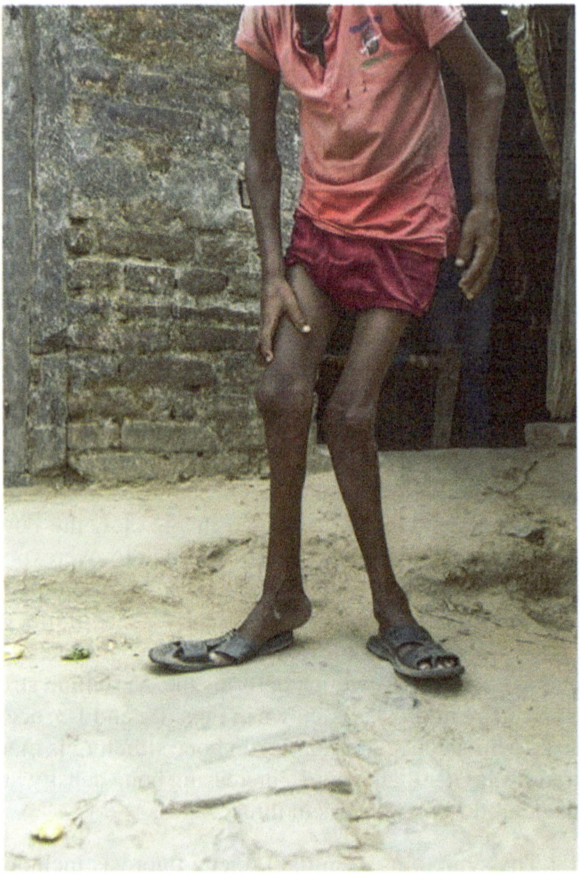

Fig. 1.2 Patient with skeletal fluorosis. Photo by Jal Seva Charitable Foundation, New Delhi. http://www.wateraidindia.in

1.3 Sources of Fluoride Contamination

Fluoride is mainly present in the Earth's crust as minerals, which are found in sedimentary, igneous, and metamorphic rocks. Some of the minerals that contain fluoride include Fluorspar (CaF_2) obtained from limestone and sandstone, Cryolite (Na_3AlF_6) sourced from granite, and Fluorapatite ($Ca_5(PO_4)_3F$) and Sellaite (MgF_2) extracted from bituminous dolomite (Jadhav et al., 2015).

Naturally, the presence of these minerals near water sources causes an increase in fluoride concentrations. It is shown that regions with high concentrations of calcium, aluminum, magnesium, fluoride are relatively higher in the natural water sources. It has also been observed that aridity is positively correlated with elevated fluoride concentration. This happens because groundwater in arid zones has a higher consumption than replenishment, allowing fluoride to leach through and concentrate in the aquifer. Furthermore, the pH of the subsoil and water affects the concentration

Fig. 1.3 This image shows a man suffering from 'Poker back,' which is a severe case of skeletal fluorosis. This image was uploaded to the India Water Portal at https://www.flickr.com/photos/indiawaterportal/9693169272

of fluoride that will be present. The higher the pH, the higher the fluoride concentration. Plate tectonics and volcanic activities can cause changes in the local rock structures, which can expose the water source, mainly the aquifer, to a fluoride-rich rock, increasing the concentration of fluoride.

Fluoride concentrations in water sources started to fluctuate after countries started to industrialize (Yadav et al., 2021). Untreated industrial effluents, when discharged into the environment, pollute the soil and water of the region. Industries related to the extraction of phosphorus for fertilizers, metal industries, semiconductor plants, aluminum plants, uranium hexafluoride production, clay industries, etc., have a high concentration of fluoride, in some cases up to thousands of mg/L, in their effluents (Yadav et al., 2021). When these effluents are disposed of without any treatment, they contaminate the soil and surface waters nearby. Fluoride accumulated in the soil percolates into the aquifer, contaminating the groundwater source.

1.4 Occurrence of Fluoride

1.4.1 Occurrence on a Global Scale

Water scarcity has become a major problem in the past decade. Arid regions have been the most affected region. The number of semi-arid and arid regions has also increased. As discussed above, compared to areas with milder climate conditions, arid regions have higher fluoride concentrations. It can be positively correlated that regions with higher fluoride-rich rocks would have higher fluoride concentrations in surrounding sources of water (Jadhav et al., 2015). Fluoride contamination in groundwater has been recorded largely in regions that are humid and in the tropical regions of the world (Yadav et al., 2021). As depicted in Fig. 1.4, the presence of fluoride-contaminated waters has been confirmed in many latitudes of Africa, Asia, South America, Oceania and the United States. India and China are the countries in which the population is most affected by fluoride contamination.

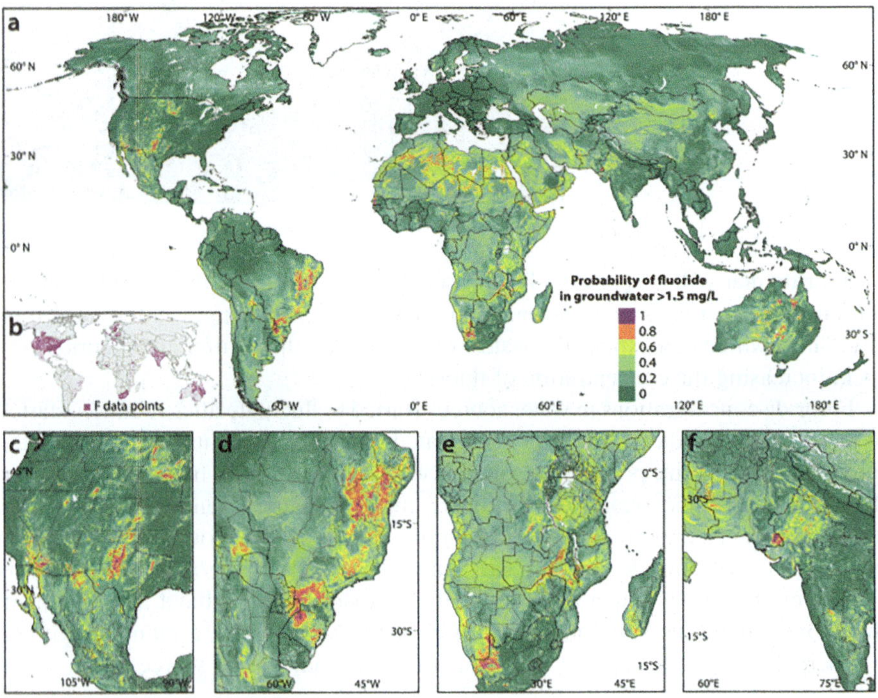

Fig. 1.4 Locations all over the world with probable fluoride concentrations greater than 1.5 mg/L, as suggested by the World Health Organization (WHO) in groundwater in the given ranges: purple: confirmed probability, 60% < red < 80% probability, 40% < yellow < 60% probability, 20% < light green < 40% probability, green < 20% probability, turquoise: regions have fluoride concentrations within the permissible limit (Podgorski & Berg, 2022)

1.4.2 *Occurrence in Africa*

In many African countries, fluorosis is endemic. Many African countries face problems with poor water quality and high fluoride concentrations are the main contributors. Almost all African countries show fluoride detections greater than the permissible limit for drinking water suggested by the WHO (2011). Countries such as Kenya, the Democratic Republic of Congo, Algeria, Ethiopia, Ghana, Malawi, Nigeria, Sudan, Tanzania, Uganda and South Africa are some of the most affected (Table 1.1 and Fig. 1.5). High-fluoride-concentration rocks are mainly found in the East African rift valley region. This also means that a large population in the countries in that region faces health problems such as dental and skeletal fluorosis. It should be noted that surface waters contain more fluoride than groundwater in the east-African rift valley. This is quite counterintuitive, as one would expect groundwater in contact with fluoride-rich rocks would expect to have a higher concentration of fluoride compared to surface waters. This was discussed by Malago et al. (2017) in fluoride levels in surface and groundwater in Africa. Countries such as Kenya, Tanzania, and Ethiopia are the most affected because they are located in the east African rift valley.

Table 1.1 Country-wise ranking of population at potential risk of fluoride contamination in groundwater

Rank	Country	Population at risk (range in millions)	Rank	Country	Population at risk (range in millions)
1	India	49(26–89)	11	Malawi	4(3.5–4.8)
2	China	22(1–50)	12	Zambia	3.4(1.4–3.6)
3	Democratic Rep. Congo	15(2–16)	13	Mozambique	2.6(1.7–3.4)
4	Ethiopia	9.6(4–13.8)	14	Angola	2.2(0.7–2.4)
5	Pakistan	7.6(4.2–8.3)	15	Afghanistan	1.7(0.5–4.8)
6	Kenya	7.5(4.2–8.3)	16	Cameroon	1.6(0.3–2.5)
7	Nigeria	7.4(1–17)	17	Madagascar	1.4(0.7–2.3)
8	Tanzania	6.9(2.7–7.9)	18	Chad	1.2(0.1–2.2)
9	Uganda	4.8(0.9–8)	19	Niger	1.2(0.2–2.6)
10	Yemen	4.3(2.6–4.4)	20	Myanmar	1.1(0.07–3.3)

Podgorski and Berg (2022)

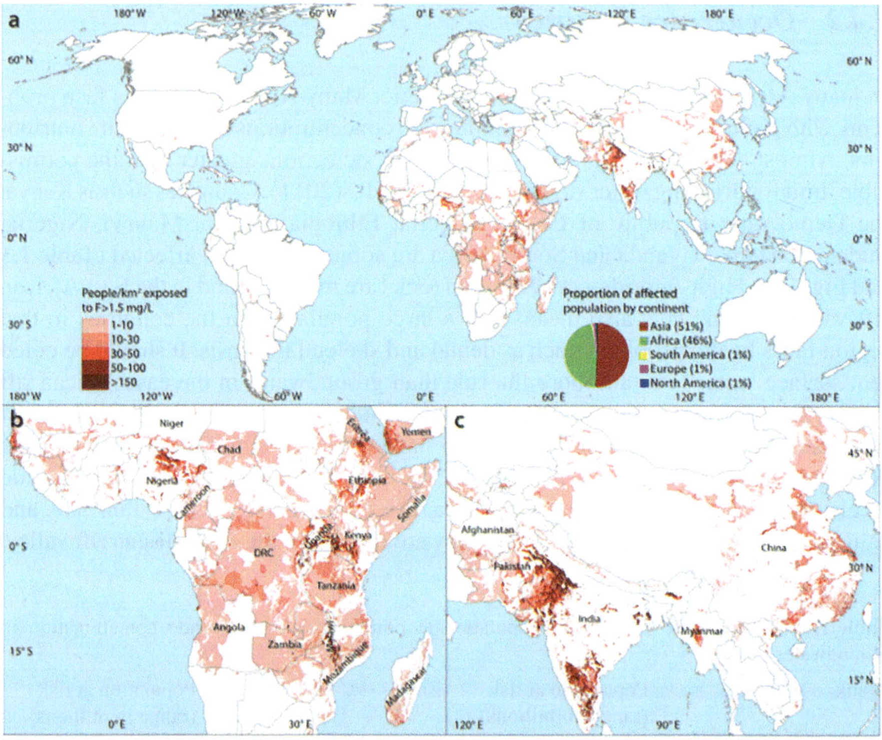

Fig. 1.5 Locations around the world with the population affected due to high concentrations of fluoride in groundwater (Podgorski & Berg, 2022)

1.4.3 Occurrence in India

India has approximately 14% of the world's fluoride deposits (Teotia & Teotia, 1984). This indicates why India is one of the countries most affected by fluoride contamination (Tables 1.1 and 1.3). The regions of southern India have high concentrations in their water sources due to the high deposits of fluoride-rich igneous rocks. Many regions in Andhra Pradesh have high fluoride contamination as many regions there have fluoride-rich granite rocks. Some data samples from Nalgonda in Andhra Pradesh have shown a fluoride concentration of up to 3100 mg/L (Rao et al., 1993). Regions in Northern and Central India have high concentrations of fluoride contamination due to soil dust deposition, an increase in crustal fluoride load, and proximity to Aluminum plants (Yadav et al., 2021). In the states of Uttar Pradesh and Madhya Pradesh in north India, the concentration of fluoride is higher than recommended by WHO, but ranges between 2 and 3 mg/L. In some places, like Sonbhadra in Uttar Pradesh (Table 1.2), samples have shown high fluoride concentrations (up to 6 mg/L) studied by Raju et al. (2009). In these regions, fluoride concentrations are elevated due to soil dust deposition (Figs. 1.6 and 1.7).

Table 1.2 Selective districts in India with high concentrations of fluoride in groundwater

State	District/Specific region	Fluoride concentration (mg/L)
Andhra Pradesh	Hyderabad	0.38–4
	Ranga Reddy	0.4–4.8
	Nalgonda	320–3100
Assam	Karbi Anlong	0.4–20.6
Bihar	Shallow	0.1–2.5
Delhi		0.2–32.5
Kerala	Palghat	0.2–5.75
Madhya Pradesh	Chandidongri	1.5–4.0
	Shivpuri	0.2–6.4
Orissa		0.1–10.1
Rajasthan		0.1–10.1
Tamil Nadu		0.5–4
	Kancheepuram	1–3.24
Gujarat	Cambay	0–10
Uttar Pradesh	Varanasi	0.2–2.1
	Sonbhadra	0.48–6.7
	Mathura	0.6–2.5

Yadav et al. (2021)

Recently, studies have been exploring predictive methods to predict a probability where a region would have a high fluoride concentration. This prediction method considers 25 independent variables such as aridity, amount of carbonate sedimentary rocks, clay fraction in the surrounding soil, irrigated area, Temperature, Sand fraction in the soil, soil pH, etc., to form a statistical model that helps predict a probability that the region having high fluoride concentration. These statistical models have shown an accuracy of 70–80% under specified conditions of the minimum fluoride concentration cut-off point and refining of the twenty-five independent variables (Podgorski et al., 2018).

From maps and data, it is quite evident that the presence of fluoride, in relatively high concentrations of fluoride in the geography, does not need to the population being affected by fluoride contamination. Most counties that have their population exposed to fluoride-contaminated water are underdeveloped nations.

Table 1.3 Percentage of the affected population and areas of high risk in India

State/Territory	Percentage of areas classified as high-risk	Percentage of population in high-risk areas in 2015
Andaman and Nicobar	0	0
Andhra Pradesh	51	14
Arunachal Pradesh	0	0
Assam	15	6
Bihar	4	1
Chandigarh	0	0
Chhattisgarh	6	2
Dadra and Nagar Haveli	0	0
Daman and Diu	20	7
Delhi	45	11
Goa	0	0
Gujarat	50	17
Haryana	81	30
Himachal Pradesh	0	0
Jammu and Kashmir	10	1
Jharkhand	9	4
Karnataka	44	14
Kerala	0	0
Ladakh	10	1
Lakshadweep	0	0
Madhya Pradesh	23	9
Maharashtra	9	3
Manipur	1	3
Meghalaya	0	0
Mizoram	0	0
Nagaland	9	8
Orissa	3	0
Puducherry	0	0
Punjab	56	17
Rajasthan	81	33
Sikkim	0	0
Tamil Nadu	41	14
Telangana	48	19
Tripura	0	0
Uttar Pradesh	13	4

(continued)

Table 1.3 (continued)

State/Territory	Percentage of areas classified as high-risk	Percentage of population in high-risk areas in 2015
Uttaranchal	0	0
West Bengal	23	9

Podgorski et al. (2018)

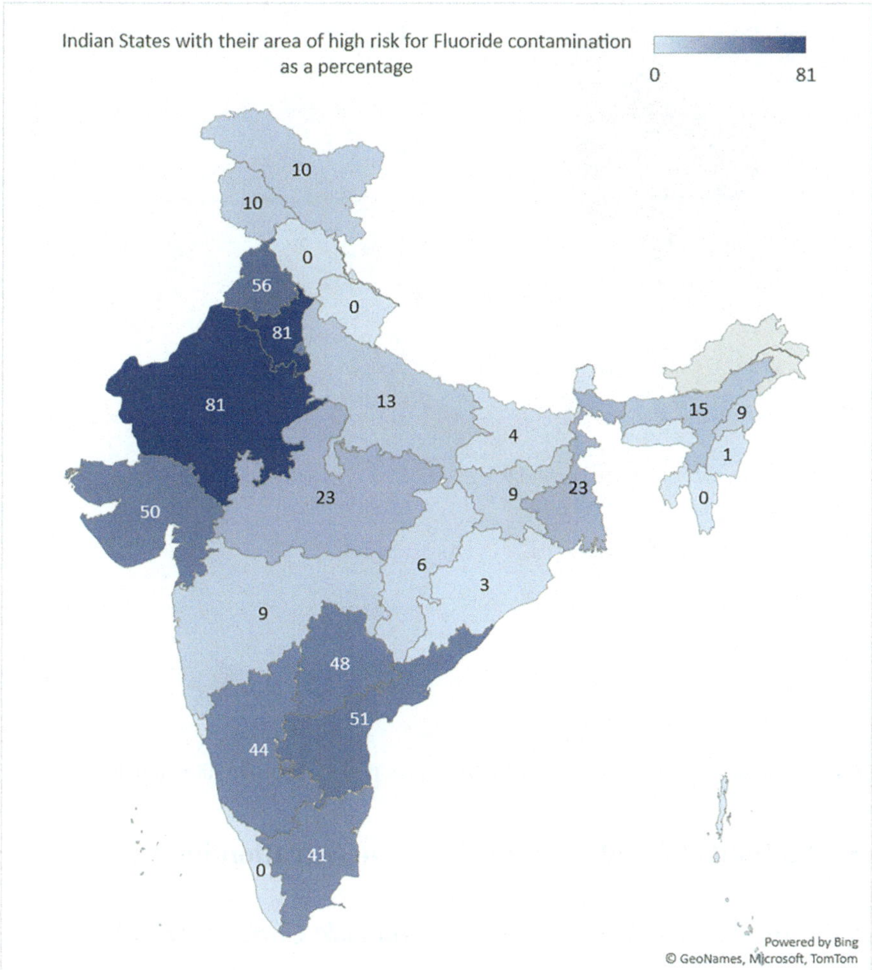

Fig. 1.6 Indian States with their area of high risk for fluoride contamination as a percentage

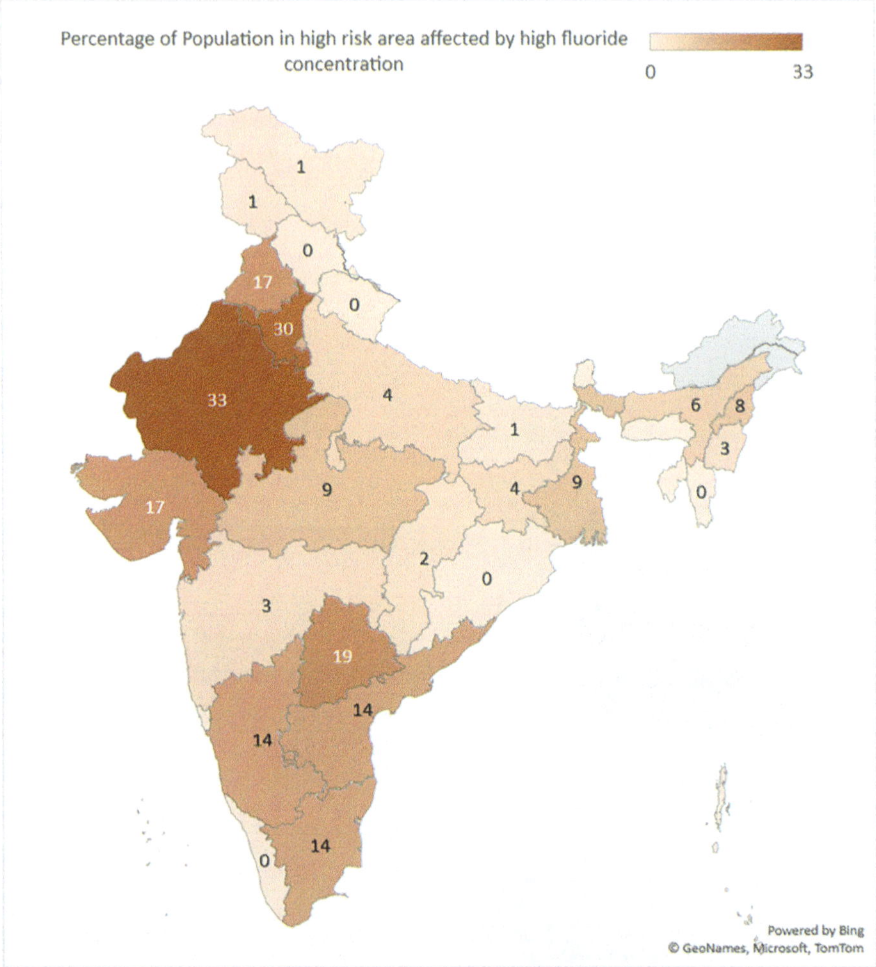

Fig. 1.7 Percentage of the population affected by high fluoride concentration in India

1.5 Methods to Reduce Fluoride Concentration in Water

1.5.1 In Situ Methods to Reduce Fluoride Concentration

In arid and semi-arid regions, fluoride concentrations in groundwaters are found to be significantly higher due to leaching and lack of replenishment of aquifers. The simplest way to reduce the fluoride concentration is by dilution. In other words, the aquifer needs to be replenished using recharging damns, percolation tanks, and recharge pits. These are in situ methods to reduce fluoride concentrations (Yadav et al., 2021).

1.5.2 Ex Situ Methods to Reduce Fluoride Concentration

1.5.2.1 Electrocoagulation

The electric current is used to settle and separate suspensions, dissolved and emulsified impurities. This method is known as electrocoagulation and Electrocoagulation/Floatation. Charged species are destabilized and neutralized, forming a mass together by passing an electric current, forcing them to settle down and come out of the solution. Fluoride forms a complex with the reduced metal/species of the anode and settles as a solid mass, becoming a separate phase. It is an effective technique for the removal of fluoride from industrial effluents as it can bring the fluoride concentration in the effluent to about 1.5 mg/L from 10–20 mg/L. It does not require any external addition of chemicals or any other sludge treatment. The only drawback of this method is that it requires electricity to be carried out, which increases the operational costs of the process, making it expensive (Jadhav et al., 2015).

1.5.2.2 Chemical Precipitation/Condensation

In this method, certain chemicals are added, which force the fluoride ions in the solution to precipitate and separate from the solution. Mainly, alum (Aluminium salts) and calcium salts are used to help precipitate fluoride from the solution/water to be purified. For this process, to achieve maximum fluoride removal, the pH of the solution/contaminated water should be at around 6–7. The pH of polluted water post-treatment may increase to 12. Calcium salt requirements are much lower than those of aluminum salt, in terms of quantity, to precipitate out the same amount of fluoride. A maximum of 96% reduction in fluoride concentration has been observed for a 109 mg/L initial concentration of 109 mg/L (Jadhav et al., 2015). This method has a low initial cost, but has an extensive requirement for downstream sludge treatment, which requires high maintenance. Treatment water has an unpleasant taste due to the addition of various salts; hence, further treatment is required if the contaminated water is intended to be consumed as potable water.

1.5.2.3 Adsorption

In this method, fluoride ions are adsorbed on a porous solid. Contaminated water is passed through a bed of this porous solid. The bed adsorbs the fluoride ions, reducing their concentration at the outlet. This method is the most reliable in the industry because it is robust, inexpensive, efficient, and does not cause harm to the environment. The adsorbent (the solid on which fluoride ions are adsorbed) can be regenerated and reused. About 75% of the adsorbent can be recovered per cycle (Jadhav et al., 2015). Materials such as activated alumina, red mud (a waste product from the Aluminum production industry), calcite, activated carbon, and activated kaolinites act as cheap and effective adsorbents. For this process, it is necessary to maintain the pH at about 5–7.5 to observe optimum adsorption. The only disadvantage is that, after a certain number of cycles, the adsorbent needs to be changed as it gets exhausted. Furthermore, the regeneration steps increase secondary pollution as fluoride-rich waste must be discarded.

1.5.2.4 Membrane Technology

In this method, membranes are used that selectively separate fluoride ions from contaminated water. Typically, three types of membrane separation techniques can be used to separate fluoride from wastewater. These are nanofiltration, reverse osmosis, and Electrodialysis. These methods can guarantee up to 75–96% reduction in fluoride concentration. These processes are advantageous in terms of the fact that no external chemical is needed to carry out the operation. Membrane selection and maintenance are the major hassles of this method.

1.6 Conclusions

Without water, life on Earth is absolutely impossible. Groundwater is one of the main sources of drinking water for people. The explanation above shows that fluoride is necessary for optimal health in tiny amounts, but at higher concentrations, it poses a health risk. For many people, fluoride poisoning of the water is a serious issue. Natural, human-made, or industrial sources all contaminate water sources. Long-term consumption of water containing a high fluoride content results in skeletal and dental fluorosis. Part III of this book discusses many more health effects. Fluoride levels in water sources can be very high in regions that are arid and have high relative temperatures, and rocks in the region have a high fluoride content. The countries of India, China, the Democratic Republic of Congo, Ethiopia, and many other countries are the most affected by the fluoride contamination of their groundwater sources. The in situ and ex situ treatment schemes of fluoride removal, such as coagulation, adsorption, precipitation, filtration, etc., are discussed in Part II of this book.

Acknowledgements The authors declare no conflict of interest. This work did not receive any specific grant from funding agencies in the public, commercial, or not-for-profit sectors. The authors acknowledge the Institute of Chemical Technology, Mumbai, for the resources provided in writing this chapter.

References

Jadhav, S. V., Gadipelly, C. R., Marathe, K. V., & Rathod, V. K. (2014). Treatment of fluoride concentrates from membrane units using salt solutions. *Journal of Water Process Engineering, 2*, 31–35.

Jadhav, S. V., Bringas, E., Yadav, G. D., Rathod, V. K., Ortiz, I., & Marathe, K. V. (2015). Arsenic and fluoride contaminated groundwaters: A review of current technologies for contaminants removal. *Journal of Environmental Management, 162*, 306–325.

Malago, J., Makoba, E., & Muzuka, A. N. N. (2017). Fluoride levels in surface and groundwater in Africa. *Journal of Environmental Health.* https://doi.org/10.11648/j.ajwse.20170301.11

Onipe, T., Edokpayi, J. N., & Odivo, J. O. (2020). A review on the potential source and health implications of fluoride in groundwater of Sub-Saharan Africa. *Journal of Environmental Health.* https://doi.org/10.1080/10934529.2020.1770516

Podgorski, J. E., & Berg, M. (2022). Global analysis and prediction of fluoride in groundwater. *Nature Communications.* https://doi.org/10.1038/s41467-022-31940-x

Podgorski, J. E., Labhesetwar, P., Saha, D., & Berg, M. (2018). Prediction modeling and mapping of groundwater fluoride contamination throughout India. *Environmental Science & Technology, 52.*

Raju, N. J., Dey, S., & Das, K. (2009). Fluoride contamination in groundwaters of Sonbhadra district, Uttar Pradesh, India. *Current Science*, 979–985

Rao, N. V. R., Rao, N., Rao, K. S. P., & Schuiling, R. D. (1993). Fluoride distribution in waters of Nalgonda district, Andhra Pradesh, India. *Environmental Geology, 21*, 84–89.

Teotia, S. P. & Teotia, M. (1984). Endemic fluorosis in India: A challenging national health problem. *The Journal of the Association of Physicians of India, 32*(4), 347–352.

WHO, G. (2011). Guidelines for drinking-water quality. World health organization

Yadav, M., Singh, G., & Jadeja, R. N. (2021). Fluoride contamination in groundwater, impacts, and their potential remediation techniques. *Groundwater geochemistry: Pollution and remediation methods* (1st ed., pp. 22–41). Wiley.

Chapter 2
Fluoride Pollution in Subsurface Water: Challenges and Opportunities

Sonam Gupta, Nivedita Mishra, Ankit Kumar, and Akhilesh Kumar Yadav

Abstract Fluoride contamination in groundwater is one of the drinking water crisis globally. Although its presence is necessary in small quantities, it is harmful to humans with intakes of more than 1.5 mg L^{-1} through contaminated drinking water due to geological factors and geochemical processes. A high fluoride content in drinking water results in skeletal fluorosis, as well as long-term liver, kidney, and brain damage. One of the most crucial challenges for drinking water safety is the management of fluoride pollution and fluorosis. To reduce the probability of fluorosis, it is essential to have a better understanding of the mechanisms underlying the presence of fluoride in the chemistry of subsurface water and the ability to identify high-risk locations using geographic data and remote sensing. The utilization of other sources of water or mixing should be prioritized. The development of stable technologies and integrated devices that are efficient, affordable, and manageable, as

S. Gupta (✉)
Department of Environmental Science, Babasaheb Bhimrao Ambedkar University, Lucknow 226025, India
e-mail: guptasonam3333@gmail.com

N. Mishra
CSIR—National Botanical Research Institute, Lucknow 226001, India
e-mail: niveditaact@gmail.com

A. Kumar
Department of Chemical Engineering, Indian Institute of Technology Gandhinagar, Gandhinagar 382355, India
e-mail: ankit.rs.che14@iitbhu.ac.in

A. K. Yadav
Department of Mining Engineering, Indian Institute of Technology (Banaras Hindu University), Varanasi 221005, India

Department of Environmental Engineering and Management, Chaoyang University of Technology, Taichung 413310, Taiwan

Environmental Science and Engineering Department, Indian Institute of Technology Bombay, Mumbai 400076, India

A. K. Yadav
e-mail: yadavbasti@gmail.com

A. K. Yadav et al. (eds.), *Advanced Treatment Technologies for Fluoride Removal in Water*, Water Science and Technology Library 125,
https://doi.org/10.1007/978-3-031-38845-3_2

well as fundamental research on defluoridation reagents and novel materials, should receive significant amount of focus. To achieve stable gains, the design, construction, operation, and monitoring of defluoridation facilities should be thoroughly evaluated and improved. This chapter highlights the extent of fluoride contamination, the effect of fluoride contamination on human health, and the available defluoridation methods.

Keywords Status of contamination · Sources of contamination · Health effects · Remediation methods

2.1 Introduction

The quality of the water we drink is correlated with our overall quality of life. One of the most important resources for supporting life on Earth is water. Groundwater is now under additional stress due to the growing population and the decreased reliance on surface water resources (Mukherjee, 2018). Groundwater is one of the most exploited water resources that fulfills the drinking water needs of around 50% of the world's population and approximately 70% of the irrigation purpose of agricultural land often known as the 'hidden sea' or "hidden ocean" (Jadhav et al., 2015). Anthropogenic activities have disrupted the cycle of heavy metals, nonmetals, and metalloids. Among them, fluoride is the most prevalent contaminant in terms of affected areas and people (Sahu, 2019). Groundwater pollution due to fluoride has been the main cause of concern for the worldwide water problem and poses many health risks from ingesting water (Bibi et al., 2017).

Although fluoride is one of the essential components for humans, an excessive amount can have harmful effects on health. According to previous reports, the intake of fluoride-contaminated groundwater from geogenic sources is the main cause of population exposure (Kumar et al., 2016). Additionally, landfills, industrial effluent, on-site sanitation systems, and the use of phosphatic fertilizers contribute to groundwater contamination (Sharma et al., 2017). Fluoride enters groundwater through a variety of sources and entry points, which could have negative impacts on human health (Ghosh et al., 2013). Rock-water interactions and contact time, pH-dependent dissolution, aqueous ionic concentrations, atmospheric deposition and mobilization through carbonate and bicarbonate are some of the factors that affect groundwater fluoride contamination (Malago et al., 2017; Raj & Shaji, 2017; Singh et al., 2020). Fluorosis, a condition caused by consuming fluoride-contaminated groundwater, affects approximately 200 million people in 25 different countries (Khatri & Tyagi, 2015). Although fluoride is found in food, air, and water, drinking water remains the main source of exposure (Vithanage & Bhattacharya, 2015).

It is well known that fluoride has an impact on the neurological system, skeletal muscles, bones, and teeth. Salivation, nausea, diarrhea, and stomach pain are the most common signs and symptoms (Chouhan & Flora, 2010). Many socioeconomic problems were brought about by the high fluoride uptake, and many wells were abandoned as a result of fluoride-contaminated groundwater (Kimambo et al., 2019). A study claims that endemic fluorosis affects at least 25 different nations around the world (Susheela et al., 2013). Fluoride remediation techniques include electrocoagulation, membrane processes, ion exchange processes, coagulation and precipitation methods, and adsorption. Due to the severity of the problem, this contamination requires prompt treatment to reduce the growing threat it poses in the near future. This chapter aims to present an overview of the fluoride contamination scenario in the context of the global and Indian subcontinents, including sources of contamination, the effects on human health, and defluoridation technologies currently in use for the treatment of contaminated water. This work will be highly beneficial for the implementation of appropriate and affordable removal technology to provide safe water to rural populations in poor countries.

2.2 Fluoride Contamination

2.2.1 Global Scenario

The occurrence of fluoride contamination in groundwater is influenced by several factors, including rock composition, topography, climate, and rainfall, which lead to varying levels of contamination between regions. Fluoride contamination is prominent on the continents of Asia and Africa, where levels range from 20 to 79.2 ppm (Raj & Shaji, 2017; Rasool et al., 2018). Several investigations reported that Afghanistan's province had the highest concentration (up to 79.2 mg/L). Fluoride concentration has been reported to range from 20 to 79.2 mg/L in various regions, including Vietnam (Ninh Hoa), South Korea, Malaysia (Jenderam Hilir), Pakistan (Kalalanwala) and Afghanistan (Province of Afghanistan) (Farooqi & Farooqi, 2015; Hayat & Baba, 2017; Shamsuddin et al., 2016). These high levels of fluoride in water were primarily due to natural geological processes, the presence of rocks containing fluoride and the occurrence of apatite minerals in groundwater (Msonda et al., 2007). Other countries with high groundwater fluoride levels include Sri Lanka, Israel, Indonesia, China, Ethiopia, Iran, Turkey, Canada, and Germany (Ali et al., 2016). The fluorosis problem in South Africa has existed since 1935 as a result of fluoride levels ranging from 0.05 to 13 ppm (Ncube and Schutte, 2005).

The Yuncheng Basin in China recorded fluoride contamination of up to 14.10 mg/L in groundwater, while the fluoride concentration in groundwater in the Muteh region of Iran ranged from 0.02 to 9.2 mg/L (Keshavarzi et al., 2010; Li et al., 2015). In the Ethiopian Rift Valley, almost 80% of the population lives in the valley (Rango et al., 2010). In various areas, including Anuradhapura in Sri Lanka (Chandrajith et al., 2012), East Java in Indonesia (Heikens et al., 2005), and sections of Anatolia in Turkey (Oruc, 2008), the fluoride level in groundwater ranged from 10 to 20 mg/L. The F content was found to range from 0.30 to 6.45 mg/L in the Gaza Strip (Jabal et al., 2014). Groundwater contamination levels of fluoride ranging from 5 to 10 mg/L have been reported in various regions, including Posht-e-Kooh-e-Dashtestan in southern Iran, the Mizunami region in Japan, Shanxi and Inner Mongolia, Taiyuan basin, Datong basin, and northern China, the Hetao Basin, Oyu Tolgoi, and Tavan Tolgoi in Mongolia, as well as NW Taiz city in Yemen and Saudi Arabia (Alabdulaaly et al., 2013; Hu et al., 2013; Nakazawa et al., 2016). In Bangladesh and Myanmar's Myingyan Township, fluoride concentrations between 1.5 and 5 mg/L have also been recorded (Bacquart et al., 2015). Groundwater fluoride pollution was below 1.5 mg/L in northern Jordan, the Sulaimani district of Iraq, Bahrain (semi-arid parts of Bahrain), and Thailand (remote places) (Khursheed et al., 2015; Rukah & Alsokhny, 2004).

2.2.2 Indian Scenario

According to studies, 14% of all geogenic fluoride is detected in India, indicating that fluoride poisoning in the country has geogenic origins (Mukherjee, 2018). According to research by the Central Groundwater Board, fluoride levels in 184 districts and 19 states of India are higher than the acceptable limit of 1.5 ppm (CGWB-2017). A surge in the prevalence of fluorosis has been noted in India, where 62 million people are at risk of fluoride poisoning (Mukherjee & Singh, 2018). For example, the presence of fluoride and nitrate above the desired levels indicated fluoride enrichment in the groundwater of the Siwani Block, western Haryana, likely due to anthropogenic activities (Ali et al., 2018). In addition, it was discovered that fluoride contamination made groundwater in the Dongargaon region of India unfit for human consumption. However, it can be used for irrigation (Sahu et al., 2017). The Andhra Pradesh region of Nalgonda has had high levels of fluoride. Subsequently, it was discovered that numerous areas of the nation had significant concentrations of fluoride, Andhra Pradesh, Rajasthan, Haryana, Punjab, Gujarat and Assam being the worst affected states (Kumar et al., 2016).

Fluoride concentrations in groundwater are known to exceed the World Health Organization drinking water standard (1.5 mg/L) in semi-arid regions surrounding Gujarat (Raza et al., 2016). Jammu and Kashmir, Maharashtra, Chhattisgarh, Karnataka, Tamil Nadu, Jharkhand, Uttar Pradesh, and Bihar have reported groundwater fluoride concentrations below 5 mg/L (Sahu et al., 2017; Subba Rao et al., 2016). Fluoride concentrations have been recorded in Andhra Pradesh, Delhi, Gujarat, Kerala, Orissa, and West Bengal between 5 and 10 mg/L. The areas of Madhya Pradesh, Punjab, Assam and Haryana all had fluoride values between 10 and 20 mg/L. According to reports, the maximum fluoride content in many areas of Rajasthan has exceeded 20 mg/L (CGWB, 2014). According to reports, Rajasthan and Gujarat are the states most severely hit (80–100%), followed by Punjab and Andhra Pradesh (60–80%). Between 21 and 40% of the areas in Maharashtra, Jharkhand, Madhya Pradesh, Tamil Nadu, Kerala, West Bengal, and Sikkim are contaminated with fluoride (Yadav et al., 2019a, 2019b). Fluoride contamination is reported in some regions of Jammu and Kashmir, Karnataka, Haryana, Uttar Pradesh, and Assam (10–20%). Approximately 70% of the districts of Andhra Pradesh have groundwater contamination with fluoride. Fluoride contamination mostly affects Delhi's North-west and West Areas. In Gujarat, up to 7.5 mg/L of fluoride concentration has been found in about 92% of the districts. In Rajasthan, up to 38 mg/L of fluoride concentration has been detected in 97% of the districts, which are primarily desert areas. 37% of the districts in West Bengal fall within the fluoride pollution zone. In Punjab, fluoride concentrations have been recorded up to 11.30 mg/L in the Sangrur and Ghorenab regions, while the possibility of high F contamination is present in approximately 77% of the other districts. 60% of the districts in Orissa fall into the high-fluoride content zone (CGWB, 2010). Figure 2.1 shows fluoride concentrations of more than 1.5 mg/L in different states of India.

2.3 Sources of Fluoride Contamination in Groundwater

The average amount of fluoride in rocks found in the crust of the Earth is 625 mg/kg (Vithanage & Bhattacharya, 2015). Elevated levels of fluoride in groundwater are caused by various geological processes and human actions (Fig. 2.2). The potential sources through which fluoride enters the environment and reaches groundwater are discussed below.

2.3.1 Geogenic Sources or Natural Sources Can Be Categorized as Follows

2.3.1.1 Fluoride-Containing Rocks and Minerals

The primary source of fluoride is found in rocks that contain minerals rich in fluoride. The fluoride content varies depending on the kind of rock, with ultramafic rocks having a value of 100 mg/kg, alkaline igneous rocks having a value of 1000 mg/kg and marine shales having a value of 1300 mg/kg (Ozsvath, 2009). Sources of fluoride in groundwater are granitic rocks with fluoride rich minerals such as muscovite, hornblende, and amphiboles (Vithanage & Bhattacharya, 2015). Hydrothermal vein deposits are linked to fluorite, also known as fluorspar (Ozsvath, 2009). Due to the substantial amount of fluoride that coal contains (295 mg/kg), which contributes to fluoride levels in the environment, coal is another source of fluoride.

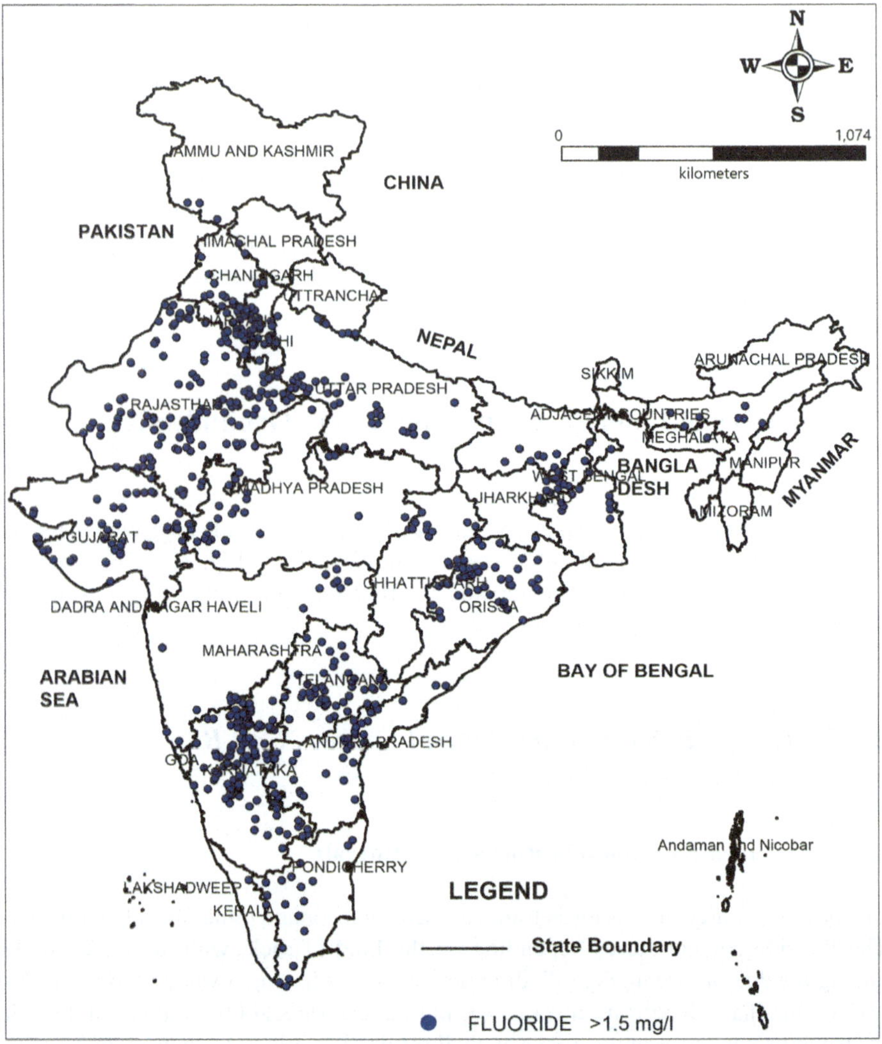

Fig. 2.1 Map showing fluoride contamination in different states of India (Source: CGWB)

2.3.1.2 Geothermal Sources

The second most important natural source is volcanic and geothermal activity. According to estimates, between 60 and 6000 kilotons of inorganic fluorides are emitted annually from volcanoes on a global scale (Camargo, 2003). Fluoride-containing minerals are commonly associated with late-stage pegmatite granites, hydrothermal vein deposits, and rocks that solidify from highly evolved magmas (Scaillet & Macdonald, 2004). At the boundary of a subduction zone, volcanic materials produce approximately 2000 mg/kg of energy (Anazawa, 2006).

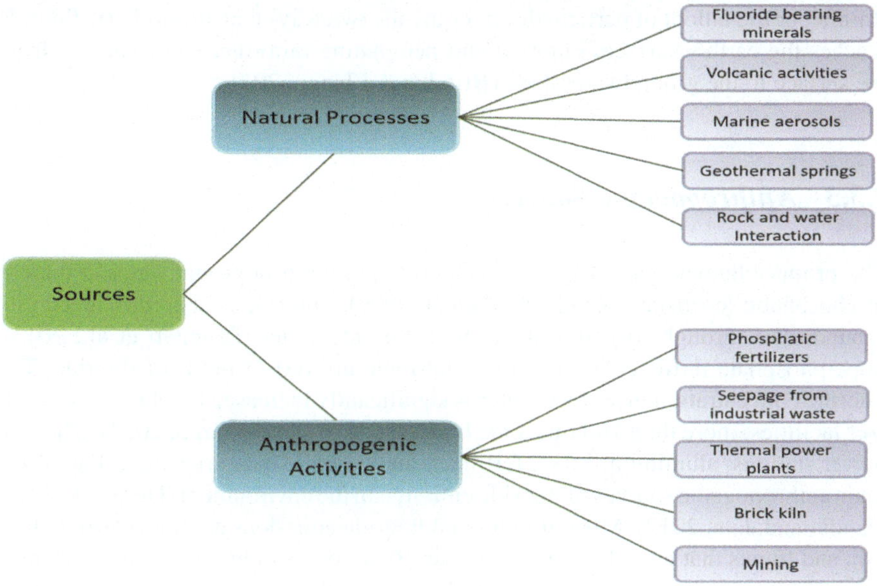

Fig. 2.2 Different sources of fluoride contamination in groundwater

Most geothermal water is alkaline or neutral in pH, creating ideal circumstances for fluoride to be desorbed from fluoride-bearing minerals such as micas and amphiboles (Garcia & Borgnino, 2015). The chemical weathering and alteration of these rocks leads to the dissolution and subsequent release of fluoride ions from minerals such as fluorite, fluorapatite, and mica. These minerals are considered the major sources of fluoride in the environment (Reddy et al., 2013). The high fluoride content of the hot spring water is mainly due to pre-Cambrian rocks composed of granite, schist and amphibolite (Marbaniang et al., 2014).

2.3.2 Atmospheric Deposition

Airborne soil dust, marine aerosols, and volcanic gas emissions are the main natural sources of fluorine in the atmosphere (Tavener & Clark, 2006). Globally, marine aerosols can produce up to 20 kilotons of inorganic fluorides each year (Camargo, 2003). According to reports, the amount of fluoride released by atmospheric air and precipitation into the environment varies from 0.01 to 0.4 $\mu g/m^3$ and 0.089 mg/L, respectively, both of which are nearly undetectable (Gupta et al., 2005). Fluorides in the atmosphere can be either gases or particulates. Mostly released in particle form are calcium fluoride, aluminium fluoride, lead fluoride, sodium hexafluorosilicate, and calcium phosphate fluoride, while hydrogen fluoride, sulfur hexafluoride, silicon tetrafluoride, and fluorosilicic acid are mostly released in gaseous form (WHO, 2002).

Rainfall or the fallout of particle fluorides are the two ways that atmospheric fluoride reaches the earth's surface. Fluoride and percolating rainwater can easily go from the surface to the groundwater zone (Brindha & Elango, 2011).

2.3.3 Anthropogenic Sources

The primary human-caused source of fluoride pollution in groundwater is the use of phosphatic fertilizers (Kundu & Mandal, 2009). Fluoride is also introduced into groundwater through irrigation using high fluoride water (Pettenati et al., 2013). Superphosphate fertilizer is shown to contribute up to 0.34 mg/L of fluoride. The fluoride concentration in surface water is significantly increased by phosphate fertilizer facilities, more than 100 times the background level (Camargo, 2003). Thermal power stations, aluminium, steel, fertilizer, and ceramic industries are a few of the main anthropogenic sources of fluoride emission in the environment (Dey et al., 2012; Nandimandalam, 2012). Most of industrial fluoride emissions are fluoride-rich dust, ash, and fumes that are released into the air. By air deposition, these emissions may contaminate water, soil and plants with fluoride in the industrial region, as well as other locations that are far from the source of the emissions (Ranjan & Ranjan, 2015).

Streams close to aluminium smelters may have fluoride levels that are more than 10 times higher than the 0.05 mg/L natural background limit. Industries that produce bricks, phosphatic fertilizer, glass and ceramics, and coal-fired power plants are the biggest contributors to the high fluoride levels (Camargo, 2003; Vike, 2005). Fluorine released into the atmosphere by human activity is very reactive and readily hydrolyzes to generate hydrogen fluoride. When coal is burned, fluoride is released into the atmosphere, which can cause particles from the surface of the soil to enter the groundwater zone (Karthikeyan et al., 2010). The amount of fluoride released into the environment as gas and particulate matter is increasing as a result of the widespread usage of fossil fuels in the industrial sectors. The brick kiln industries also contribute to the elevated fluoride concentration. Clay is used in brickmaking to create bricks with extremely high fluoride concentrations (Cape et al., 2003). Increased groundwater consumption and mining activities speed up fluoride dissolution (Kumar et al., 2016).

2.4 Health Effects

Consumption of fluoride higher than the permissible level has acute and chronic effects on humans. Exposure to excessive fluoride can lead to various health complications in both genders in different age groups, including dental fluorosis, skeletal fluorosis, osteoporosis, kidney damage, bone deformities, reproductive organ dysfunction, and nerve and muscle degeneration, among others (Yadav et al., 2019a,

2019b). Several effects due to the consumption of contaminated groundwater in different human body systems have been described below.

2.4.1 *Dental Fluorosis*

Humans who consume excessive amounts of fluoride-contaminated drinking water while their teeth are developing, often between the ages of six and eight, suffer dental fluorosis. Dental fluorosis is caused by the excessive absorption of fluoride into these tissues. It is also known as hypoplasia or hypomineralization of dental enamel. In particular in the first seven years of life, teeth are particularly sensitive to fluorosis during the tooth formation or calcification period. Dental enamel is composed of nearly 87% crystalline calcium phosphate in the form of hydroxyapatite (Arlappa et al., 2013). Excessive consumption of fluoride through drinking water can cause the displacement of hydroxide ions (OH–) from hydroxyapatite, resulting in the formation of fluorapatite. This process is responsible for the appearance of dental fluorosis. Long-term ingestion of fluoride-contaminated water causes teeth to become hard and brittle. Enamel also develops pits, discoloration (from yellow to brown to black), and mottling. Surface discoloration may appear in the form of spots or horizontal streaks. Dental fluorosis can take various forms such as normal, mild, moderate, and severe, depending on how much fluoride is consumed and how severely the water is contaminated (Fig. 2.3). For moderate forms of dental fluorosis, horizontal white striations emerge on the surface of the tooth, or opaque areas of chalky white discolorations may also appear (Pini et al., 2015; Susheela, 2003). Osmotic patches can change color from yellow to brown or even black in moderate to severe cases of dental fluorosis (advanced forms), and eventually, increased porosity of the teeth can result in structural deterioration such as pitting or chipping (Pini et al., 2015).

2.4.2 *Skeletal Fluorosis*

It is a medical condition marked by the accumulation of excessive amounts of fluoride in the bones, resulting in increased bone density and mass. This may cause a variety of symptoms related to the skeletal and joint systems, including joint pain, stiffness, and restricted movement, as well as calcification and hardening of ligaments and tendons around the joints. North Rajasthan, Gujarat, and Andhra Pradesh in India are the three states where the problem is most prevalent. States like Madhya Pradesh, Maharashtra, Punjab, and Haryana are marginally affected, while West Bengal, Bihar, Assam, Tamil Nadu, and Uttar Pradesh are barely affected (Arlappa et al., 2013). Skeletal fluorosis (which can be acute or chronic) is caused by prolonged exposure

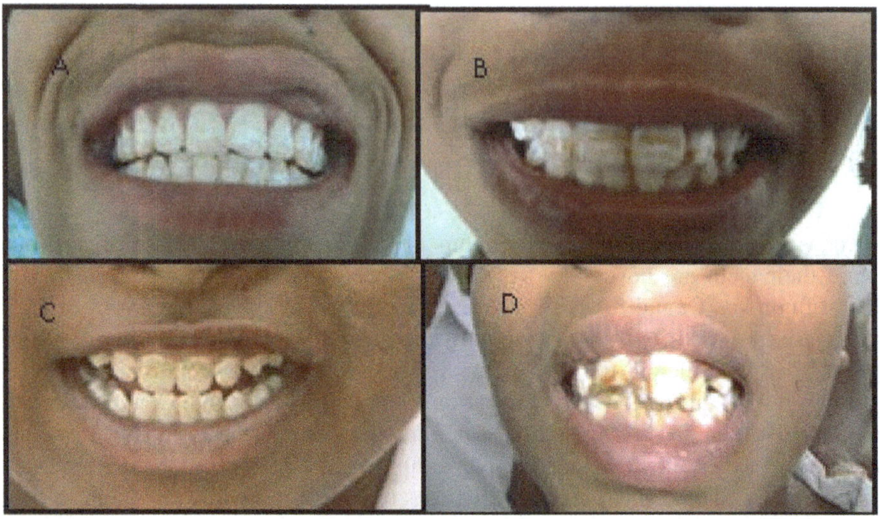

Fig. 2.3 Variation in degrees of dental fluorosis in children aged 5–11 very mild (**a**); mild (**b**); moderate (**c**); severe (**d**)

to high amounts of F^{-} (> 4 mg/L), whether from eating or breathing. Fluorosis, which can be severely debilitating, can be caused by long-term exposure (more than 10 years) to high fluoride concentrations (> 10 mg/L) (Arlappa et al., 2013).

In the early stages of skeletal fluorosis, people can experience mild symptoms such as muscle weakness, joint pain, stiffness, and occasional bone pain, as well as chronic fatigue. As the condition progresses to intermediate stages, osteosclerosis and bone calcification may occur, which can eventually lead to hardening and tightening of the joints. The condition known as "poker back" occurs when skeletal fluorosis is severe enough to cause bones to grow increasingly rigid over time. Ultimately, this leads to crippling skeletal fluorosis that begins with restricted joint movements, bone abnormalities, acute calcification of ligaments, muscle atrophy, and neurological impairments (Itai et al., 2010).

2.4.3 Reproductive System

Fluoride is believed to be one of the key contributing elements to the global infertility crisis that currently plaguing society. Fluoride exposure is associated with an increase in follicle stimulating hormone levels, a luteinizing hormone and reduction in thyroid hormone and estrogen levels, and changes in the estrogen receptor to the androgen receptor (ER/AR) ratio (Wang et al., 2009). There was a decrease in the level of circulating testosterone in a male patient with skeletal fluorosis (Long et al., 2009).

2.4.4 Kidney and Liver

The kidney, liver and heart all exhibit histopathological and functional alterations when exposed for an extended period of time to excessive fluoride concentrations (Kumari & Kumar, 2011). A study revealed that chronic kidney disease progresses more quickly when fluoride is continuously consumed through food (CKD) (Dharmaratne, 2015). Even if they ingest the permitted amount of fluoride, someone with renal disease is more likely to develop osteoporosis due to a decreased ability to eliminate fluoride through urine.

2.4.5 Brain

Excess fluoride increases the risk of neurotoxicity, which can impair one's capacity for learning and remembering. According to recent studies, children in high-fluoride locations have relatively weaker mental abilities than children in low-fluoride areas (Zhang et al., 2015). The level of lipid peroxidation increases due to the increased fluoride content, which also inhibits several key brain enzymes. Fluoride can obstruct the brain's ability to function by directly affecting neurotransmitters, myelin, and neurons (Kabir et al., 2020).

2.4.6 Cytotoxicity

In excess consumption of fluoride has the potential to cause chromosomal abnormalities and genetic alterations in mammalian cells (Zuo et al., 2018). Even at low doses, fluoride has a considerable effect on the synthesis of DNA and proteins. Furthermore, excessive fluoride has been associated with nucleosome DNA damage and modifications in cell structure (Zhang et al., 2006).

2.5 Methods for the Remediation of Fluoride

People living in fluoride-contaminated areas contaminated with fluoride can suffer from various health problems. To alleviate these issues, several preventive and remedial measures can be adopted. These include providing alternative sources of water and improving the nutritional status of the affected population. Additionally, various defluoridation techniques can be used to remove excess fluoride from drinking water. Table 2.1 provides a detailed comparison of the advantages and disadvantages of different defluoridation techniques. It is important to note that each technique has its own limitations, and the most suitable method can vary depending on factors such as cost, feasibility, and the specific needs of the affected community. However, with proper implementation and management, defluoridation techniques can be an effective means of mitigating the health risks associated with fluoride exposure.

2.5.1 *Ion Exchange*

This procedure uses a substance known as an ion exchanger, where the unwanted ions are removed by allowing water to pass through its bed. In the resin, the fluoride ion swaps places with the chloride ion, and this process continues until all the sites in the resin are filled. The residue is regenerated by washing it with water that has sodium chloride salt dissolved in it, leading to the replacement of fluoride with chloride and begins to function as an ion exchanger (Ojekunle et al., 2016). A unique driving reason behind the replacement of chloride ions in the increased electronegativity of the resin is the fluoride ion. The removal of fluoride from aqueous solutions by the ion exchange method has a high potential (up to 95%). Although the resins can be easily replenished, the treatment is unprofitable due to the high cost of the resins. Another drawback of the method is that the regeneration process produces a lot of waste that contains fluoride (Jadhav et al., 2015).

2.5.2 *Coagulation/Precipitation*

Researchers tried to create an economical and sustainable method to remove fluoride from water in the early 1930s. In terms of impurities from the water, this procedure reduces turbidity. A specific chemical known as a coagulant is used during the coagulation process to destabilize the tiny particles present in water. Aluminium, alum, iron, lime, sodium aluminate, zeolites, ferric chloride, and silica gel are some of the coagulants used to remove fluoride from water (Ahmad et al., 2022). The most frequently utilized of them are alum and lime. The Nalgonda method is the greatest illustration of how fluoride can be removed from water using the coagulation/precipitation method. This process involves the addition of aluminium salts, bleaching powder, and lime to water that has been contaminated with fluoride over the course of six successive processes, including flocculation, filtration, rapid mixing, sedimentation, and sludge concentration. The following steps are taken to complete the entire process. Insoluble flocs of aluminium hydroxide form, followed by settling of sediments in the bottom, and fluoride and bleaching powder co-precipitate (Bhatnagar et al., 2011). However, because the Nalgonda process for defluorizing water only deters around 70% of fluoride, it is not effective for treating water with high fluoride levels. Also, this procedure is expensive because of the high cost of alum.

Table 2.1 Advantages and disadvantages of different defluoridation techniques

Advantages	Disadvantages
Ion Exchange	
• High efficiency • Removal efficacy up to 95% • Retains colour and taste of water • Highly effective and efficient process • Low running cost	• Costly • Easy exposed to interfering ions (bicarbonate, sulfate, chloride, etc.) • Requires replacement of resin after multiple recycle, disposal of resin needs separate treatment • Treated water becomes acidic • Treated water exhibits greater chloride concentration
Coagulation/Precipitation	
• Efficiency is high • Chemicals used are available commercially • Work at the domestic and community level • Simple design • Simple operation process	• Costly and largely depends on the co-ion and pH of water • High amount of residual aluminium was obtained • Sludge obtained contain toxic aluminium fluoride complex • Large amount of water is retained and hence dewatering is done before disposal • High chemical dosages
Electrocoagulation	
• Simple process and design • Easily adjustable • Low sludge production, non-hazardous • Targets multiple contaminants	• pH adjustment is required • Sacrificial electrodes • Electricity use may be expensive • Disposal of brine is a problem
Reverse Osmosis	
• Quick regeneration of membrane • N > 90% helps in eliminating other dissolved solids and also a wide range of pH • Chemical process is not used • Work under a wide pH range • No chemical assist	• High cost • Saline solution is used to treat water • Salt may be creating an issue • Valuable mineral is needed after treatment for demineralization • Needs pH enhancement
Adsorption	
• Cost effective • Higher accessibility • Easy operation • Availability of number of adsorbents • Locally available and cheap	• To achieve good and adjustment of pH is required • Presence of other ion hinders the adsorption • Reduction in removal % of regenerated adsorbent • Interference of other inorganic ions • Treated water comprises greater total dissolved solids (TDS)

(continued)

Table 2.1 (continued)

Advantages	Disadvantages
Nano-filtration	
• No chemical is used • Production is very high • No ion interference • Divider for the organic micropollutant • Inorganic toxic suspended solid	• High cost • Issue of membrane degradation scaling and fouling • Requires remineralization of treated water • Discards all vitals minerals

2.5.3 *Electrocoagulation*

It is a rapid and effective technique commonly used for removing flocculating agents generated during the electro-oxidation process, which typically involves the use of a sacrificial anode made of iron or aluminum. This method helps to reduce the amount of sludge that needs to be disposed of because it does not use any chemical coagulants or flocculants. This approach uses three basic processes: coagulation, hydrodynamics, and electrochemistry. The electrocoagulation reactor consists of an electrolytic cell with a cathode and an anode (Mollah et al., 2001). Ghosh et al. (2013) proposed the electrocoagulation procedure to remove fluoride from drinking water with a fluoride concentration between 2 and 10 mg L^{-1} using monopolar and bipolar connections. They discovered that the bipolar connection preferred fluoride removal over the monopolar connection. The final recommended breaking point of fluoride (1 mg L^{-1}) under bipolar connection was reached at 625 A m^{-2} in less than 30 min. Additionally, the operational expenses for mono and bipolar connections are 0.38 and 0.62 US \$ m^{-3}, respectively.

2.5.4 *Reverse Osmosis*

A semipermeable membrane divides a tank into two during the reverse osmosis process. A semipermeable membrane allows contaminated water to pass from one side to the other with the help of hydraulic pressure. Salts and several other contaminants cannot travel through the semipermeable barrier, although water and tiny impurities can (Dubey et al., 2021). When the pressure on the contaminated side of the semipermeable is elevated, the process of osmosis is accelerated, and this increases the rate at which water can pass through the membrane, from the impure side to the clean side (Sehn, 2008). In the absence of interference from other ions, the removal efficiency of this process approaches 90% or higher (Assefa, 2006). The efficacy of membrane fluoride removal is influenced by the two crucial variables of pH and temperature. The membrane for water purification is chosen depending on the nature, cost, temperature, pressure, recovery, and salt rejection (Velazquez-Jimenez et al., 2015). The primary disadvantage of reverse osmosis is the removal of all the ions from the water. Minerals are required for various metabolic processes as well as for normal growth. The treated water must be remineralized, which adds to its cost, in order for the minerals to fully dissolve.

2.5.5 *Nanofiltration*

Nanofiltration is recognized as the best membrane method for eliminating fluoride among all membrane procedures due to its high specific membrane selectivity (Tahaikt et al., 2007). Nanofiltration membranes have somewhat smaller pores than reverse osmosis membranes, which makes it easier for the solvent and solute to flow across the membrane. As a result, nanofiltration requires little pressure and has a quick flow (Hu & Dickson, 2006). Nanofiltration was discovered to outperform other membrane approaches in the removal of fluoride and the desalination of some brackish water (Diawara et al., 2011). Two commercial nanofiltration membranes, NF-270 and NF-90, were compared and a reduction in fluoride level was observed from 10 to 1.5 mg L^{-1} in NF-270 while 20 to 0.5 mg L^{-1} in NF-90 (Hoinkis et al., 2011).

2.5.6 *Adsorption*

Adsorption is the process of creating a material layer on an adsorbent's surface. As a result, the adsorption process can be described using the stages below. Initially, the surface of the adsorbent particles from the heterogeneous solution developed a layer of fluoride followed by the adsorption of fluoride ions onto the surface of fluoride. Finally, intraparticle diffusion occurs wherein adsorbed fluoride is transferred to the inner surfaces of porous adsorbent materials (Obulapuram et al., 2021). In particular, the amount of fluoride adsorbed on the adsorbent surface per unit mass of the adsorbent shows the degree of removal of fluoride pollution from the water. The initial fluoride concentration, pH of the water, nature of adsorbent utilized, period of contact, and presence of interfering ions all affect the efficacy of the adsorbent in removing fluoride (Waghmare & Arfin, 2015a). Adsorption is regarded as the most efficient method for smaller populations due to the clear design, simple operating procedure, variety of adsorbent availability, and low set up cost (Waghmare & Arfin, 2015b). Active alumina is extensively employed as an adsorbent at the domestic and communal levels. The usage of active alumina as an adsorbent was discovered to entail both the adsorption process and the ion exchange approach (Ghorai & Pant, 2004). Fluoride may be removed from water using active alumina; however, doing so can have harmful effects on health. Furthermore, the complex chemical formed when aluminum and fluoride combine is known to cause Alzheimer's disease (Yadav et al., 2018). Granular ferric hydroxides have been successful in eliminating fluoride from drinking water (Shams et al., 2010).

2.6 Conclusion

Fluoride contamination in drinking water is a persistent issue in many regions of the world, particularly in developing countries of Asia. The presence of fluoride in geological formations makes it a challenge to address through source control, necessitating remediation methods to reduce exposure. Groundwater is particularly vulnerable to fluoride contamination due to the prevalence of fluoride-bearing parent bedrock, with high concentrations often found in areas with granitic compositions. The impact of fluorosis is significant, with water-stressed regions bearing the greatest burden. Effective defluorination techniques are essential in areas where fluorine water is the primary drinking source, and efforts must be made to increase public awareness of risks and promote the adoption of safe drinking water practices. While many initiatives have been taken to address the issue, more needs to be done to provide affordable, adaptable, and acceptable defluoridation technologies. Regular monitoring of groundwater is required to identify suitable defluoridation techniques, and communities must be encouraged to actively participate in the process. It is essential to establish and maintain more community defluoridation plants and educate households about domestic defluoridation techniques to reduce the risk of fluorosis in fluoride-contaminated areas.

Acknowledgements The corresponding author is grateful to the DAIC, Ministry of Social Justice and Empowerment, Govt. of India for providing financial support.

References

Ahmad, S., Singh, R., Arfin, T., & Neeti, K. (2022). Fluoride contamination, consequences and removal techniques in water: A review. *Environmental Science: Advances.*

Alabdulaaly, A. I., Al-Zarah, A. I., & Khan, M. A. (2013). Occurrence of fluoride in ground waters of Saudi Arabia. *Applied Water Science, 3*, 589–595.

Ali, S., Shekhar, S., Bhattacharya, P., Verma, G., Chandrasekhar, T., & Chandrashekhar, A. K. (2018). Elevated fluoride in groundwater of Siwani Block, Western Haryana, India: A potential concern for sustainable water supplies for drinking and irrigation. *Groundwater for Sustainable Development, 7*, 410–420.

Ali, S., Thakur, S. K., Sarkar, A., & Shekhar, S. (2016). Worldwide contamination of water by fluoride. *Environmental Chemistry Letters, 14*, 291–315. https://doi.org/10.1007/s10311-016-0563-5

Anazawa, K. (2006). Fluorine and coexisting volatiles in the geosphere: The role in Japanese volcanic rocks. *Fluorine and the Environment: Atmospheric Chemistry, Emissions & Lithosphere*, 187.

Arlappa, N., Qureshi, A. A. T. I. F., & Srinivas, R. (2013). Fluorosis in India: An overview. *International Journal of Research and Development Health, 1*(2), 97–102.

Assefa, B. (2006). Defluoridation of Ethiopian rift valley region water using reverse osmosis membranes. *Zede Journal, 23*, 1–6.

Bacquart, T., Frisbie, S., Mitchell, E., Grigg, L., Cole, C., Small, C., & Sarkar, B. (2015). Multiple inorganic toxic substances contaminating the groundwater of Myingyan Township, Myanmar:

Arsenic, manganese, fluoride, iron, and uranium. *Science of the Total Environment, 517*, 232–245.

Bhatnagar, A., Kumar, E., & Sillanpää, M. (2011). Fluoride removal from water by adsorption—A review. *Chemical Engineering Journal, 171*(3), 811–840.

Bibi, S., Kamran, M. A., Sultana, J., & Farooqi, A. (2017). Occurrence and methods to remove arsenic and fluoride contamination in water. *Environmental Chemistry Letters, 15*, 125–149.

Brindha, K., & Elango, L. (2011). Fluoride in groundwater: Causes, implications and mitigation measures. *Fluoride Properties, Applications and Environmental Management, 1*, 111–136.

Camargo, J. A. (2003). Fluoride toxicity to aquatic organisms: A review. *Chemosphere, 50*(3), 251–264. https://doi.org/10.1016/S0045-6535(02)00498-8

Cape, J. N., Fowler, D., & Davison, A. (2003). Ecological effects of sulfur dioxide, fluorides, and minor air pollutants: Recent trends and research needs. *Environment International, 29*(2–3), 201–211. https://doi.org/10.1016/S0160-4120(02)00180-0

Central Ground Water Board (CGWB). (2014). *Ground water year book 2012–13 Rajasthan State.* Government of India, Ministry of Water Resources, Regional Office Data Centre Western Region Rajasthan.

CGWB. (2010). *Groundwater quality in shallow aquifers of India.* Ministry of Water Resource, Faridabad. http://cgwb.gov.in/

Chandrajith, R., Padmasiri, J. P., Dissanayake, C. B., & Prematilaka, K. M. (2012). Spatial distribution of fluoride in groundwater of Sri Lanka. *Journal of the National Science Foundation of Sri Lanka, 40*, 4.

Chouhan, S., & Flora, S. J. (2010). Arsenic and fluoride: Two major ground water pollutants. *Indian Journal of Experimental Biology, 48*(7), 666–678.

Dey, R. K., Swain, S. K., Mishra, S., Sharma, P., Patnaik, T., Singh, V. K., Dehury, B. N., Jha, U., & Patel, R. K. (2012). Hydrogeochemical processes controlling the high fluoride concentration in groundwater: A case study at the Boden block area, Orissa, India. *Environmental Monitoring and Assessment, 184*, 3279–3291. https://doi.org/10.1007/s10661-011-2188-2

Dharmaratne, R. W. (2015). Fluoride in drinking water and diet: The causative factor of chronic kidney diseases in the North Central Province of Sri Lanka. *Environmental Health and Preventive Medicine, 20*(4), 237–242.

Diawara, C. K., Diop, S. N., Diallo, M. A., Farcy, M., & Deratani, A. (2011). Performance of nanofiltration (NF) and low pressure reverse osmosis (LPRO) membranes in the removal of fluorine and salinity from brackish drinking water. *Journal of Water Resource and Protection, 3*(12), 912.

Dubey, S., Agarwal, M., & Gupta, A. B. (2021). Fluoride removal using Alum & PACl in batch & continuous mode with subsequent microfiltration. *Membrane and Water Treatment, 12*(2), 83–93.

Farooqi, A., & Farooqi, A. (2015). Status of As and F^- groundwater and soil pollution in Pakistan. *Arsenic and Fluoride Contamination: A Pakistan Perspective*, 21–33.

Garcia, M. G., & Borgnino, L. (2015). Fluoride in the context of the environment. In V. R. Preedy (Ed.), *Fluorine: Chemistry, analysis, function and effects* (pp. 3–18). The Royal Society of Chemistry.

Ghorai, S., & Pant, K. K. (2004). Investigations on the column performance of fluoride adsorption by activated alumina in a fixed-bed. *Chemical Engineering Journal, 98*(1–2), 165–173.

Ghosh, A., Mukherjee, K., Ghosh, S. K., & Saha, B. (2013). Sources and toxicity of fluoride in the environment. *Research on Chemical Intermediates, 39*, 2881–2915. https://doi.org/10.1007/s11164-012-0841-1

Gupta, S. K., Deshpande, R. D., Agarwal, M., & Raval, B. R. (2005). Origin of high fluoride in groundwater in the North Gujarat-Cambay region, India. *Hydrogeology Journal, 13*, 596–605. https://doi.org/10.1007/s10040-004-0389-2

Hayat, E., & Baba, A. (2017). Quality of groundwater resources in Afghanistan. *Environmental Monitoring and Assessment, 189*, 1–16.

Heikens, A., Sumarti, S., Van Bergen, M., Widianarko, B., Fokkert, L., Van Leeuwen, K., & Seinen, W. (2005). The impact of the hyperacid Ijen Crater Lake: Risks of excess fluoride to human health. *Science of the Total Environment, 346*(1–3), 56–69.

Hoinkis, J., Valero-Freitag, S., Caporgno, M. P., & Pätzold, C. (2011). Removal of nitrate and fluoride by nanofiltration—A comparative study. *Desalination and Water Treatment, 30*(1–3), 278–288.

Hu, K., & Dickson, J. M. (2006). Nanofiltration membrane performance on fluoride removal from water. *Journal of Membrane Science, 279*(1–2), 529–538.

Hu, S., Luo, T., & Jing, C. (2013). Principal component analysis of fluoride geochemistry of groundwater in Shanxi and Inner Mongolia, China. *Journal of Geochemical Exploration, 135*, 124–129.

Itai, K., Onoda, T., Nohara, M., et al. (2010). Serum ionic fluoride concentrations are related to renal function and menopause status but not to age in a Japanese general population. *Clinica Chimica Acta, 411*(3–4), 263–266.

Jabal, M. S. A., Abustan, I., Rozaimy, M. R., & Al-Najar, H. (2014). Fluoride enrichment in groundwater of semi-arid urban area: Khan Younis City, southern Gaza Strip (Palestine). *Journal of African Earth Sciences, 100*, 259–266.

Jadhav, S. V., Bringas, E., Yadav, G. D., Rathod, V. K., Ortiz, I., & Marathe, K. V. (2015). Arsenic and fluoride contaminated groundwaters: A review of current technologies for contaminants removal. *Journal of Environmental Management, 162*, 306–325. https://doi.org/10.1016/j.jenvman.2015.07.020

Kabir, H., Gupta, A. K., & Tripathy, S. (2020). Fluoride and human health: Systematic appraisal of sources, exposures, metabolism, and toxicity. *Critical Reviews in Environmental Science and Technology, 50*(11), 1116–1193.

Karthikeyan, K., Nanthakumar, K., Velmurugan, P., Tamilarasi, S., & Lakshmanaperumalsamy, P. (2010). Prevalence of certain inorganic constituents in groundwater samples of Erode district, Tamilnadu, India, with special emphasis on fluoride, fluorosis and its remedial measures. *Environmental Monitoring and Assessment, 160*, 141–155. https://doi.org/10.1007/s10661-008-0664-0

Keshavarzi, B., Moore, F., Esmaeili, A., & Rastmanesh, F. (2010). The source of fluoride toxicity in Muteh area, Isfahan, Iran. *Environmental Earth Sciences, 61*, 777–786.

Khatri, N., & Tyagi, S. (2015). Influences of natural and anthropogenic factors on surface and groundwater quality in rural and urban areas. *Frontiers in Life Science, 8*(1), 23–39. https://doi.org/10.1080/21553769.2014.933716

Khursheed, D. A., Abdulateeef, D. S., Fatah, A. O., & Rauf, A. M. (2015). Fluoride concentration of well water in different areas of Sulaimani province. *Sulaimani Dental Journal, 2*, 67–71.

Kimambo, V., Bhattacharya, P., Mtalo, F., Mtamba, J., & Ahmad, A. (2019). Fluoride occurrence in groundwater systems at global scale and status of defluoridation–state of the art. *Groundwater for Sustainable Development, 9*, 100223. https://doi.org/10.1016/j.gsd.2019.100223

Kumar, M., Das, A., Das, N., Goswami, R., & Singh, U. K. (2016). Co-occurrence perspective of arsenic and fluoride in the groundwater of Diphu, Assam, Northeastern India. *Chemosphere, 150*, 227–238. https://doi.org/10.1016/j.chemosphere.2016.02.019

Kumari, S., & Kumar, A. (2011). Fluoride toxicity enhances phagocytic activity of macrophages in spleen of rats. *Asian Journal of Experimental Biological Sciences, 2*(2), 283–287.

Kundu, M. C., & Mandal, B. (2009). Assessment of potential hazards of fluoride contamination in drinking groundwater of an intensively cultivated district in West Bengal, India. *Environmental Monitoring and Assessment, 152*, 97–103. https://doi.org/10.1007/s10661-008-0299-1

Li, C., Gao, X., & Wang, Y. (2015). Hydrogeochemistry of high-fluoride groundwater at Yuncheng Basin, northern China. *Science of the Total Environment, 508*, 155–165.

Long, H., Jin, Y., Lin, M., Sun, Y., Zhang, L., & Clinch, C. (2009). Fluoride toxicity in the male reproductive system. *Fluoride, 42*(4), 260–276.

Malago, J., Makoba, E., & Muzuka, A. N. (2017). Fluoride levels in surface and groundwater in Africa: A review. *American Journal of Water Science and Engineering, 3*(1), 1–17. https://doi.org/10.11648/j.ajwse.20170301.11

Marbaniang, D. G., Diengdoh, D. F., & Nongpiur, C. L. (2014). Preliminary investigation of the physico-chemical properties of a geothermal spring (hot spring) located at Jakrem, Meghalaya, India. *Keanean Journal of Science, 3*, 35–42.

Mollah, M. Y. A., Schennach, R., Parga, J. R., & Cocke, D. L. (2001). Electrocoagulation (EC)—Science and applications. *Journal of Hazardous Materials, 84*(1), 29–41.

Msonda, K. W. M., Masamba, W. R. L., & Fabiano, E. (2007). A study of fluoride groundwater occurrence in Nathenje, Lilongwe, Malawi. *Physics and Chemistry of the Earth, Parts A/B/C, 32*(15–18), 1178–1184.

Mukherjee, A. (2018). *Overview of the groundwater of South Asia* (pp. 3–20). Springer. https://doi.org/10.1007/978-981-10-3889-1_1

Mukherjee, I., & Singh, U. K. (2018). Groundwater fluoride contamination, probable release, and containment mechanisms: A review on Indian context. *Environmental Geochemistry and Health, 40*(6), 2259–2301.

Nakazawa, K., Nagafuchi, O., Okano, K., Osaka, K. I., Hamabata, E., Tsogtbaatar, J., & Choijil, J. (2016). Non-carcinogenic risk assessment of groundwater in South Gobi, Mongolia. *Journal of Water and Health, 14*(6), 1009–1018.

Nandimandalam, J. R. (2012). Evaluation of hydrogeochemical processes in the Pleistocene aquifers of middle Ganga Plain, Uttar Pradesh, India. *Environmental Earth Sciences, 65*, 1291–1308. https://doi.org/10.1007/s12665-011-1377-1

Ncube, E. J., & Schutte, C. F. (2005). The occurrence of fluoride in South African groundwater: A water quality and health problem. *Water SA, 31*(1), 35–40. https://doi.org/10.4314/wsa.v31i1.5118

Obulapuram, P. K., Arfin, T., Mohammad, F., Khiste, S. K., Chavali, M., Albalawi, A. N., & Al-Lohedan, H. A. (2021). Adsorption, equilibrium isotherm, and thermodynamic studies towards the removal of reactive orange 16 dye using Cu (I)-polyaninile composite. *Polymers, 13*(20), 3490.

Ojekunle, O. Z., Ojekunle, O. V., Adeyemi, A. A., et al. (2016). Evaluation of surface water quality indices and ecological risk assessment for heavy metals in scrap yard neighbourhood. *Springerplus, 5*(1), 1–16.

Oruc, N. (2008). Occurrence and problems of high fluoride waters in Turkey: An overview. *Environmental Geochemistry and Health, 30*(4), 315–323.

Ozsvath, D. L. (2009). Fluoride and environmental health: A review. *Reviews in Environmental Science and Bio/technology, 8*, 59–79. https://doi.org/10.1007/s11157-008-9136-9

Pettenati, M., Perrin, J., Pauwels, H., & Ahmed, S. (2013). Simulating fluoride evolution in groundwater using a reactive multicomponent transient transport model: Application to a crystalline aquifer of Southern India. *Applied Geochemistry, 29*, 102–116. https://doi.org/10.1016/j.apgeochem.2012.11.001

Pini, N. I. P., Sundfeld-Neto, D., Aguiar, F. H. B., Sundfeld, R. H., Martins, L. R. M., Lovadino, J. R., & Lima, D. A. N. L. (2015). Enamel microabrasion: An overview of clinical and scientific considerations. *World Journal of Clinical Cases: WJCC, 3*(1), 34.

Raj, D., & Shaji, E. (2017). Fluoride contamination in groundwater resources of Alleppey, southern India. *Geoscience Frontiers, 8*(1), 117–124. https://doi.org/10.1016/j.gsf.2016.01.002

Rango, T., Bianchini, G., Beccaluva, L., & Tassinari, R. (2010). Geochemistry and water quality assessment of central Main Ethiopian Rift natural waters with emphasis on source and occurrence of fluoride and arsenic. *Journal of African Earth Sciences, 57*(5), 479–491. https://doi.org/10.1016/j.jafrearsci.2009.12.005

Ranjan, R., & Ranjan, A. (2015). *Fluoride toxicity in animals. SpringerBriefs in Animal Sciences* (1st ed.). Springer.

Rasool, A., Farooqi, A., Xiao, T., Ali, W., Noor, S., Abiola, O., Ali, S., & Nasim, W. (2018). A review of global outlook on fluoride contamination in groundwater with prominence on the Pakistan current situation. *Environmental Geochemistry and Health, 40*, 1265–1281.

Raza, M., Farooqi, A., Niazi, N. K., & Ahmad, A. (2016). Geochemical control on spatial variability of fluoride concentrations in groundwater from rural areas of Gujrat in Punjab, Pakistan. *Environmental Earth Sciences, 75*, 1–16.

Reddy, D. V., Nagabhushanam, P., & Ramesh, G. (2013). Turnover time of Tural and Rajvadi hot spring waters, Maharashtra, India. *Current Science*, 1419–1424.

Rukah, Y. A., & Alsokhny, K. (2004). Geochemical assessment of groundwater contamination with special emphasis on fluoride concentration, North Jordan. *Geochemistry, 64*(2), 171–181.

Sahu, B. L., Banjare, G. R., Ramteke, S., Patel, K. S., & Matini, L. (2017). Fluoride contamination of groundwater and toxicities in dongargaon block, Chhattisgarh, India. *Exposure and Health, 9*, 143–156. https://doi.org/10.1007/s12403-016-0229-3

Sahu, P. (2019). Fluoride pollution in groundwater. In *Groundwater development and management: Issues and challenges in South Asia* (pp. 329–350).https://doi.org/10.1007/978-3-319-75115-3_14

Scaillet, B., & Macdonald, R. (2004). Fluorite stability in silicic magmas. *Contributions to Mineralogy and Petrology, 147*, 319–329. https://doi.org/10.1007/s00410-004-0559-1

Sehn, P. (2008). Fluoride removal with extra low energy reverse osmosis membranes: Three years of large scale field experience in Finland. *Desalination, 223*(1–3), 73–84.

Shams, M., Nabizadeh Nodehi, R., Hadi Dehghani, M., Younesian, M., & Hossein Mahvia, A. (2010). Efficiency of granular ferric hydroxide (GFH) for removal of fluoride from water. *Fluoride, 43*(1), 61.

Shamsuddin, M. K. N., Suratman, S., Ramli, M. F., Sulaiman, W. N. A., & Sefie, A. (2016, July). Hydrochemical assessment of surfacewater and groundwater quality at bank infiltration site. In *IOP Conference Series: Materials Science and Engineering* (Vol. 136, No. 1, p. 012073). IOP Publishing.

Sharma, S., Jha, P. K., Ranjan, M. R., Singh, U. K., Kumar, M., & Jindal, T. (2017). Nutrient chemistry of River Yamuna, India. *Asian Journal of Water, Environmental Pollution, 14*, 61–70. https://doi.org/10.3233/AJW-170016

Singh, A., Patel, A. K., & Kumar, M. (2020). Mitigating the risk of arsenic and fluoride contamination of groundwater through a multi-model framework of statistical assessment and natural remediation techniques. In *Emerging issues in the water environment during Anthropocene: A South East Asian perspective* (pp. 285–300). https://doi.org/10.1007/978-981-32-9771-5_15

Subba Rao, N., Dinakar, A., Surya Rao, P., Rao, P. N., Madhnure, P., Prasad, K. M., & Sudarshan, G. (2016). Geochemical processes controlling fluoride-bearing groundwater in the granitic aquifer of a semi-arid region. *Journal of the Geological Society of India, 88*, 350–356. https://doi.org/10.1007/s12594-016-0497-3

Susheela, A. K. (2003). *A treatise on fluorosis, revised second edition* (pp. 13–14). Fluorosis Research and Rural Development Foundation.

Susheela, A. K., Mondal, N. K., & Singh, A. (2013). Exposure to fluoride in smelter workers in a primary aluminium industry in India. *The International Journal of Occupational and Environmental Medicine, 4*(2), 61–72.

Tahaikt, M., El Habbani, R., Haddou, A. A., et al. (2007). Fluoride removal from groundwater by nanofiltration. *Desalination, 212*(1–3), 46–53.

Tavener, S. J., & Clark, J. H. (2006). Fluorine: Friend or Foe? A green chemist's perspective. In A. Tressaud (Ed.), *Fluorine and the environment: Agrochemicals, archaeology, green chemistry and water (Chapter 5)* (pp. 177–202). Elsevier.

Velazquez-Jimenez, L. H., Vences-Alvarez, E., Flores-Arciniega, J. L., Flores-Zuniga, H., & Rangel-Mendez, J. R. (2015). Water defluoridation with special emphasis on adsorbents-containing metal oxides and/or hydroxides: A review. *Separation and Purification Technology, 150*, 292–307.

Vike, E. (2005). Uptake, deposition and wash off of fluoride and aluminium in plant foliage in the vicinity of an aluminium smelter in Norway. *Water, Air, and Soil Pollution, 160*, 145–159. https://doi.org/10.1007/s11270-005-3862-1

Vithanage, M., & Bhattacharya, P. (2015). Fluoride in the environment: Sources, distribution and defluoridation. *Environmental Chemistry Letters, 13*, 131–147.

Waghmare, S. S., & Arfin, T. (2015a). Fluoride removal by clays, geomaterials, minerals, low cost materials and zeolites by adsorption: A review. *International Journal of Science, Engineering and Technology Research, 4*(11), 3663–3676.

Waghmare, S. S., & Arfin, T. (2015b). Fluoride removal by industrial, agricultural, and biomass wastes as adsorbents. *International Journal of Advance Research and Innovative Ideas in Education, 1*, 628–653.

Wang, H., Yang, Z., Zhou, B., Gao, H., Yan, X., & Wang, J. (2009). Fluoride-induced thyroid dysfunction in rats: Roles of dietary protein and calcium level. *Toxicology and Industrial Health, 25*(1), 49–57.

World Health Organization (WHO). (2002). *Fluorides, I. P. C. S. International Programme on Chemical Safety (Environmental Health Criteria 227).* Geneva.

Yadav, K. K., Kumar, S., Pham, Q. B., Gupta, N., Rezania, S., Kamyab, H., ... Cho, J. (2019a). Fluoride contamination, health problems and remediation methods in Asian groundwater: A comprehensive review. *Ecotoxicology and Environmental Safety, 182*, 109362.

Yadav, K. K., Kumar, V., Gupta, N., Kumar, S., Rezania, S., & Singh, N. (2019b). Human health risk assessment: Study of a population exposed to fluoride through groundwater of Agra city, India. *Regulatory Toxicology and Pharmacology, 106*, 68–80.

Yadav, N., Rani, K., Yadav, S. S., Yadav, D. K., Yadav, V. K., & Yadav, N. (2018). Soil and water pollution with fluoride, geochemistry, food safety issues and reclamation—A review. *International Journal of Current Microbiology and Applied Sciences, 7*, 1147–1162.

Zhang, S., Zhang, X., Liu, H., Qu, W., Guan, Z., Zeng, Q., Jiang, C., Gao, H., Zhang, C., Lei, R., Xia, T., Wang, Z., Yang, L., Chen, Y., Wu, X., Cui, Y., Yu, L., & Wang, A. (2015). Modifying effect of COMT gene polymorphism and a predictive role for proteomics analysis in children's intelligence in endemic fluorosis area in Tianjin, China. *Toxicological Sciences, 144*(2), 238–245.

Zhang, Y., Sun, X., Sun, G., Liu, S., & Wang, L. (2006). DNA damage induced by fluoride in rat osteoblasts. *Fluoride, 39*(3), 191–194.

Zuo, H., Chen, L., Kong, M., et al. (2018). Toxic effects of fluoride on organisms. *Life Sciences, 198*, 18–24.

Part II
Fluoride Removal Techniques

Chapter 3
Fluoride Pollution Control Techniques and Principles

Divyadeepika, Krishna Yadav, and Jyoti Joshi

Abstract The lesser and excessive amounts of F^- ions are both unhealthy for human health. Due to an insufficient amount of F^-, the formation and decay of teeth is observed, but excessive fluoride is linked to the diagnosis of fluorosis leading to hyperactivity, Musculoskeletal abnormalities, and brain damage due to excessive fluoride exposure during the development of tooth enamel. Before opting for the method of treatment for the fluoride-contaminated water, the chemical composition of surface and subsurface water is imperative. The presence of volcanic ash and some fertilizers in the soil also leads to an increase in the concentration of F^-. Sometimes a low concentration of F^- also has effective health benefits, but F^- at concertation > 1 ppm can lead to several health hazards. Excessive use of it over a long period of time can cause changes in DNA structure. In order to produce usable water, numerous technologies are being used for the removal of F^- and its derivatives. The major technologies used for the removal of F^- from waste water are green nanomaterials, capacitive deionization (CDI), membrane technology, and electrocoagulation. Apart from F^- removal from waste water, techniques used for natural water are equally important, viz. adsorption technology using various adsorbents. Some effective adsorbents are namely zeolites, alumina, organic based, Shell based including carbon-based nanotubes and graphite, metallopolymers and variety of microspheres. The defluoridation of water using modified activated alumina, chitosan derivatives, clays and muds, composites, and various separation techniques having merits and demerits are in use. Mechanisms used in the fluoride removal techniques are generally Adsorption, Nano-adsorption, Reverse-osmosis, Coagulation-Precipitation, Electrodialysis, Electrocoagulation, Nanofiltration, Ion exchange, Membrane dialysis etc.

Divyadeepika · K. Yadav · J. Joshi (✉)
Organic & Medicinal Chemistry Laboratory, Department of Chemistry, Malaviya National Institute of Technology Jaipur, Jaipur 302017, India
e-mail: jjoshi.chy@mnit.ac.in

Divyadeepika
e-mail: 2020rcy9615@mnit.ac.in

K. Yadav
e-mail: 2022rcy9015@mnit.ac.in

A. K. Yadav et al. (eds.), *Advanced Treatment Technologies for Fluoride Removal in Water*, Water Science and Technology Library 125,
https://doi.org/10.1007/978-3-031-38845-3_3

Keywords Adsorption · Coagulation · Dialysis · Nanofiltration · Capacitive deionization

3.1 Introduction

To lead a healthy life, access to clean and safe is absolutely necessary. Due to specific geographical, economic, and financial obstacles, safe drinking water is inaccessible in many regions of the world. Poor water quality areas cause significant social and health problems. Due to this, the provision of clean drinking water is regarded as a priority in many countries, and developing and impoverished nations are not able to meet some drinking water requirements imposed by regulatory bodies (Poonia et al., 2021; Shirin & Yadav, 2014). Groundwater constitutes a sizeable amount of fresh water that is acceptable for use and intake by both humans as well as animals because of its higher microbial activity in comparison to surface water. However, several chemical compounds and elements have increased in concentration and polluted groundwater as a result of various geological processes. Furthermore, the only supply of water in such regions has been chemically polluted due to the unauthorized disposal of effluent from agricultural, industrial, and urban operations. Both people and aquatic life are affected by these polluted water sources (Shirin et al., 2022; Zhang et al., 2016). Therefore, measures to improve drinking water quality would significantly improve health (Hayat & Baba, 2017). Consuming polluted water causes waterborne infections, which have demonstrable and serious detrimental impacts on human well-being and society's economy (Onipe et al., 2020).

Inorganic contaminants such as fluoride, chromium, lead, mercury, arsenic, copper, cyanide, and antimony, in addition to carbonate and non-carbonated hardness, harm water resources. They may enter drinking water through industrial operations, natural sources, or plumbing methods.

Fluoride (F^-) is relatively more studied groundwater contaminants compared to other because of its detrimental effects on the pervasiveness of people across nations. The maximum permitted level for F^- for drinking purposes is 1.5 mg L^{-1}, which is higher than the recommended safe limit of 1.0 mg L^{-1} (WHO, 2011). Water with F^- levels of 5 mg L^{-1} or above may induce stiffness in the bones or joints, which can result in skeletal fluorosis. Water with F^- values of 1.5–4 mg L^{-1} could cause dental fluorosis when ingested over time (WHO, 2017). Table 3.1 (López-Guzmán et al., 2019) summarizes the health impacts.

The main sources of F^- in water bodies are minerals such as fluorapatite, fluorspar, hydroxyapatite and cryolite which are present in surrounding rocks, sediments, and clay minerals. One of the most important natural groundwater quality issues affecting India's arid and semi-arid regions is F^- pollution (Reddy et al., 2016). Groundwater flow rates are higher near a river because there are fewer residence durations and less opportunities for water to contact F^- bearing rocks. As a result, the level of F^- in the groundwater has decreased (Ali et al., 2019).

Table 3.1 Health risk effects associated with F^- ingestion*

Fluoride concentration (mg L^{-1})	Health effects
< 0.5	More cases of tooth caries
0.5–1.5	Strengthening the molars and bones
1.5–4	Oral fluorosis in kids (mottling in teeth)
> 4	Fluorosis in the skeleton and teeth (deformities in the bone)
> 10	Fluorosis cases, thyroid problems, female infertility, cancer, and Alzheimer's disease cases

*Classification is based on Dissanayake (1991)

To support dental health, fluoride ions are typically added to drinking water. However, the misuse of this chemical poses a major threat to the environment and all living beings due to its tremendous toxicity. Fluorosis of the teeth and bones, acute stomach difficulties, DNA damage, kidney failure, Alzheimer's disease, infertility, thyroid trouble, cancer, and even brain damage are all serious health problems that can result from excessive exposure to fluoride (Lacson et al., 2021; Yadav et al., 2014).

3.2 Defluoridation Techniques

In the developing world, adsorption/ion-exchange and coagulation procedures continue to be the most popular fluoride removal techniques. Many nations have adopted bone char and the Nalgonda approach, or a mix of the two, at both the domestic and community levels. Adsorption, a very old technique, uses adsorbents such as clays, soils, calcium- and carbon-based materials (Getachew et al., 2015), synthetic compounds, and materials based on alumina or aluminum. It has also been investigated how to remove F^- ions from aqueous solutions using a range of zeolites, such as reversed and modified zeolites, as well as specific ion exchange resins made of cross-linked polystyrene. Although membrane use has effectively decreased the necessary levels, because of its great accessibility and minimal impact on the pocket, surface adsorption continues to be a crucial component of defluoridation research. That is the reason why researchers continue to hunt for alternative ways to enhance inexpensive adsorbents in hopes of improving their general efficacy, even as new strategies gain in popularity (Mohan et al., 2007), (Biswas et al., 2007).

3.2.1 Adsorption Technique

Adsorption is thought to be a desirable technique for removing fluorine from water. Adsorption is useful, easy to create and use, effective, and most importantly, affordable. In general, temperature, pH, sorbent/sorbate concentrations, sorption kinetics, coexisting ions, sorbent particle size, and adsorbent surface area regulate fluoride adsorption from water (Loganathan et al., 2013). Often, pH is discovered to be the essential component. The pH range between 4 and 9 is typically regarded as the range in which water loses the most fluoride. Several adsorbent substances, such as clay and layered double hydroxide, metal oxide and hydroxide, biosorbent, carbon-based adsorbent, and several others, have been tried over the years to find a defluoridating agent that is efficient and reasonably priced. Because they are affordable and generally available, biomass-based adsorbents are frequently used. Chemically treated activated carbon, such as peach stones treated with H_3PO_4, rice straw and water hyacinth treated with H_3PO_4, and cotton stems treated with KOH, have also been used as adsorbents to remove harmful dyes. For water defluoridation, activated carbons made from various biomasses have been tested. The adsorption capacity of activated carbons made from biomass, such as banana peels, coffee husks, moringa indica leaves, and coconut shells, has been found to be less than 0.5 mg/g. Despite reports of large surface area in some cases and the fact that the adsorption efficiency varies depending on a variety of parameters, it is often quite low. Only after intercalation with other metals such as Zr or Al is an increase in adsorption capability observed.

Numerous adsorbent substances have been reported in the literature, such as activated alumina coated silica gel and alumina, calcite, shell carbon and activated coconut, fly ash, and sawdust. Other adsorbent substances include rice husk, groundnut shell serpentine, magnesia, tricalcium phosphate, activated soil sorbent, bone charcoal, defluoron1, defluoron-2, carbion (Vinati et al., 2015) (Table 3.2).

Table 3.2 Advantages and disadvantages related to the adsorption technique

Advantage	Disadvantage
High efficiency	Strong pH dependence and limited application
Cost effective	Adsorption is hampered by competing ions such as phosphate, nitrate, bicarbonates, etc
Availability of a wide range of adsorbents	Repetition of regeneration and decreased adsorbent effectiveness with subsequent regeneration
Simple operation	The adsorbent that has been used up could be deemed hazardous waste
Highly eco-friendly	The adsorbent leaks into the water systems, which might lower the quality of the water

3.2.1.1 Bone and Bone Charcoal

Smith and Smith were the ones who initially identified and reported on the fluoride-scavenging ability of bone (1937). But the high price of bone became a deterrent to its broader use. However, it quickly became clear that bone char, which is produced by carbonizing bone between 1100 and 1600 °C, had greater scavenging ability than raw bone. Subsequently, bone char took the role of bone as a defluoridizing agent. In essence, tricalcium phosphate and carbon make up bone char, which is created from burnt, pulverized animal bones that contain no biological substance. This substance was created to discolor cane syrups used to make sugar and was found to be more affordable than bone. One hypothesis is that, by means of the ion exchange process, fluoride ions bound to bone char where the phosphate in the bone char exchange with fluoride ions. The following is how fluoride is absorbed by hydroxyapatite, according to Fan et al., (2003):

$$Ca_{10}(PO_4)_6(OH)_2 + nF^- = Ca_{10}(PO_4)_6(OH)_{2-n}F_n + nOH^- \tag{3.1}$$

The first known defluoridation agent is bone charcoal, which was widely available commercially due to its extensive use in the sugar business and was in use in the US during the 1940s and 1960s.

One of the drawbacks of the bone char technique is that subsequent regenerations result in a reduction in capacity; as a result, the medium is highly expensive abandoned with high prices rather than refreshed. However, the bone char procedure has the ability at remove fluoride to incredibly low levels (Shirin & Yadav, 2021; Ayoob et al., 2008).

3.2.1.2 Clays and Soils

A thorough investigation on the adsorption of fluoride onto rocks and soils was published in 1967 by Bower and Hatcher. Since then, additional investigations have already been carried out employing a variety of resources.

Clay

Bower and Hatcher in 1967 asserted that OH^- ions are released concurrently with fluoride adsorption on rocks and soils. Furthermore, it was discovered that the Langmuir adsorption isotherm accurately describes concentration-dependent fluorine adsorption. The surface charge and the ion exchange in solution are highly affected by the structure of the clay (Puka, 2004). For negatively charged ions such as fluoride, the greater the more positive the surface. Given that pH has an impact on the charges at the edges of phyllosilicates and minerals with changeable charges, such as haematite, goethite, and gibbsite, adsorption capacity is highly pH dependent.

In general, charges are positive in an acidic atmosphere and negative in an alkaline one.

Soils

Numerous study teams have investigated fluoride adsorption on soils (Wang & Reardon, 2001). Examined how different exchangeable sodium concentrations affected the deposition of fluoride onto sodic soils. Fluoride adsorption was found to decrease with an increase in the proportion of soil exchangeable sodium at equilibrium fluoride concentration. Omueti and Jones investigated how Illinois sediments absorbed fluoride. They claimed that Freundlich and the Langmuir isotherms were used to characterize fluoride adsorption onto soils at low concentrations. Furthermore, it was proposed that the existence of amorphous aluminum hydroxides was the cause of fluoride adsorption on soils. Used earth samples from Ethiopia to study the defluorination of water. Studies have determined that the use from Ando soils of Kenya to defluoridate drinking water appeared to be a cost-effective and effective technique.

3.2.1.3 Nanosorbants for Defluoridation

Nanotechnology has become one of many promising methods for removing fluoride in recent years. Nanoparticles (NPs) have been demonstrated to be excellent fluoride adsorbents, because of some of their unique characteristics. Some of the properties include compact size, strong reactivity, ease of separation, high surface area, excellent catalytic activity, potential reactivity, and a large number of active sites for adsorption. Surface energies, free active oxidation states, and high adsorption ability of NPs are the results of the aforementioned properties (Dhillon & Kumar, 2015).

There has been several research on the development and use of NPs or nanoparticles for the removal of fluoride. The surfaces have been modified with functionalized polymers to add surface reaction. Because of its effective fluoride removal, the nanocomposite of polyaniline and iron (NC) (conducting polymer) produced using the chemical co-precipitation method has drawn the attention of researchers.

Metal oxide-based sorbents, such as those made of aluminum or iron oxide (Chai et al., 2013), are effective sorbents for fluoride uptake due to their cost-effectiveness, chemical uniformity, excellent magnetic separability, and good effects on the environment, having a limited pH range for application.

A member of the IVB group of phosphates, which also contains Ti, Zr (He & Chen, 2014), and Hf phosphate, Zr-phosphate (ZrP) exhibits outstanding water insolubility, chemical consistency, great resistance to strong acids, great resistance to organic reagents, and environmental friendliness (Wang et al., 2012). The Donnan membrane concept has frequently been used to create effective water-remediation adsorbents.

3.2.1.4 Calcium-Based Adsorbent

The strength of calcium's attraction to fluoride ions is well-recognized. As a result, significant studies on the removal of fluoride have been conducted using various calcium-based sorbents. Calcite has been used as a sorbent in studies on fluoride absorption from acid-treated water. Adsorbent treatment was performed prior to acid treatment to produce significant amounts of Ca ions for fluoride ion uptake. The precipitation and absorption techniques worked together to defluoridate the water. A 7 g/L adsorbent dose was used to remove the most fluoride.

3.2.1.5 Carbon-Based Defluoridation

When properly modified and handled, carbon materials can be used as effective adsorbents. In many nations, activated carbon is an expensive substance because it is made from costly, nonrenewable resources like coal. As a result, inexpensive carbon sources such as sugarcane bagasse and natural bamboo sawdust have been used as adsorbents for wastewater purification. In this manner, work on creating less expensive and renewable carbon adsorbents have continued (Pehlivan et al., 2013). Delonix regia pod carbon was recently used by Ajisha and Rajagopal to defluoridate effluent. Without applying any chemicals, the authors created Delonix regia capsules' activated carbon and showed that adsorption is improved at higher temperatures. Although the method has the drawback that it requires a lower pH to remove fluoride ions from polluted or wastewater (Ajisha & Rajagopal, 2015). When Ce-containing particles are dispersed in carbon material, the drawbacks of employing affordable carbon materials for defluoridation are eliminated.

Graphite

Graphene and its variants (Liu et al., 2016), with their large specific surface area of 2630 m^2/g have proven to be efficient wastewater adsorbents. Abe et al., (2004) reported the fluoride absorption capacity of several carbon-based adsorbents in the following order.

$$\begin{aligned} &bone\ char > coal\ charcoal > wood\ charcoal \\ &> carbon\ black > petroleum\ coke. \end{aligned}$$

Fishbone charcoal was studied for fluoride adsorption by Bhargava and Killedar (1992). Furthermore, chemically converted graphene (CCG) has been used to remove fluorine from water because it has several kinds of oxygen-containing groups on its carbon backbone (Li et al., 2011). However, little interaction between fluoride ions and CCG results in a limitation of limited fluoride adsorption capacity. Graphene with a large surface area and Al $(OH)_3$ with strong fluoride affinity were combined by Chen et al., (2013) to form a highly effective adsorbent.

Alumina-Impregnated Graphitic Carbon

The adsorption of fluoride on impregnated and unimpregnated alumina activated carbons was studied by Ramos. Stirring the solution of Al nitrate at a steady pH and calcination at 300 °C under N_2 gas produced these carbons which have 3–5 times higher adsorption capacity than the regular one. The adsorption of fluoride on coated carbon was affected by both the pH of the impregnating fluid and the calcination temperature. Increasing the level of calcination from 300 to 1000 °C decreased fluoride absorption.

Carbon Nanotubes

To investigate how carbon nanotubes might be used to adsorb fluoride, a group headed by Li et al. (2003) created aligned carbon nanotubes (ACNTs) through the ferrocene-catalyzed breakdown of xylene (Li et al., 2003). Adsorption isotherms for g-Al_2O_3, activated carbon, carbon nanotubes, and common soil produced under identical circumstances revealed that *carbon nanotubes* > *soil* > *g*-Al_2O_3 > *activated carbon* was the order of adsorption. Due to the outstanding one-dimensional (1D) nanostructure and enhanced physical, chemical, and electronic characteristics of carbon nanotubes (CNTs), researchers are keenly interested in the study (Gupta et al., 2009; Loganathan et al., 2013; Sathish et al., 2007). Furthermore, their typically high sorption capacity with a broad surface area is provided by their hollow and layered nanostructure. (Cruz et al., 2012). These CNTs' characteristics were used by Sankararamakrishnan et al. to create the CGCNT absorbent. The uptake capacity of CGCNTs was higher than that of oxidized CNTs (11.1 mg/g), at 20 mg/g. To stop using harsh chemicals in experiments to regenerate used CGCNTs, more research is still needed.

Alumina-Impregnated Carbon Nanotubes

The use of carbon nanotubes to aid in fluoride adsorption on alumina was also investigated by Li's team in 2001 and 2003. By pyrolysis of the mimic catalyst formed by combination of propylene-hydrogen and Ni particles, the nanotubes were created. Alumina that resembled a sponge and was sustained by carbon nanotubes was crushed and sieved to the proper particulate matter. Little amounts of NaOH or HNO_3 were used to alter the pH before shaking a 0.2 g/100 L solution of adsorbent for 12 h with NaF solutions. It was discovered how pH affected the adsorption isotherms. Al_2O_3/carbon nanotubes were found to have an adsorption capability that was > 13.5 times that of AIC-300 carbon, > 4 times that of g-Al_2O_3, and higher than that of polymeric resin IRA-410.

3.2.1.6 Alumina and Aluminum-Based Adsorbent

Researchers have long recognised the significance of sorbents based on alumina and aluminum used for defluoridation of water and wastewater (Vithanage & Bhattacharya, 2015). The formation of hydroxide in aqueous media by alkaline earth metal (IIA) oxide adsorbents (MgO and CaO) is known to raise the pH of the solution. Similar to this, because of their low quantity, adsorbents for precious metal such as rare earth metals oxide are quite costly. Consequently, for fluoride absorption in a broad pH range, Bayerite/Boehmite NCs are used so that the drawbacks of adsorbents' above-stated are solved.

Alumina

Alumina needs to be heated to be a useful adsorbent. This normally involves pyrolysis of $Al(OH)_3$, gibbsite, or other materials, either rapidly at a high temperature (flash calcination) or slowly to obtain a mostly crystalline product (Rozic et al., 2006). The final product, transitional alumina (between Al_2O_3 and gibbsite), was less crystalline compared to either alumina or gibbsite made by slow calcination. The less crystalline product has a surface area of approximately 200 m^2/g of alumina, making it the fastest and most efficient fluoride adsorbent.

The order of selectivity according to Johnston and Heijnen (2002) for anion adsorption on activated alumina in the pH range of 5.5–8.5 is as follows:

$$\begin{aligned} &OH- > H_2AsO_4^- > Si(OH)_3O^- > HSeO_3^- > F^- > SO_4^{2-} \\ &> CrO_4^{2-} > HCO_3^- > Cl^- > NO_3^- > Br^- > I^- \end{aligned}$$

Alumina Plus Iron Oxide

After aging and dehydrating, the combined hydroxide that results from the equimolar coprecipitation of iron (III) and aluminium hydroxides from a chloride mixture is a more effective fluoride adsorbent than either of the individual hydroxides (Biswas et al., 2007). The result that there was a new form of compound was supported by physical measurements of the materials indicating the independent nature of hydroxides from any type of bonding. Equivalent moles of Al and Fe (III) chloride by the addition of ammonia-precipitated adsorbent and then the capacity of that adsorbent was investigated by Chubar et al. (2005). Different anions such as fluoride, chloride, bromide, and bromate were also examined. Investigations into the binding of anionic species of As (III) and As (V) have revealed similar findings. According to Dey et al. (2004), hydrous ferric oxides can remove fluoride from polluted water.

Alumina Plus Calcium Minerals

The Nalgonda technique is based on the adsorption of fluoride on flocs of aluminum hydroxide as they are generated in solution. The technique is named after the Indian town of Andhra Pradesh where it was developed; the fluoride-contaminated water is quickly combined with two chemicals, lime (CaO) and alum (aluminum sulfate or potassium aluminum sulphate). Formed aluminum hydroxides generate "cotton wool"-like flocs after mild stirring, which transport the majority of the dissolved fluoride and are eventually removed. Many nations, including Tanzania, Kenya, Kenya, Senegal, and India, have adopted the Nalgonda method. Dahi et al. have described straightforward, low-cost setups that are appropriate for households in third-world countries. Despite claims that the Nalgonda procedure is the best method for removing fluoride, detractors have pointed out various drawbacks.

Red Mud

Due to its availability and affordability, red mud, which mostly contains iron oxides and aluminum oxides, has received substantial research as an adsorbent for anionic pollutants, mostly As^{3+} and As^{5+}. Red mud was cleaned, acid-activated, and dried to create an adsorbent that could prevent the requirement for general water acidification (Cengeloglu et al., 2002) to be defluoridated.

Natural Polymers

Natural polymers such as chitin and chitosan with potential functional groups are essential components of any bioadsorbent used to remove different kinds of aquatic contaminants.

Chitin is a polymer that can be isolated from crustacean shells and contains hydroxyl functional groups and amino acids. Chitosan, which is often accessible as shellfish-processed waste. According to reports, the adsorption of F^- using chitosan from synthetic water samples occurred extremely quickly.

3.2.2 Precipitation and Coagulation Technique

The first tests to create a method to eliminate F^- from water were conducted in the early 1930s. Scientists had been working hard at the time to develop a sustainable and economically viable technique to lower the levels of F^- ion concentration in water (Kut et al., 2016). Water turbidity can be reduced by using chemical coagulants to remove a variety of contaminants, such as colloidal particles and dissolved organic substances (El-Gohary et al., 2010). Coagulation is a chemical procedure that destabilizes tiny particles in water by using specialized chemicals called coagulants. The most common coagulants are trivalent atoms, such as iron and aluminum. This procedure is used for the removal of F^- ion from water. F^- can be removed from water using a variety of compounds, including alum, Na silicate, zeolites, Na aluminate, Cao, silica gel, $Al_2O_3 \cdot H_2O$, and $FeCl_3$. None of the materials listed above, with the exception of lime and alum, has been found to be particularly useful. F^- can be eliminated from water by using both lime and alum at the same time. However, the removal of fluoride is negatively affected by the existence of other organic and inorganic compounds as coagulant. Essentially, two processes take place: the assimilation of impurities into noncrystalline metal OH^- precipitate, also known as "sweep flocculation," and the charge neutralization of negatively charged colloids by cationic hydrolysis products. Numerous variables, particularly pH and coagulant dose, affect the relative significance of these two mechanisms (Gregory & Duan, 2001). When cations from solution, charge neutralization is a straightforward process for destabilizing negatively charged particles. Positively charged species may bind to negatively charged surfaces simply through electrostatic attraction or through the creation of specific surface complexes. Small amounts of hydrolyzing coagulants are generally responsible for the destabilization of colloids and these results in the neutralization of the charge particle. A net positive charge due to charge reversal brought about by significant quantities of coagulants leads to restabilization of the particles. Even micromolar concentrations of the coagulant can cause destabilization and restabilization of particles. Therefore, careful dosage management of the coagulant is necessary for the best destabilization by charge neutralization. In contrast to the theory of charge neutralization, higher coagulant doses show more impact on coagulation. This is due to sweep flocculation and rapid and extensive hydroxide precipitation, both of which occur at greater coagulant dosages.

Nevertheless, it is evident that impurity-assisted particles are entangled in the developing noncrystalline precipitate of OH^- and can be easily eliminated from water by the sedimentation process. Exact processes of sweep flocculation are not entirely understood. As a result, "sweeping away" particles from water using an OH^- precipitate is termed "sweep flocculation". This method often results in better particle removal than when particles are simply destabilized by charge neutralisation. Because of the higher solid concentration in sweep flocculation, the rate of aggregation is greatly enhanced. It is also likely that the ability of precipitated hydroxides to attach to the particles may result in stronger aggregates. Hence, both charge neutralization and sweep flocculation are combined in process of coagulation by hydrolyzing coagulants (Gregory & Duan, 2001). There are two types of coagulation-based defluoridation processes:

1. Fluoride precipitation caused by an appropriate reagent and a chemical reaction.
2. Fluoride coprecipitation, which entails its simultaneous precipitation from the same solution with a macro component.

3.2.2.1 Precipitation of Fluoride

Lime

The cheapest and most popular defluoridation method is the addition of lime ($Ca(OH)_2$) or other calcium salts ($CaSO_4$, $CaCl_2$), which precipitates fluoride as insoluble CaF_2 (Ksp = 3 × 10 − 11 at 25 °C). This method is used mainly for wastewaters with high concentrations of fluorides.

The following describes the lime precipitation process:

$$Ca(OH)_2 + 2F^- \rightarrow CaF_2 \downarrow + 2OH^- \quad (3.2)$$

The limit of fluoride solubility in the CaF_2 system should be affected by precipitation, according to theory. pH and residual Ca^{2+} concentration affect total soluble fluoride concentration. If $Ca(OH)_2$ is used as the source of lime, liming has a significant drawback in that the pH will increase with calcium dose, as stated in Eq. 3.2. As a result, it is crucial to use the right amount of lime to keep the pH of drinking water within acceptable ranges (Reardon & Wang, 2000).

Magnesium Oxide

In addition to lime, magnesium oxide has also been used to remove F^- ions from drinking water. When magnesium oxide is added to water containing fluoride, the magnesium oxide is converted to magnesium hydroxide as follows:

$$MgO + H_2O \rightarrow Mg(OH)_2 \quad (3.3)$$

When magnesium hydroxide and fluoride ions mix in reaction (Eq. 3.3), nearly insoluble magnesium fluoride is created as

$$Mg(OH)_2 + 2F^- \rightarrow MgF_2 \downarrow + 2OH^- \quad (3.4)$$

Based on this procedure, Rao and Mamatha (2004) created an easy-to-use DDU in India. The addition of small quantities (0.15–0.20 g/L) of $NaHSO_4$ raises the pH of the treated water samples using magnesium oxide to 10–11, which is then corrected in appropriate ranges (6.5–8.5).

Calcium and Phosphate Compound

According to Eqs. 3.5–3.8, Precipitation of fluoride from drinking water is theoretically possible using the substances monosodium phosphate ($NaH_2PO_4 \cdot H_2O$) and calcium chloride ($CaCl_2 \cdot 2H_2O$) as fluorapatite and/or calcium fluoride (Fig. 3.1).

(a) (b)

Fig. 3.1 Comparison of **a** aluminum precipitates and **b** aluminum precipitates with F. Adopted by Lawler et al. (2016)

$$CaCl_2 \cdot 2H_2O = Ca^{2+} + 2Cl^- + 2H_2O \tag{3.5}$$

$$NaH_2PO_4 \cdot H_2O = PO_3^{-4} + Na^+ + 2H^+ + H_2O \tag{3.6}$$

$$Ca^{2+} + 2F^- = CaF_2 \tag{3.7}$$

$$10Ca^{2+} + 6PO^{4-} + 2F^- = Ca_{10}(PO_4)_6F_2 \tag{3.8}$$

3.2.2.2 Coprecipitation of Fluoride

Alum

Several studies have looked at the ability to defluoridate water. The chemistry of the alum's elimination of fluoride is highly complex, despite the fact that it is frequently utilised in the purification of water. Al^{3+} is hydrated to create $Al(H_2O)_6{}^{3+}$ when enough aluminum salts (Al) are in contact with water to exceed the limit of solubility of the metal hydroxide. This aquometal ion then goes through additional hydrolysis to form hydroxo complexes that are mononuclear, dinuclear, and even polynuclear like $Al_7(OH)_{17}{}^{4+}$, $Al_8(OH)_{20}{}^{4+}$, $Al_{13}(OH)_{34}{}^{5+}$, and $Al_6(OH)_{15}{}^{3+}$, eventually inducing the hydroxide of metal $Al(OH)_3$. Although a number of polymeric species put forth, their presence has not proven conclusively established; as a result, they may be viewed as a transitional form in the transformation salts of soluble metals into crystalline precipitates. According to researchers, Using hydrolytic processes can be expressed as follows:

$$Al^{3+} + H_2O \leftrightarrow Al(OH)^{2+} + H^+ \quad (3.9)$$

$$Al^{3+} + 2H_2O \leftrightarrow Al(OH)^{2+} + 2H^+ \quad (3.10)$$

$$7Al^{3+} + 17H_2O \leftrightarrow Al_7(OH)_{17}^{4+} + 17H^+ \quad (3.11)$$

$$Al(H_2O)_3 + 6H_2O \leftrightarrow Al(H_2O)_5(OH)_2 + H_3O^+ \quad (3.12)$$

$$Al^{3+} + 3H_2O \leftrightarrow Al(OH)_3(s) + 3H^+ \quad (3.13)$$

$$Al^{3+} + 4H_2O \leftrightarrow Al(OH)^4 + 4H^+ \quad (3.14)$$

These equations (Aksu & Gönen, 2004) make it clear where the acidity of the metal ions comes from. The amount of alum added will therefore affect the pH of the entire system, as will the alkalinity (buffering ability). In addition to the basic aquometal ions, the hydroxometal complexes generated are easily adsorbed at interfaces, and this adsorption is what causes colloids in salts of aluminum-treated water to become unstable by neutralization of charges. Heavy molecular mass polymers are capable of simultaneously adhering to two or more particles and "polymer bridging," or joining them.

Fluoride ions were eliminated from solutions in the coprecipitation process using Al salts by being entangled on the gelatinous $Al(OH)_3$ flocs. The most recent scientific studies in this area indicate that the process by which fluoride is removed from water when an aluminum salt is added may involve either adsorption/ligand exchange or by the coprecipitation of hydroxide ions and fluorine with aluminium ions to generate a precipitate with the structural formula $Al_nF_m(OH)_{3n-m}$ (Eqs. 3.15 and 3.16). (Hu et al., 2003). The data of Mekonen et al. also support the elimination of fluoride through the bulk solution phase's adsorption complexation or adsorption with $Al(OH)_3(s)$ (2001).

$$nAl_3(aq) + (3n - m)OH^-(aq) + mF^- \rightarrow Al_nF_m(OH)_{3n-m}(s) \quad (3.15)$$

$$Al_n(OH)_{3n}(s) + mF^-(aq) \rightarrow Al_nF_m(OH)_{3n-m}(s) + mOH^+(aq) \quad (3.16)$$

Due to the large dose needs, the high pH of the potable water, problems with sludge disposal, and the simultaneous discharge of silica, alum treatment is not economically practical and cannot be viewed or advised as a method of defluoridation method.

Lime and Alum (Nalgonda Technique)

Alum may remove F from water in two different ways: (a) by adhering to the surface of an aluminum precipitate and (b) by incorporating F into the structure of an aluminium precipitate (Lawler et al., 2016). The Indian National Environmental Engineering Research Institute (NEERI) developed the Nalgonda defluoridation process in 1975. The Nalgonda process is a common domestic procedure carried out in communities near Nalgonda where fluorosis is rampant (Andhra Pradesh, India). Although its origins in India, the Nalgonda method is now widely used throughout Asia. Coagulation/flocculation, rapid mixing, disinfection, filtration sedimentation, and concentration of sludge are the first six phases of the Nalgonda process. Alum is the main coagulant used to remove F^- from water, also known as hydrated aluminum sulfate $Al_2(SO_4)_3 \cdot 3H_2O$. When a saline environment for improved F^- elimination, lime is introduced. Although various methods are proposed, the Nalgonda method of purifying water to remove the F^- ion is still not fully understood (Malik et al., 2010). By applying Nalgonda procedures, the following reactions occur:

$$Al_2(SO_4)_3 \cdot 18H_2O \rightarrow 2Al^{3+} + 3SO_4^{2-} + 18H_2O \quad (3.17)$$

$$2Al^{3+} + 6H_2O \rightarrow 2Al(OH)_3 + 6H^+ \quad (3.18)$$

$$F^- + Al(OH)_3 \rightarrow Al - F\ \text{Complex} + \text{undefined product} \quad (3.19)$$

$$6Ca(OH)_2 + 12H^+ \rightarrow 6Ca^{2+} + 12H_2O \quad (3.20)$$

According to the equation, $Ca(OH)_2$ can react with F^- to produce CaF_2, which is insoluble in water (3.21).

$$Ca(OH)_2 + 2F^- \rightarrow CaF_2 + 2OH^- \quad (3.21)$$

The Nalgonda procedure cannot be used to treat water polluted with large amounts of F^- since its accuracy for F^- removal is less than 70%. Because alum is costly, using the Nalgonda technique is expensive. The water is cleaned of colloidal particles via coagulation in addition to F^- (Table 3.3).

3.2.3 *Membrane Techniques*

Modernization of water purification is necessary to expand access to pure water. Nanofiltration (NF), reverse osmosis (RO), and methods for electric-driven membranes, such as ED have all been developed as cutting-edge and suitable methods for defluorizing water.

Table 3.3 Advantages and disadvantages related to precipitation techniques

Advantage	Disadvantage
Widely used in both households and society	It removes some fluoride (18–33%) as a precipitate and transforms a significant amount into hazardous soluble aluminum fluoride ions
Less water waste and minimal disposal issues	Using aluminum sulfate significantly increases the concentration of sulphate ions, occasionally to levels over the permitted limit of 400 mg/L
Minimal energy needed	Maintenance is highly expensive
The taste and smell of the water don't alter	The procedure is not totally automated, so adequate care must be taken when adding chemicals and managing the process
Eliminate turbidity, bacteria, and organic contaminants in addition	Compared to other methods, it produces a lot of sludge
Produces water of consistent and acceptable quality	Diseases, as well as neurobehavioral and pathophysiological alterations in the human body, are caused by high levels of residual aluminum detected in treated water

3.2.3.1 Nanofiltration

A low-pressure technique is called nanofiltration NF to remove larger dissolved particles in descending order of nanometers. The procedure is regarded a middle step between RO and ultrafiltration. As NF membranes contain holes wider than RO ones, both solute and solvent ions may travel across them with little resistance. Table 3.4 shows the pros and drawbacks associated with NF (Elazhar et al., 2013). Compared to other membrane techniques for water treatment, such as reverse osmosis and ED, NF is an effective defluoridation method. The process is far more energy efficient, thanks to the technique's significantly lower operating pressure than that of other pressure-driven approaches. The membranes also have a number of benefits over conventional purification methods, organic contaminants, high flux and high multivalent anion salt retention, and comparatively inexpensive operating and maintenance expenses. In a comparison investigation on the commercial NF membranes efficacy of the NF9 and NF5, at fluoride removal by Nasr et al. (2013). The authors have also looked at how fluoride ions are rejected when there are other ions present. Groundwater experiments have revealed that the hydration of the ions affects their ability to be retained. Due to their higher solvation energy, it was found that fluoride ions were more easily sustained than chloride ions (Nasr et al., 2013). Ceramic membranes exhibit efficient defluoridation, but their high cost prevents their practical use. High-temperature and non-fired clay mixtures are typically factors that affect the price of the ceramic membrane. Thermal treatment and uncooked clay combinations typically determine how much the ceramic membrane costs. The price is also increased by the technique of manufacture. Nandi et al. created affordable ceramic membranes by combining sodium silicate, kaolin, quartz, carbonates, and carbon (Nandi et al.,

Table 3.4 Advantages and disadvantages related to nanofiltration

Advantage	Disadvantage
Extremely effective elimination of fluoride	Membrane fouling at the nanoscale limits the membrane's ability to be reused
Efficient rejection of microorganisms, organic materials, and suspended solids	To achieve the desired separation, single-membrane separation is insufficient
A broad pH range is applicable	Membrane lifespan reduction and decreased chemical resistance
There is no demand for chemically consistent and high-quality water	Certain chemicals have not been completely eliminated

2008). Chakrabortty et al. (2013) used a module with a flat-layer cross-flow NF membrane to model and simulate the removal of fluoride and evaluate its economic feasibility. With the help of the membrane componant, a pure water flux of 158 L/m^2 h was successfully created, lowering the water fluoride content below the allowable barrier at just 14 kgf/cm^2 of trans-membrane pressures and 750 L/h of volumetric bend. The membrane mechanisms bring the high pH down to an allowable level. As a result, the membrane filtering system offered a practical and affordable method of cleaning up fluoride groundwater.

3.2.3.2 Reverse Osmosis

Using ionic exclusion and molecular sieving, the membrane-based reverse osmosis also removes up to 99% of divalent ions and 90–96% of monovalent ions. The process uses hydrostatic force to push water through a partially permeable barrier while maintaining salt residue (Meenakshi & Maheshwari, 2006). The parameters of raw water, maintenance, temperature, pressure, etc., affect the technique's efficiency. It runs at a higher pressure and rejects more dissolved solids. Recently, the World Health Organization (WHO) and one of the best processes for treating water, according to the US Environmental Protection Agency (USEPA), is RO.

3.2.3.3 Dialysis and Electrodialysis

Although a membrane to keep solutes in place, the principle of dialysis is the isolation of substances by means of passage through a membrane. Using DC voltage, ions are electrochemically separated using ion-exchange membranes (Piddennavar, 2013). Ionic elements from the source water are transferred through the cathode (charged particles) and anode under the action of a driving force in order to pay attention to the polluted water and produce more diluted and cleaned water (negatively charged species).

3.2.4 Ion-Exchange Techniques

Ion-exchange materials used in the ion-exchange process to treat water. This method removes the main ionic pollutant (fluoride) from water while softening the water (Chaudhary & Prasad, 2015). Calcium DOWEX G-26(4) can be used for fluoride-specific purification (resin for cation exchange in strong acids). The resin form's Ca (II) ions precipitate with fluoride obtained from water. Furthermore, fluoride may be removed using quaternary ammonium moiety in extremely basic anion exchange resin when equilibrium is reached in following order:

$$\text{Matrix} - R_3N^+ - Cl^- + F^- \leftrightarrows \text{Matrix} - R_3N^+ - F^- + Cl^- \qquad (3.22)$$

Polymeric materials have a number of benefits, including being simple to handle, being widely accepted by users, producing high-quality water, and being quick and easy to renew. In order to remove harmful ions from water, polymeric ion exchange resins have been applied frequently on a modest scale. It has been discovered that, when some resins are carefully chosen, compared to those for anion exchange, cation exchange resins (Ku et al., 2002) are more discriminating. Unfortunately, the chosen resin has a reduced selectivity, and its performance is additionally hampered by the high sodium salt concentration. Consequently, it has always been possible to create polymeric resins for the selective removal of fluoride from groundwater sources (Li et al., 2011).

3.2.5 Electrochemical Techniques

3.2.5.1 Electrocoagulation

A filtering method called electrocoagulation (EC) removes suspended particles from water to subm levels (for example, F). In the electrolytic process known as electrocoagulation, metallic cations are synthesized when an electric potential is supplied by an additional source of power, sacrificial anodes (Kobya et al., 2016). The use of electrocoagulation has grown over the last ten years. Recently, electrocoagulation has been suggested to be a reliable method for removing F^- from groundwater. Using electrocoagulation, a variety of potentially hazardous pollutants including pigment, oil, F^- and toxic metals may be successfully removed. While defluoridating water, electrocoagulation retains valuable components already present in raw water without creating secondary contaminants. According to Sinha et al. (2012), electrocoagulation using aluminum electrodes and 230 V DC was very effective in simultaneously eliminating both F and aluminium. In drinking water F^- removal was also achieved by increasing the retention duration. The most crucial factor in EC is charge loading (electron density that is a passage of electricity transmitted through a solution); this regulates the EC response rate, which affects the coagulation speed. They discovered

Table 3.5 Advantages and disadvantages related to electrocoagulation

Advantage	Disadvantage
That is a productive method	Aluminum can leak into water when electrocoagulation is used
With this method, a variety of pollutants can be eliminated	Powerful electrical output is required
Secondary pollutants are not produced by electrocoagulation	
Using this method prevents the removal of useful ions	
This approach results in high F^- elimination effectiveness (greater than 80%)	
This method can help eradicate certain microorganisms	

that there is no linear link between F-depletion performance and charge loading. Although it is the most important factor in EC on its own, charge loading is not thought to be a crucial factor in the elimination of F^-. The amount of F^- present in treated wastewater is initially reduced by an increase in charge loading, but a crucial feature is that F-concentration has decreased, but not significantly. In comparison to the active alumina process and Nalgonda technology, the electrocoagulation method may release less aluminum from water (Table 3.5).

3.2.5.2 Electrosorption

Lately, it has proven possible to increase the sorption capacity of conventional systems as well as to treat a variety of polluted fluids (Lounici et al., 2004). The performance of the adsorbent was greatly improved by the electric field-based activation method of Lounici et al. for alumina. In every instance, this method proved to be more effective than the standard activation method. Although the performance of traditional alumina was not significantly impacted by the ionic strength, the electro activated adsorbent's capacity to absorb fluoride increased as the ionic strength increased to around 55%. When choosing and evaluating an adsorbent's economic value for a given application, regeneration is a key factor. Any novel method that may lower the cost of the regeneration process will help increase the effectiveness and allure of column bed adsorption (Lounici et al., 2001). However, in this instance, it is discovered that electrode sorption approaches are superior. The electrical field production promotes the regeneration of alumina by increasing the mobility of the hydroxide ions or making the pores' active sites more accessible. Without any loss in sorption capacity, three successive regenerations could be performed. In addition, NaOH, a cleaning chemical, was used with 90% less waste than with traditional regeneration methods. The electrosorption system achieved 95% of the adsorption efficiency was restored. Furthermore, compared to existing regeneration methods, water was substantially less needed, to about 6% of the treated water volume to replenish the saturated bed.

3.2.6 Capacitive Deionization (CDI)

The last ten years have seen a lot of interest in the CDI brackish water desalination method because it offers promising benefits such as cost effectiveness, environmental friendliness, chemical-free operation, and slow pressure.

In general, commercially available activated carbon is used to material made of porous carbon with many particular surfaces for CDI electrodes. However, it is crucial to note that significant efforts are being made to provide superior large surface area carbon electrodes with features such as extreme ion removal capacity and low transport in a range of MOPS-derived carbon electrodes, hierarchical porous carbon architectures (Zhang et al., 2013), and graphene-based electrode materials.

There are no potentially irreversible chemical reactions, only electrostatic interactions, in CD processes. In CD cells, activated carbon may experience a remarkable cycle between charging and discharging. Although, this is true provided that the electrode potentials do not reach a certain point when a certain voltage is applied to the cells at which the system's stability is threatened (water-splitting carbon oxidation etc.)

3.3 Comparison of Different Techniques for F Removal

The following conclusions on fluoride removal techniques were drawn from the review of the literature:

- Table 3.2 shows that although adsorption has been shown to be promising from a financial, operational, and technical point of view, disposing of the sludge created by this phenomenon is a significant issue.
- The coagulation/precipitation method can only remove 33% of F^- and produces a large amount of residue with a lot of remaining aluminum.
- The ion-exchange process is quite efficient in removing fluoride, but it is costly and results in water with a high chloride level.
- Although nanofiltration techniques and reverse osmosis are effective in removing fluoride to a greater extent, they are also quite costly and remove some vital ions from water.
- If the aforementioned issues are taken into account, the electrocoagulation approach is suitable for removing fluoride; however, this process requires constant power supply.

- Last but not least, it would be important to remember that various Asian nations have different concentrations of fluoride and other ion competitors in their groundwater and that each strategy has drawbacks (Table 3.6).

3.4 Conclusions

Due to urbanisation and the rising global population, F^- pollution in water has recently increased substantially. The human body's teeth, bones and skeleton contain about 96% of the total F^-, which could result in excessive F^- levels that could lead to a number of health problems, including dyspnea, salivation, sweating, anorexia, stiffness, muscle weakness, severe gastroenteritis restlessness, and tachycardia. To purge fluoride from drinking water, several procedures had been devised. Three main approaches to defluoridation can be grouped together: (a) utilization of alternate water sources; (b) improvement of one's diet; and (c) defluoridating water. According to a survey of the literature, a number of techniques, including precipitation/coagulation, RO, electrocoagulation, and nanofiltration, have been proposed. Ion-exchange procedures and adsorption have been employed to remove F^- from water. The best course of action may be to defluoridate the water because an improved diet and alternative water sources are not always feasible for individuals living in Asia. Ion exchange and adsorption, coagulation, membrane techniques, and electrochemical techniques are important approaches. Many substances can effectively remove F^- during the precipitation/coagulation process, but they cannot increase the concentration of fluoride to the necessary levels. The compound that seems to be used the most frequently is alum. Furthermore, compared to the normal precipitation/coagulation process, electrocoagulation produces substantially less waste sludge. Compared to RO, nanofiltration requires less pressure, which reduces energy expenditures. Nevertheless, RO membranes are less costly than nanofiltration membranes.

On a large scale, extensively used methods include adsorption or ion exchange. Precipitation methods are less successful than sorption methods in lowering fluoride concentration, which is often considerably below acceptable levels. Only activated alumina and bone char were confirmed to be efficient at the operational level, despite the fact that numerous adsorbents with exceptionally high potential had been discovered as indicated in this article. Because pH controls the entire process chemistry, it is crucial to the sorption process and the most significant element. Membrane separations consume less energy since they do not need additives and may be carried out under isothermal circumstances at low temperatures. However, they cost a little to install and are maintain and run and vulnerable to membrane breakdown, scaling, and fouling. Electrochemical processes have high installation and maintenance costs. High costs of technology may be a major obstacle to adoption, particularly in underdeveloped nations. It seems that the social and environmental acceptability of the techniques also merits consideration. Hence, the "most suited" alternative need not

Table 3.6 An overview of the defluoridation techniques

Defluoridation technique	State of strengths/development	Limitations
Precipitation calcium oxide	Coagulation is the most affordable, well-known, and often utilised technique	High levels of sludge are produced, and treated water has a high pH. Precipitation had poor settling properties. The fluoride content in wastewater is high
Precipitation by magnesium oxide	Existing technologies Reasonable price	The amount of sludge produced and the pH of the treated water are both high
Precipitation calcium chloride and monosodium phosphate	New technology with consistent operating costs. No health risks associated with chemical misuse or overdose, unlike traditional precipitation methods	Too little contact time allows contaminants to leak into treated water more often. Calcium phosphates may precipitate in the upper portions of the filter bed as a result of prolonged contact time. The removal efficiency will be reduced by both of these methods
Coprecipitation by alum	Reputable and proven technique. Often used in places with a high population of F^-	The pH of the treated water is too low. Higher fluoride concentrations demand high dosage requirements. Sulfate and Al concentrations in treated water were expected, especially at high pH levels
Coprecipitation by alum and lime (Nalgonda)	Well-known and often used technology for small-scale pilot projects at the community and home levels. Easily accessible chemicals. Simple to use and maintain	Controlling doses for various raw water sources with varied concentrations of alkalinity and fluoride can be challenging. The treated water has significant levels of hardness, pH, and residual aluminum
Adsorption/ion exchange	Long-standing method for use in the community	Give water a taste Low levels of social acceptance
Bone and Bone Char	Well-known and reputable technology. Excellent potential. Local applications benefit from local availability and processing resources. Removing fluorine to extremely low levels	With repeated regenerations, capacity decreases dramatically. Higher in cost than in coagulation methods. Religious beliefs in many countries and communities place restrictions on usage limited acceptability in society

(continued)

Table 3.6 (continued)

Defluoridation technique	State of strengths/development	Limitations
Clays	Economical. Very few local applications	The likelihood of defluoridation is typically minimal. It is quite tough to regenerate
Activated alumina	Really well-established method. One of the greatest technologies today in use. pH 5 yields the best results. Little disruption from counterions of constant potential. Several uses	Expensive compared to bone char and coagulation procedures. The potential is reduced by high pH. Regeneration causes a material decrease of 5–10% and a capacity reduction of 30–40% when there is an increase in the concentration (> 0.2 mg/L)
Electro-coagulation	A new technique Comparing the EC method with the conventional coagulation process reveals how efficient it is	Other anion interference, such as sulfate. A requirement for routine sacrificial electrode replacement. Expensive due to high electric power usage
Electro-sorption	A new technique The absorbent capacity was increased by almost 50%. Great opportunity for regeneration	Expensive due to high electric power usage
Reverse osmosis	Developed and well-researched technology Several applications in the market. Prevalent in many wealthy nations. Little footprint. Salts and organics are also eliminated	Polarization phenomena are sensitive. possibilities for bacterial and mineral fouling The proper mineral balance may not be present in treated water. Inadequate water recovery. High price
Nanofiltration	The membrane separation that is widely used. Lower transmembrane pressures and larger water flows than RO	Greater sensitivity to pH and ionic strength than RO. Leaves significant amounts of retained fraction. Expensive method Need for skilled operators

(continued)

Table 3.6 (continued)

Defluoridation technique	State of strengths/development	Limitations
Electrodialysis	Good method for simultaneously desalination and defluoridation Commercially established. More affordable than RO More resistant to fouling	A significant level of pre-treatment is necessary. Low-molecular-mass noncharged chemical removal is ineffective. Membrane scaling. The quality of treated water is lower than that of RO
Donnan dialysis	Treatment of fluoride has recently attracted interest. However, the concentration gradient serves as the driving force in electro membrane processes. A permanent division between solutions that cannot be undone, even after the system is shut off from the outside world	A so-called driving counter-ion must be added to the stripping solution for operation. Decreased effectiveness in highly salinized waters. Expensive method

be the "best available technology." Ultimately, despite substantial progress in defluoridation research, finding a long-term, all-encompassing end of this dilemma is still out of reach.

Acknowledgements The Department of Chemistry of MNIT Jaipur provided infrastructural assistance for the work presented, and the University Grants Commission (UGC) awarded junior research fellowships. The authors are grateful to both organizations.

References

Abe, I., Iwasaki, S., Tokimoto, T., Kawasaki, N., Nakamura, T., & Tanada, S. (2004). Adsorption of fluoride ions onto carbonaceous materials. *Journal of Colloid Interface Sciience, 275*, 35–39.

Ajisha, M. A. T., & Rajagopal, K. (2015). Fluoride removal study using pyrolyzed Delonix regia pod, an unconventional adsorbent. *International Journal of Environmental Science and Technology, 12*, 223–236.

Aksu, Z., & Gönen, F. (2004). Biosorption of phenol by immobilized activated sludge in a continuous packed bed: Prediction of breakthrough curves. *Process Biochemistry, 39*, 599–613.

Ali, S., Thakur, S. K., Sarkar, A., & Shekhar, S. (2019). Worldwide contamination of water by fluoride. *Environmental Chemistry Letters, 14*, 291–315.

Ayoob, S., Gupta, A. K., & Bhat, V. T. (2008). A conceptual overview on sustainable technologies for the defluoridation of drinking water. *Journal of Environmental Science and Technology, 38*, 401–470.

Bhargava, D. S., & Killedar, D. J. (1992). Fluoride adsorption on fishbone charcoal through a moving media adsorber. *Water Research, 26*, 781–788.

Biswas, K., Saha, S. K., & Ghosh, U. C. (2007). Adsorption of fluoride from aqueous solution by a synthetic iron(III)–aluminum(III) mixed oxide. *Industrial Engineering Chemistry Research, 46*, 5346–5356.

Cengeloglu, Y., Kir, E., & Ersoz, M. (2002). Removal of fluoride from aqueous solution by using red mud. *Seperation and Purification Technology, 28*, 81–86.

Chai, L., Wang, Y., Zhao, N., Yang, W., & You, X. (2013). Sulfate-doped Fe_3O_4/Al_2O_3 nanoparticles as a novel adsorbent for fluoride removal from drinking water. *Water Research, 47*, 4040–4049.

Chakrabortty, S., Roy, M., & Pal, P. (2013). Removal of fluoride from contaminated groundwater by cross flow nanofiltration: Transport modeling and economic evaluation. *Desalination, 313*, 115–124.

Chaudhary, V., & Prasad, S. (2015). Rapid removal of fluoride from aqueous media using activated dolomite. *Analytical Methods, 7*, 8304–8314.

Chen, Y., Zhang, Q., Chen, L., Bai, H., & Li, L. (2013). Basic aluminum sulfate@graphene hydrogel composites: Preparation and application for removal of fluoride. *Journal of Materials Chemistry, 1*, 13101–13110.

Chubar, N. I., Samanidou, V. F., Kouts, V. S., Gallios, G. G., Kanibolotsky, V. A., Strelko, V. V., & Zhuravlev, I. Z. (2005). Adsorption of fluoride, chloride, bromide, and bromate ions on a novel ion exchanger. *Journal of Colloid and Interface Science, 291*, 67–74.

Cruz, E. F. D. L., Zheng, Y., Torres, E., Li, W., Song, W., & Burugapalli, K. (2012). Zeta potential of modified multi-walled carbon nanotubes in presence of poly (vinyl alcohol) hydrogel. *International Journal of Electrochemical Science, 7*, 3577–3590.

Dey, S., Goswami, S., & Ghosh, U. C. (2004). Hydrous ferric oxides: A scavenger of fluoride from contaminated water. *Water Air and Soil Pollution, 158*, 311–323.

Dhillon, A., & Kumar, D. (2015). Nanocomposite for the detoxification of drinking water: Effective and efficient removal of fluoride and bactericidal activity. *New Journal of Chemistry, 39*, 9143–9154.

Dissanayake, C. B. (1991). The fluoride problem in the groundwater of Sri Lanka: Environmental management and health. *International Journal of Environmental Studies, 38*, 137–155.

El-Gohary, F., Tawfik, A., & Mahmoud, U. (2010). Comparative study between chemical coagulation/precipitation (C/P) versus coagulation/dissolved air flotation (D/DAF) for pre-treatment of personal care products (PCPs) wastewater. *Desalination, 252*, 106–112.

Elazhar, F., Tahaikt, M., Zouahri, A., Taky, M., Hafsi, M., & Elmidaoui, A. (2013). Defluoridation of moroccan groundwater by nanofiltration and electrodialysis: Performances and cost comparison. *Journal of World Applied Science, 22*, 844–850.

Fan, X., Parker, D. J., & Smith, M. D. (2003). Adsorption kinetics of fluoride on low-cost materials. *Journal of Water Research, 37*, 4929–4937.

Getachew, T., Hussen, A., & Rao, V. (2015). Defluoridation of water by activated carbon prepared from banana (Musa paradisiaca) peel and coffee (Coffea arabica) husk. *International Journal of Environmental Science and Technology, 12*, 1857–1866.

Gregory, J., & Duan, J. (2001). Hydrolyzing metal salts as coagulants. *Pure and Applied Chemistry, 73*, 2017–2026.

Gupta, A. K., Deva, D., Sharma, A., & Verma, N. (2009). Adsorptive removal of fluoride by micro nanohierarchal web of activated carbon fibers. *Industrial & Engineering Chemistry Research, 48*, 9697–9707.

Hayat, E., & Baba, A. (2017). Quality of groundwater resources in Afghanistan. *Environmental Monitoring Assessment, 189*, 318.

He, J., & Chen, J. P. (2014). A zirconium-based nanoparticle: Essential factors for sustainable application in treatment of fluoride containing water. *Journal of Colloid Interface Science, 416*, 227–234.

Hu, C. Y., Lo, S. L., & Kuan, W. H. (2003). Effects of co-existing anions on fluoride removal in electrocoagulation (EC) process using aluminium electrodes. *Water Research, 37*, 4513–4523.

Johnston, R., & Heijnen, H. (2002). *Safe water technology for arsenic removal*. Report. World Health Organization (WHO).

Kobya, M., Demirbas, E., & Ulu, F. (2016). Evaluation of operating parameters with respect to charge loading on the removal efficiency of arsenic from potable water by electrocoagulation. *Journal of Environmental Chemical Engineering, 4*, 1484–1494.

Ku, Y., Chiou, H. M., & Wang, W. (2002). The removal of fluoride ion from aqueous solution by a cation synthetic resin. *Seperation Science and Technology, 37*, 89–103.

Kut, K. M. K., Sarswat, A., Srivastava, A., Pittman, C. U., & Mohan, D. (2016). *Groundwater for Sustainable Development, 2–3*, 190–212.

Lacson, C. F. Z., Lu, M. C., & Huang, Y. H. (2021). Fluoride containing water: A global perspective and a pursuit to sustainable water defluoridation management-an overview. *Journal of Cleaner Production, 280*, 124236.

Lawler, D., Katz, L., Gee, I., Yeo, S., & Herrboldt, J. (2016). *Sub-project A2: Simultaneous removal of inorganic contaminants, DBP precursors, and particles in alum and ferric coagulation.* University of Texas at Austin.

Li, W., Cao, C. Y., Wu, L. Y., Ge, M. F., & Song, W. G. (2011). Superb fluoride and arsenic removal performance of highly ordered mesoporous aluminas. *Journal of Hazardous Material, 198*, 143–150.

Li, Y. H., Wang, S. G., Zhang, X. F., Wei, J. Q., Xu, C. L., Luan, Z. K., & Wu, D. H. (2003). Removal of fluoride from water by carbon nanotube supported alumina. *Environmental Technology, 24*, 391–398.

Liu, L., Cui, Z., Ma, Q., Cui, W., & Zhang, X. (2016). One-step synthesis of magnetic iron-aluminum oxide/graphene oxide nanoparticles as a selective adsorbent for fluoride removal from aqueous solution. *RSC Advances, 6*, 10783–10791.

Loganathan, P., Vigneswaran, S., Kandasamy, J., & Naidu, R. (2013). Defluoridation of drinking water using adsorption processes. *Journal of Hazardous Materials, 248–249*, 1–19.

López-Guzmán, M., Alarcón-Herrera, M., Irigoyen-Campuzano, J., Torres-Castañón, L., & Reynoso-Cuevas, L. (2019). Simultaneous removal of fluoride and arsenic from well water by electrocoagulation. *Science of the Total Environmental, 678*, 181–187.

Lounici, H., Adour, L., Belhocine, D., Elmidaoui, A., Bariou, B., & Mameri, N. (2001). Novel technique to regenerate activated alumina bed saturated by fluoride ions. *Journal of Chemical Engineering, 81*, 153–160.

Lounici, H., Belhocine, D., Grib, H., Drouiche, M., Pauss, A., & Mameri, N. (2004). Fluoride removal with electro-activated alumina. *Desalination, 161*, 287–293.

Malik, A. H., Nasreen, S., Mahmood, Q., Khan, Z. M., Sarwar, R., Jilani, G., & Khan, A. (2010). Strategies for low-cost water defluoridation of drinking water—A review of progress. *Journal of Chemical Society of Pakisthan, 32*, 550–558.

Meenakshi, & Maheshwari, R. C. (2006). Fluoride in drinking water and its removal. *Journal of Hazardous Material, 137*, 456–463.

Mohan, S. V., Ramanaiah, S., Rajkumar, B., & Sarma, P. (2007). Removal of fluoride from aqueous phase by biosorption onto algal biosorbent Spirogyra s IO2: Sorption mechanism elucidation. *Journal of Hazardous Material, 141*, 465–474.

Nandi, B. K., Uppaluri, R., & Purkait, M. K. (2008). Preparation and characterization of low-cost ceramic membranes for micro-filtration applications. *Applied Clay Science, 42*, 102–110.

Nasr, A. B., Charcosset, C., Amar, R. B., & Walh, K. (2013). Defluoridation of water by nanofiltration. *Journal of Fluorine Chemistry, 150*, 92–97.

Onipe, T., Edokpayi, J. N., & Odiyo, J. O. (2020). A review on the potential sources and health implications of fluoride in groundwater of Sub-Saharan Africa. *Journal of Environmental Science and Health, 55*, 1078–1093.

Pehlivan, E., Tran, H. T., Wki, O., Schmidt, C., Zachmann, D., & Bahadir, M. (2013). Sugarcane bagasse treated with hydrous ferric oxide as a potential adsorbent for the removal of As(V) from aqueous solutions. *Food Chemistry, 138*, 133–138.

Piddennavar, R. (2013). Review on defluoridation techniques of water. *International Journal of Engineering Science, 2*, 86–94.

Poonia, T., Singh, N., & Garg, M. (2021). Contamination of arsenic, chromium and fluoride in the Indian groundwater: A review, meta-analysis and cancer risk assessment. *International Journal of Environmental Science and Technology, 18*, 2891–2902.

Puka, L. R. (2004). *Fluoride adsorption modelling and the characterization of clays for defluoridation of natural waters*. MSc dissertation, Faculty of Science, Rand Afrikaans University.

Rao, M. S., & Mamatha, P. (2004). Water quality in sustainable water management. *Current Science, 87*, 942–947.

Reardon, J. E., & Wang, Y. (2000). A limestone reactor for fluoride removal from wastewaters. *Environmental Science and Technology, 34*, 3247–3253.

Reddy, A. G. S., Reddy, D. V., Kumar, M. S., & Naik, P. K. (2016). Evaluation of fluoride enrichment processes in groundwater of Chimakurthy granitic pluton complex in Prakasam District India. *African Journal of Environmental Science and Technollogy, 10*, 350–379.

Rozic, L., Novakovic, T., Petrovic, S., Cupic, Z., Grbavcic, Z., & Rosic, A. (2006). The sorption and crystallographic characteristics of alumina activated in a reactor for pneumatic transport. *Journal of Serbian Chemical Society, 71*, 1237–1246.

Sathish, R. S., Raju, N., Raju, G., Nageswara Rao, G., Kumar, K. A., & Janardhana, C. (2007). Equilibrium and kinetic studies for fluoride adsorption from water on zirconium impregnated coconut shell carbon. *Journal of Seperation Science and Technology, 42*, 769–788.

Shirin, S., Jamal, A., Emmanouil, C., Singh, V. P., & Yadav, A. K. (2022). Assessment and characterization of waste material used as backfilling in an abandoned mine. *International Journal of Coal Preparation and Utilization*, 1–9. https://doi.org/10.1080/19392699.2022.2118259

Shirin, S., & Yadav, A. K. (2014). Physico chemical analysis of municipal wastewater discharge in Ganga River, Haridwar district of Uttarakhand, India. *Current World Environment, 9*, 536–543. https://doi.org/10.12944/CWE.9.2.39

Shirin, S., & Yadav, A. K. (2021). Assessment and suitability for the dumping of flyash in an abandoned mine. *Environmental Challenges, 4*, 100111. https://doi.org/10.1016/j.envc.2021.100111

Sinha, R., Khazanchi, I., & Mathur, S. (2012). Fluoride removal by a continuous flow electrocoagulation reactor from groundwater of Shivdaspura. *International Journal of Engineering and Research and Application, 2*, 1336–1341.

Vinati, A., Mahanty, B., & Behera, S. (2015). Clay and clay minerals for fluoride removal from water: A state-of-the-art review. *Journal of Applied Clay Science, 114*, 340–348.

Vithanage, M., & Bhattachary, P. (2015). Fluoride in the environment: Sources, distribution and defluoridation. *Environmental Chemical Letters, 13*, 131–147.

Wang, L., Wu, X. L., Xu, W. H., Huang, X. J., Liu, J. H., & Xu, A. W. (2012). Stable Organic–inorganic hybrid of polyaniline/a–zirconium phosphate for efficient removal of organic pollutants in water environment. *ACS Applied Materials & Interfaces, 4*, 2686–2692.

Wang, Y., & Reardon, E. J. (2001). Activation and regeneration of a soil sorbent for defluoridation of drinking water. *Journal of Applied Geochemistry, 16*, 531–539.

WHO. (2017). *Guidelines for drinking-water quality* (4th ed.). World Health Organization.

WHO, G. (2011). Guidelines for drinking-water quality. *World Health Organization, 216*, 303–304.

Yadav, A. K., Sahoo, S. K., Mahapatra, S., Kumar, A. V., Pandey, G., Lenka, P., & Tripathi, R. M. (2014). Concentrations of uranium in drinking water and cumulative, age-dependent radiation doses in four districts of Uttar Pradesh, India. *Toxicological and Environmental Chemistry, 96*, 192–200. https://doi.org/10.1080/02772248.2014.934247

Zhang, Q., Du, Q., Jiao, T., Zhang, Z., Wang, S., Sun, Q., & Gao, F. (2013). Rationally designed porous polystyrene encapsulated zirconium phosphate nanocomposite for highly efficient fluoride uptake in waters. *Scientific Reports, 3*, 2551–2559.

Zhang, Z., Wang, J. J., Ali, A., & DeLaune, R. D. (2016). Heavy metal distribution and water quality characterization of water bodies in Louisiana's Lake Pontchartrain Basin, USA. *Environmental Monitoring Assessment, 188*, 1–14.

Chapter 4
Fluoride Removal from Aqueous Solution Using Iron-Based Materials: Preparation, Characterization, and Applications

Divya Patel, Mridu Kulwant, Saba Shirin, Ramita Varshney, Govind Pandey, and Akhilesh Kumar Yadav

Abstract Fluoride is a common element found in many minerals and rocks, and its widespread distribution and high concentration in groundwater have raised serious concerns. Adsorption is one of the most efficient approaches that have been suggested for removing it from aquatic environments. Iron-based materials, such as nanoscale zerovalent iron (nZVI), modified nZVI, spinel ferrites, Fe-based metal–organic frameworks, and their composites, have recently gained a lot of attention because of their even greater fluoride removal efficiencies from an aqueous solution compared

D. Patel (✉) · M. Kulwant
Department of Environmental Studies, The Maharaja Sayajirao University of Baroda, Vadodara 390002, India
e-mail: pateldivya9030@gmail.com

M. Kulwant
e-mail: mridu.k-envphd@msubaroda.ac.in

D. Patel
National Sugar Institute, Kalyanpur, Kanpur 208017, India

S. Shirin (✉)
Department of Environmental Science, School of Vocational Studies and Applied Sciences, Gautam Buddha University, Greater Noida 201312, India
e-mail: sabashirin83@gmail.com

S. Shirin · R. Varshney · A. K. Yadav
Department of Mining Engineering, Indian Institute of Technology (Banaras Hindu University), Varanasi 221005, India
e-mail: ramita25j@gmail.com

G. Pandey
Department of Civil Engineering, Madan Mohan Malaviya University of Technology, Gorakhpur 273010, India

Rajkiya Engineering College, Gonda 271002, India

A. K. Yadav et al. (eds.), *Advanced Treatment Technologies for Fluoride Removal in Water*, Water Science and Technology Library 125,
https://doi.org/10.1007/978-3-031-38845-3_4

to those from other adsorbents. The most popular iron-based substances used in fluoride elimination are described in this chapter. Their methods of preparation and characterization were categorised and presented in groups. The authors also explained how these materials could be used to effectively remove fluoride from water. For a comprehensive understanding of the fluoride removal mechanisms of fluoride, commonly used adsorption isotherms, kinetic models, and thermodynamic analysis are completely described. Subsequently, research on regeneration and recycling was addressing these issues. According to the results, fluoride may be effectively removed from water using materials with an iron base. However, several current obstacles need to be addressed further to enable their actual application in fluoride removal.

Keywords Fluoride · Iron based-adsorbents · Adsorption · Water · Removal mechanism

4.1 Introduction

In countries around the world, the spread of pollutants into surface and groundwater has become a significant problem. Water sources are polluted by a variety of contaminants, including organic, inorganic, heavy metals, microbiological and radioactive species. These contaminants might be in the form of suspended, dissolved, or dispersed elements. Fluoride, one of the main water contaminants, causes a number of illnesses in both humans and animals. It also has an adverse impact on insects. Figure 4.1 outlined the main effects of fluoride poisoning (Zuo et al., 2018). Additionally, fishes are negatively affected by the direct release of fluoride-loaded waste into water bodies. In many regions of India, fish is one of the most consumed foods on a daily basis. Not just in India, but also in many other nations throughout the world, fish is consumed in large quantities on a daily basis. Fluoride, which accumulates in fish bodies, enters the human system through the consumption of fluoridated fish (Ghosh and Ghosh, 2019).

Numerous investigations have revealed substantial fluoride levels in groundwater, including India (Ali et al., 2016, 2019; Patel et al., 2022; Yadav et al., 2014). One of the nation's most badly affected by fluoride on the globe is Tanzania, which is home to some of the most fluoridated rift valley regions (He et al., 2020). One of the

G. Pandey
e-mail: pandey_govind@rediffmail.com

A. K. Yadav (✉)
Environmental Science and Engineering Department, Indian Institute of Technology Bombay, Mumbai 400076, India
e-mail: yadavbasti@gmail.com

Department of Environmental Engineering and Management, Chaoyang University of Technology, Taichung 413310, Taiwan

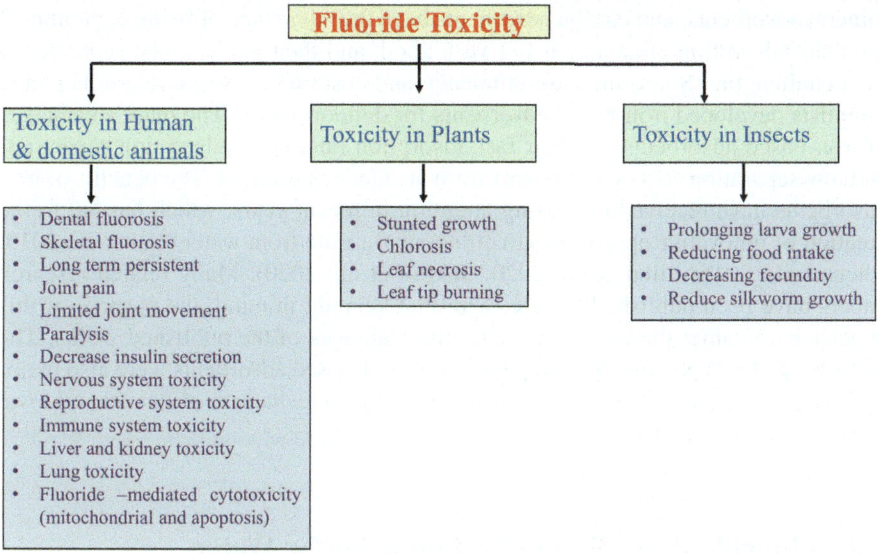

Fig. 4.1 Showing the toxicity effect of excess fluoride

key geogenic causes of fluoride enrichment in water is the weathering of fluoride-wearing minerals (fluorite, cryolite, topaz, amphibole, sellaite, etc.). Fluoride in water is also triggered by anthropogenic factors, such as brick ovens, mining, and the use of pesticides (Mukherjee & Singh, 2018). Excessive fluoride consumption can result in skeletal fluorosis and dental fluorosis. As a result, various national and international organizations have established fluoride levels in drinking water that are acceptable within certain limits. The permissible limits for fluoride in drinking water have been established at 1.5 mg/L by the World Health Organization (WHO) and 1 mg/L by the Bureau of Indian Standards (BIS), respectively (WHO, 2011). For the purpose of eliminating fluoride from water, researchers and scientists have developed a variety of methods, including precipitation/coagulation, membrane filtration, ion exchange, electrochemical treatments, and adsorption (Hegde et al., 2020). Although most of these technologies demonstrated good removal efficiency, several barriers, such as high personnel costs, high capital costs, the need to produce secondary products, and other operational difficulties, preclude their widespread use (Mukherjee & Halder, 2018). The previous study found that the adsorption technique is the most efficient approach among all to remove fluoride from water due to its wide range of cost- and environmentally-friendly approaches (He et al., 2020). The adsorbent is the fundamental component of the adsorption process, and its ability to remove fluoride is easily impacted by external and internal conditions including temperature, adsorbent pore size, coexisting ions, pH, etc.

Biosorbents, mineral adsorbents, zeolites, carbon adsorbents, and metal-based adsorbents are some prominent fluoride adsorbents (Bonyadi et al., 2019; Gai et al., 2021; Gao et al., 2021; Sadhu et al., 2021). Among them, biological adsorbents, mineral adsorbents, and carbon adsorbents have the advantage of being economical, but their adsorption efficiency is not very good, and their ability to be separated is a bit challenging. So, to increase efficiency and cost effectiveness, researchers and scientists developed iron-based adsorbents for defluoridation. The main advantages of iron-based adsorbents are their fast adsorption rate, large adsorption capacities, and low separation cost of adsorbents from an aqueous solution. The benefits of iron adsorbents have received increasing attention in recent years, which has led to the creation of innovative approaches to eliminate fluoride from water (Cai et al., 2015; Chen et al., 2018; Pillai et al., 2020; Takmil et al., 2020). Many related research papers have been published to date; with that keeping in mind, the purpose of this chapter is to gather the advantages and disadvantages of the published works. The history, synthesis process, and preparation of iron-based adsorbents were also major topics of this chapter. This chapter also covered the mechanism of fluoride removal by magnetic adsorbents.

4.2 Fluoride Remediation Technologies for Water

If the water fluoride concentration is continuously and visibly higher than the permitted limit, consider remedial fluoride treatment methods. The various categories of defluoridation procedures can be broadly categorized (Fig. 4.2).

Membrane filtration: From an industrial perspective, the membrane separation technique is more well known for sea water desalination, wastewater treatment, and groundwater defluoridation (Chufa et al., 2021). Particles are isolated in a membrane separation procedure on the assumptions about their molecular size and form using an extraordinarily well-designed semipermeable membrane. A thin polymeric film that can be porous or not and is often formed of ceramic, metal, or even liquid is the semi-permeable membrane. Reverse osmosis, nanofiltration, Donnan-dialysis, and electrodialysis are the four most popular membrane separation techniques to remove fluoride from water (Arfin & Waghmare, 2015).

Ion-exchange: Ion exchange is widely used in both municipal and industrial water treatment systems. Compared to alternative treatment options, the technique offers a number of benefits. It provides low maintenance costs, high flow rate of treated water, and is ecologically sound. Ion exchange also has several disadvantages, such as chlorine contamination, bacterial contamination, organic matter adsorption, iron fouling, calcium sulfate fouling, and organic resin contamination (Ahmad et al., 2022).

Precipitation: A widely used technique for treating water with high fluoride concentrations is precipitation. Lime and calcium chloride are used to precipitate calcium fluoride down to its solubility limit. After that, the fluoride is further reduced by coagulation using aluminum to reach the low discharge limit. Calcium has been used as a cheap source of precipitating fluoride for the treatment of drinking water

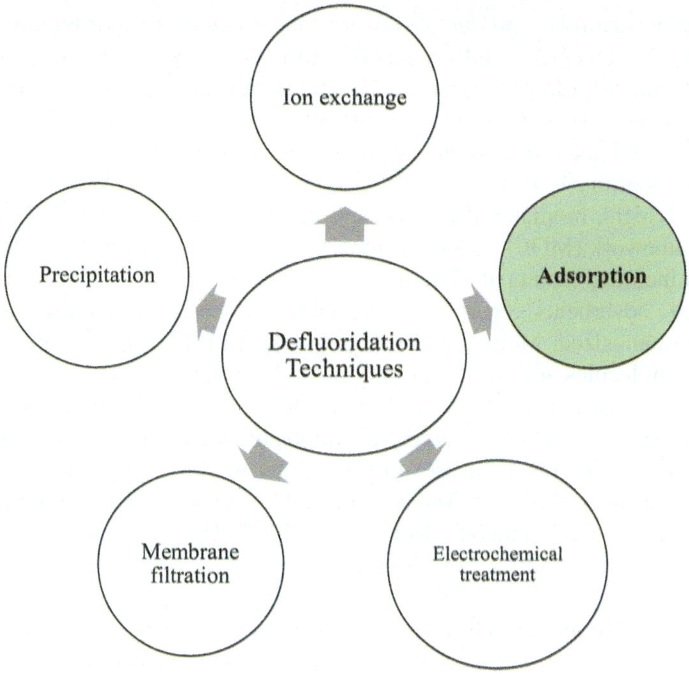

Fig. 4.2 Depicting different remediation techniques of defluoridation

for a very long time. This technique has recently been used to remediate wastewater that has high levels of fluoride by adding calcium-containing minerals. Traditional chemical precipitation, however, generates a significant amount of sludge with a high-water content (between 95 and 99). As a result, management of the financial, environmental and post-disposal aspects has been difficult due to the bulk volume (Shirin & Yadav, 2021; Lacson et al., 2021).

Electrochemical treatment: Another method of treating water is electrochemical, which has been shown to be effective in removing fluoride while requiring only moderate amounts of energy. This process uses sacrificial Al or Fe electrodes to create trivalent metal species, water electrolysis at the cathode to produce hydrogen and hydroxide ions, and electromigration of the various charged species (Lacson et al., 2021).

4.3 History and Types of Magnetic Adsorbents

The Federal Environmental Protection Agency standardized the maximum levels of contamination and the anthropogenic origin of known heavy metals (Khan et al., 2020). One such method for cleaning polluted water at a cheaper cost and without the production of harmful by-products is adsorption. Magnetic composites have received a lot of interest recently as adsorbents due to their reusability, which makes them sustainable and affordable (Sharma et al., 2022). For the treatment of contaminants including metals, dyes, and pharmaceuticals in solutions, many adsorbents have been created and put to use. Magnetic adsorbents have excellent potential to remove different contaminants from wastewater (Mehta et al., 2015). Various substances, such as polymers, biomaterials, clays, and magnetic composites based on a metal organic framework (MOF), as well as their use in wastewater treatment procedures, have been included (Liu et al., 2022).

Magnetic adsorbents generally have a large surface area since they are usually micro- and nanosized. Modifications to the surface of the solid phase, such as the binding of molecules and functional groups to the adsorbent, may increase the efficacy of the adsorptive process (da Silva et al., 2021). Many substances, such as polymers, biomaterials, clays and magnetic composites based on metal organic frameworks (MOF), as well as their uses in wastewater treatment procedures, have been covered (Sharma et al., 2022; Yadav et al., 2022). Oxana V. Kharissova reviewed micro and nanosized magnetic adsorbents in 2015. These materials are supported by organic (macromolecules, polysaccharides, polymers, and biomolecules) or inorganic (carbon, graphene, silica, and zeolites) components. Heavy metal removal, the separation, oxidation, and adsorption of oil, dyes, hazardous organic compounds, various biomolecules, pharmaceuticals, and a variety of catalytic activities are some of the primary uses for magnetic adsorbents (Kharissova et al., 2015). Spark discharge, low-temperature reaction, pulsed laser ablation, photothermal, spray pyrolysis, inert gas condensation, ion sputtering, thermal plasma, pulsed laser ablation, and flame spray pyrolysis are widely used to create these nanoparticles. Several methods are used to characterize nanoparticles, including scanning electron microscopy, X-ray photoelectron spectroscopy, transmission electron microscopy, X-ray diffraction, and energy-dispersive X-ray spectroscopy (Khan et al., 2020).

Moreover, comprehensive studies of metal oxide-based nanocomposites have produced promising results with improved metal-removal capabilities. For example, the Al_2O_3–Fe_3O_4 nanocomposite was identified as the optimal nanosized metal oxide for the removal of cadmium based on its adsorption capacity (Prasanna et al., 2019). The adsorption process was further investigated using theoretical calculations and adsorption fitting models, which showed that the connections between the adsorbent and adsorbate are mostly based on secondary noncovalent contacts and hydrophobic interactions (Hao et al., 2021). A cost-effective and environmentally friendly magnetic core material was produced using the industrial waste coal-fly-ash magnetic sphere (CMS). An extrusion-dripping technique was used to create a composite bioadsorbent called CMS@CS, which has chitosan (CS) as the shell and CMS as the core. According to structural analyzes, the resulting CMS@CS samples are either porous solid microspheres or hollow microspheres with a solid wall, depending on the preparation circumstances (Zhang et al., 2020).

4.4 Magnetic Adsorbent Preparation and Synthesis Method

The various synthesis techniques can be roughly divided into five groups (Fig. 4.3). These groups include using magnetic particles directly as adsorbents, attaching preprepared adsorbents and preprepared magnetic particles, synthesising magnetic particles on pre-prepared adsorbents, synthesising adsorbents on pre-prepared magnetic particles, and cosynthesizing adsorbents and magnetic particles (Baig et al., 2021). The seven factors that were compared were the properties of the materials, the initial concentration, the dosage of the adsorbent, the pH, the temperature, the time frame of the reaction, and the maximum adsorption capacity. Composite adsorbents are simple to modify and significantly more effective at removing fluoride because of the synergistic interaction of the many adsorbent types. Metal composite adsorbents can be created using simple coprecipitation, hydrothermal, or impregnation modification processes (Wei et al., 2022).

Coprecipitation, thermal disintegration, micro emulsion, and hydrothermal synthesis were shown to be the four most prevalent methods for creating magnetic materials in the realms of catalysis and medicine. Furthermore, the surface modification of magnetic particles was investigated on the type of coating materials (Phouthavong et al., 2022). One of several magnetic materials, magnetite (Fe_3O_4), is being extensively researched as a potential adsorbent, sensor, and for magnetic resonance imaging. In general, MNPs are made up of a magnetic core, a coating, and in some cases surface-active additives. This technique generates nanoparticles between 4 and 45 nm in size. Under specific synthesis conditions, the quality of magnetite nanoparticles is very repeatable (Roto, 2018). One such method for cleaning polluted

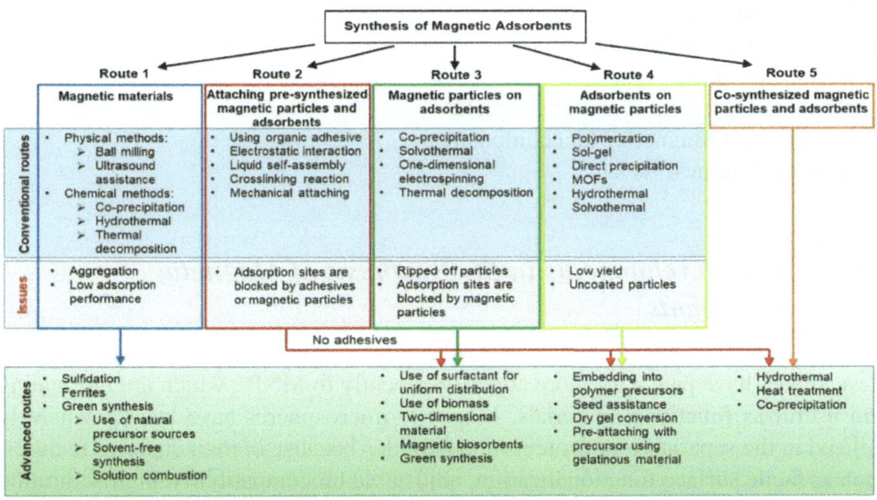

Fig. 4.3 Traditional techniques for making magnetic adsorbents

water at a cheaper cost and without the production of harmful by-products is adsorption. Magnetic compounds have received a lot of interest recently as adsorbents due to their reusability, making them sustainable and affordable (Deletic & Wang, 2019). According to a recent review, magnetic biosorbent production frequently uses iron (Fe), copper (Cu), titanium (Ti) and zinc (Zn) metal nanoparticles in the form of nitrates, carbonates, oxides, or sulfates, such as iron (III) oxide (Fe_3O_4), iron (III) chloride and zinc (Zn) oxide ($FeCl_3$) (Tee et al., 2022). The elimination of mixed pollutants is emphasized, as well as the links between adsorbent functional groups, pollutant properties, and adsorption mechanisms. In addition, the synthesis, modification, and applications of such an approach are discussed. All of these adsorbents have been studied with regard to their cost-effectiveness, potential regeneration, and environmental consequences. The stability, surface morphology, and adsorption capability of modified magnetic polysaccharide-based adsorbents were found to be very good (Shirin et al., 2019, 2021, 2022; Wang & You, 2021).

4.5 Improvements in Magnetic Particle Synthesis on Adsorbents

Advanced magnetic adsorbents are used to remove various organic and inorganic impurities from water according to their classified categories. On the basis of synthesis processes, adsorption capacities, and magnetic performances, some magnetic adsorbents are listed below.

- S-nZVI
- Ferrite (MFe_2O_4)
- Surfactant Modification
- Green Synthesis
- Uniform Distribution of Nanoparticles
- Bio-Derived Magnetic Nanocomposite
- Matrice-Confined NPs.

4.5.1 New Developments in the Synthesis of Magnetic Adsorbents

Researchers have paid a lot more attention recently to MNPs, which are fascinating and useful as functional materials. Magnetic microspheres have been extensively utilized in the separation of proteins and peptides because of their superior features, such as facile surface functionalization, adjustable biocompatibility, high-saturation magnetization, etc. (Wang et al., 2023). Polymers, as contrasted with low molecular weight (MW) compounds, are composed of a range of chain-like molecules of various molecular weights. MW of polymers are typically larger than 5000 g/mol, or on average more than 100 repeat units per molecule (Hagnauer, 1986). On the basis of the synthesis method and magnetic concerts, some roughly listed magnetic adsorbents (Phouthavong et al., 2022) are discussed below.

- Embedding into Polymer Precursors
- Application of Seed Crystals
- Pre-Mixing into Precursor Gel
- Pre-Attaching with Precursor Using Gelatinous Material
- Co-Synthesis of Magnetic Particles and Adsorbents
- Co-Synthesis via Hydrothermal Synthesis
- Co-Synthesis via Heat Treatment.

Through co-precipitation, co-synthesis Kharissova et al. (2015) published a study on micro- and nanosized magnetic adsorbents based on elemental metals, iron oxides, and ferrites, supported by inorganic materials (carbon, graphene, silica, zeolites) or organic ones (macromolecules, polysaccharides, polymers, and biomolecules) (Kharissova et al., 2015). Despite the fact that several techniques have been developed to create magnetic adsorbents with efficient adsorption performance, Phouthavong et al. (2022) reviews that concentrate on the synthesis techniques of magnetic adsorbents for wastewater treatment and their material structures have not been described. In order to remove contaminants from water streams, magnetic adsorbents have undergone recent improvements in their synthesis. This article gives a full description and discussion of these developments (Phouthavong et al., 2022).

The maximum adsorption capacities of modified biochars for Cr (VI) and AO7 were found to be 80.96 and 110.27 mg g^{-1}, respectively, according to Santhosh et al. (2020). After five cycles of adsorption/desorption, magnetic biochar showed a high efficacy in removing pollutants. The best alternative for treating water has been found to be carbon-based adsorbents because of their distinct physical and chemical characteristics. The pricey and labor-intensive catalytic reaction of activated carbon, even if it has a high capacity for sorption when employed as a commercial adsorbent, occasionally limits its application (Santhosh et al., 2020). Green synthesis of NPs uses a variety of biological substances, including plants, bacteria, algae, and fungi, but recent plant-based green synthesis of NPs has garnered more attention from researchers around the world. Numerous characterization techniques and green synthetic NPs are used in the treatment of water and wastewater due to their high efficacy and biocompatibility (Goutam et al., 2019). As a result of their benefits, magnetic nanocomposite adsorbents are intriguing candidates for possible applications in the domains of wastewater purification. This chapter used common examples to highlight current developments in magnetic nanocomposite adsorbents for wastewater purification (Liu et al., 2019). This chapter basically provides a comprehensive look at different types of adsorbents used for defluoridation, including those made from agricultural waste, carbon, industrial by—products, biopolymers, algae, fungi, nanoparticles, and nanocomposites. Also, the most common ways of desorbing and regenerating adsorbents and how well they work in aspects of their adsorption capacity have been summed up (Yadav & Jagadevan, 2021).

4.6 Regeneration and Recyclability of Magnetic Adsorbents and Their Performance

In evaluating an adsorbent's productivity, regenerability is a critical factor from an environmental, economic, and practical perspective. Regeneration is the rapid recycling or recovery of used adsorbents using procedures that are both economically and technically sustainable. Cost is one of the most important considerations while creating novel adsorbents. Many studies have concentrated on employing magnetic-based adsorbents to remove fluoride from aqueous solutions. However, the recycling, regeneration, or disposal of adsorbents after adsorption has gained little attention from studies (Dehghani et al., 2021; Kyzas et al., 2022). Thus, this aspect needs to be thoroughly researched in order to look at potential solutions for magnetic-based adsorbents. Therefore, research on the desorption, regeneration, and recovery of magnetic-based adsorbents for fluoride removal requires an assessment to compile the findings. Therefore, it is essential to regenerate magnetic-based adsorbents in order to extract fluoride from the aqueous solution and improve adsorbent sustainability. However, it is crucial to consider the type and nature of the adsorbent and adsorbate, as well as the cost and processing circumstances, when choosing an appropriate technique for regenerating adsorbents. Therefore, the technique used to regenerate the adsorbent must be economical, easy to use, efficient, and allow the adsorbent to be reused in water treatment. The adsorbent may be successfully recovered for use in the adsorption process by applying a suitable desorption agent for regenerating a magnetic-based adsorbent. Additionally, this substance must be able to practically restore the physical and adsorption properties of adsorbents to their original conditions without being expensive, damaging the environment, or damaging magnetic-based adsorbents. Therefore, the desorption and regeneration of the used adsorbents is made easier by selecting an eluent with the right desorption characteristics. Table 4.1 lists the various desorption agents reported by different researchers.

Zhou et al. used bone-derived biochar to remove fluoride, and achieved it over the duration of four cycles with a significant amount of fluoride removal (Zhou et al., 2019). The regenerability and reusability of the adsorbent up to a certain number of cycles were the focus of some other investigations, which similarly revealed a

Table 4.1 List of some regenerative chemicals for fluoride recyclability

S. No	Desorbents	Conc./Molar/Normal	References
1	NaOH	1 N	Roy (2021)
2	NaOH	0.1 N	Tan et al. (2020)
3	$NaAlO_2$	2 mmol	Wang et al. (2017)
4	Methanol	95% pure	Massoudinejad et al. (2016)
5	Ethanol	Pure	Kyzas et al. (2022)

considerable amount of fluoride removal (Dehghani et al., 2020; Jing et al., 2020; Shirin & Yadav, 2014).

4.7 Application and Advantages of Magnetic Adsorbent Technologies in Defluoridation

Magnetic adsorbents have a substantial advantage over conventional adsorbents due to their capacity to be isolated from an aqueous solution by magnetic attraction (Mehta et al., 2015). Numerous pollutants, including organic, inorganic, heavy metals, microbial, and radioactive species, deteriorate sources of water. These pollutants can exist as particles that are suspended, dissolved, or dispersed particles. The only solution to the fluoride problem in drinking water is defluoridation. Defluoridation processes, according to Senapati et al., include membrane separation, coagulation and precipitation, and adsorption and ion exchange (Senapati et al., 2018). Although it relies on the adsorbent, including its adsorption capability, contact time, dose, change in pH when added to water, and other considerations, adsorption is one of the best ways to remove fluoride from drinking water. Effective water treatment is now possible due to recent developments in nanotechnology. Carbon nanotubes, nanoscale metal oxides, nanofibers, and other nanoparticle-based products are used for water filtration (Senapati et al., 2018). An adsorbent encompasses the benefits of magnetic nanoparticles and hydrous aluminum oxide floc with magnetic separability and high affinity for fluoride, offering distinctive advantages such as simple preparation, a large capacity for adsorption, and simple separation from sample solutions by using an external magnetic field, according to a team of researchers in 2010. They discovered that Fe(3)O(4)@Al(OH)(3) NPs with strong fluoride affinity could be useful adsorbents for the treatment of fluoride-contaminated water (Zhao et al., 2010).

Metal-based adsorbents combine the benefits of fast adsorption, a large capacity for adsorption, and strong selectivity in order to efficiently remove fluoride from water bodies and have the potential to accomplish environmental sustainability goals. The metal-based adsorbents covered in this article are those made of iron, aluminum, lanthanum, cerium, titanium, zirconium, and multimetal composite materials. It explores the potential and challenges associated with the development and use of metal-based adsorbents while focusing on adsorption conditions and removal efficiency capacities (Ni et al., 2022). Many biomass products, including chitosan, microbes, lignocellulose plant materials, animal attribute materials, biologically carbonized materials, and biomass-like organic materials similar to biomass, have been described and investigated, according to Huang et al. (2022). After studying the adsorption performance with the adsorbent processes for fluoride removal, it is discovered that carbonizing materials and altering adsorbents with metal ions are more beneficial to improving the adsorption efficiency, and definitely the views of these approaches are varied (Huang et al., 2022).

The technology to remove fluoride from the water sample was shared by Sapna Nehra and her colleagues; many adsorbents were made from natural raw materials. Less adsorption sites, limited adsorption performance, and metal-ion leaching are issues with the observed adsorbents. As a result, the Langmuir and Freundlich models successfully describe the adsorption and kinetics of the pseudo-second-order model. The maximum monolayer adsorption capabilities of LC-Ce and LC are 212 mg/g and 52.63 mg/g, respectively, according to the Langmuir model (Nehra et al., 2020). Adsorption is considered to be one of the most effective methods for the removal of fluoride. Table 4.2 lists some effective adsorbents for fluoride removal by various researchers. The main disadvantages of different adsorption techniques include limited adsorption capacities, extended contact times, high dose, and very low or high pH. Many surface-modified magnetic nanoparticles have been developed in recent years to enable efficient fluoride removal from water (Senapati et al., 2018).

4.8 Mechanism of Fluoride Adsorption

Fluoride adsorption is a complex procedure that involves numerous chemical and physical interactions between the adsorbent and fluoride (Alhassan et al., 2020). Removal of fluoride by adsorption typically employs the use of specific mechanisms, such as ligand exchange, ion exchange, electrostatic contact, and Lewis acid–base interaction, as shown in Fig. 4.4. (Huang et al., 2020; Wei et al., 2022).

Energy-dispersive X-ray spectroscopy (EDS), X-ray photoelectron spectroscopy (XPS), Fourier transform infrared spectroscopy (FTIR), zeta potential tests, etc. are instruments that can confirm these mechanisms. Metal oxide or hydroxide adsorbents go through a hydration reaction when subjected to an aqueous solution, which causes a coating of hydroxyl groups to develop on their surfaces. Fluoride and the hydroxyl group exchange ligands when one metal has a stronger affinity for fluoride than the other. A trimetallic (Fe–Al–Ce) oxide modified pectin-alginnism-based bioadsorbent for the removal of fluoride was described by Raghav and his colleagues. This approach successfully eliminated fluoride due to ion exchange, complexation, and electrostatic attraction (Raghav et al., 2019). To understand how fluoride is removed, Han et al. researched Magnetic Core–Shell Fe_3O_4@LaCe. They discovered that the primary processes were the ligand exchange action and the Lewis acid–base reaction between the fluoride and the adsorbent (Han et al., 2019) and plausible mechanism depicted in Fig. 4.5.

Table 4.2 List of some reported magnetic adsorbents and their defluoridation capacity

S. No.	Adsorbents	Qmax(mg/g)	pH	References
1	Hydrous aluminum oxide embedded with Fe_3O_4 nanoparticle (Fe_3O_4@$Al(OH)_3$ NPs)	88.48	6.5	Zhao et al. (2010)
2	Polypyrrole (PPy)/Fe_3O_4 magnetic nanocomposite	22.3	6.5	Bhaumik et al. (2011)
3	Magnetic Mg–Al-layered double hydroxides composite	47.7	7	
4	The calcination product of Mg–Al–Fe hydrotalcite compound	14	6	Ma et al. (2011)
5	A novel Mn–Ce oxide adsorbent			
6	Iron–aluminum oxide nanoparticles anchored on graphene oxide (IAO/GO)	64.72	6.5	Liu et al. (2016)
7	A novel magnetic composite of La–Zr	88.5	≈ 3	Chen et al. (2016)
8	Calcinated magnetic Fe_3O_4@Fe–Ti core–shell nanoparticle slurry	41.8	3	Zhang et al. (2016)
9	A novel lanthanum-loaded magnetic cationic hydrogel (MCH-La)	136.78	7	Dong and Wang (2016)
10	Magnetic alumina aerogel (MAA)	32.1	5	Yang et al. (2016)
11	A novel adsorbent of γ-AlOOH @CS (Pseudoboehmite and chitosan shell) magnetic nanoparticles (ACMN)	67.5	7	Wan et al. (2015)
12	A novel magnetite–chitosan composite	9.43	7	Mohseni-Bandpi et al. (2015)
13	Magnetic nickel/polypyrrole (Ni/PPy) nanostructures	67.7	7	Srivastava et al. (2016)
14	MgO–$MgFe_2O_4$/GO	34.2	6	Sahoo and Hota (2018)
15	Mixed-rare earth mixed rare-earth magnetic chitosan beads (MCLRB)	22.35	5	Liang et al. (2018)
16	Ce-Zn double-layer layered Ce–Zn double hydroxide cellulose (Fe_3O_4@LDH/poly) nanocomposite adsorbent	167.62	3	Ammavasi and Mariappan (2018)
17	Synthesized magnetic core–shell Ce–Ti@Fe_3O_4 nanoparticle	44.37	7	Abo Markeb et al. (2017)

(continued)

Table 4.2 (continued)

S. No.	Adsorbents	Qmax(mg/g)	pH	References
18 ara>	Magnetic iron oxide fabricated hydrotalcite/chitosan (Fe_3O_4@HTCS) composite	5.03	7	Pandi et al. (2017)
19	Amorphous Fe oxide	60.8	3.5–10.3	Zhang et al. (2017)
20	α-Fe_2O_3 and Fe_3O_4 dispersed on in a high surface area (663 m^2/g) Douglas fir biochar (BC)	7.81	3	Bombuwala Dewage et al. (2018)
21	A novel new magnetic bio adsorbent beads composed of Fe_3O_4, chitosan, chitosan and $Al(OH)_3$ (Fe_3O_4/CS/$Al(OH)_3$)	76.63	5	Hu et al. (2018)
22	Magnetic iron oxide encrusted oxide-encrusted hydrocalumite-chitosan (Fe_3O_4@HCCS) composite	6.62	3	Pandi et al. (2019)
23	Al/Fe (oxyhydr)oxide-coated magnetite (Mag@Al_2Fe)	26.5	6.8	Fu et al. (2020)
24	Carboxylated chitosan/Fe_3O_4	0.288	3	Mohammadi et al. (2019)
25	Magnetic carbon-based adsorbents derived from agricultural biomass (PAC-Fe_3O_4)	2.74	3	Dehghani et al. (2021)
26	Magnetic activated carbon from canola stalks and then Fe_3O_4 (ACCS-Fe_3O_4)	161.2	5	Kyzas et al. (2022)
27	La_2O_3–CeO_2–Fe_3O_4 composite nanofibers (LCF NFs)	216.45	3	Jian et al. (2022)

4.9 Conclusion and Recommendation

This chapter tried to cover a variety of magnetic adsorbents that had been used in the process of removing fluoride from aqueous solutions. This chapter evaluated the benefits and drawbacks of magnetic adsorbents for the removal of fluoride based on the basis of the reviewed literature. Because of their magnetic properties, magnetic adsorbents make it easy and affordable to separate adsorbents from the aqueous solution without filtration. Iron oxides can be blended with agricultural products and other reasonably priced and easily accessible for the simple removal of fluoride from water and wastewater. Chitosan and products derived from it were used to remove fluoride from the contaminated solution because they are safe for humans.

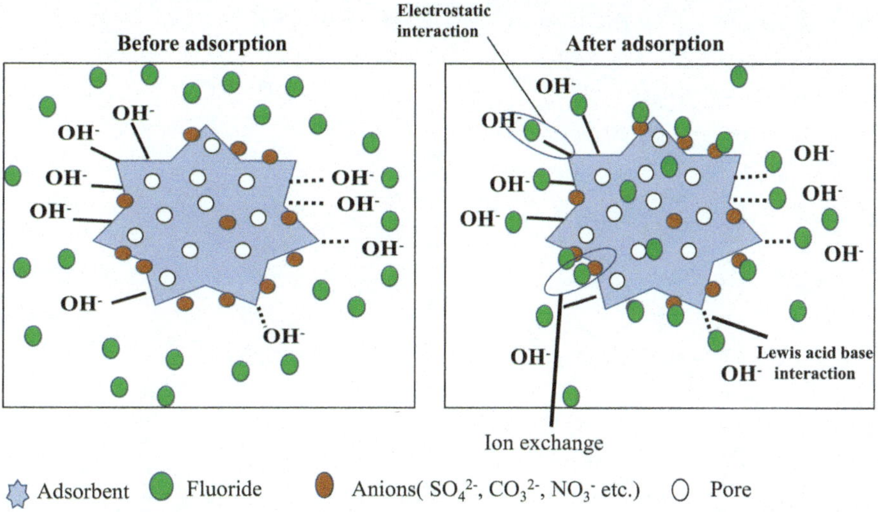

Fig. 4.4 Possible mechanism of fluoride adsorption on an adsorbent in an aqueous solution

In this chapter, we attempt to explain the mechanisms for fluoride removal mechanisms on the adsorbents. Although several magnetic adsorbents performed well in removing fluoride, additional research is required before the method can be applied on an industrial scale. Additionally, more extensive regeneration experiments were required to recover the adsorbent, improving the technique's economic viability. In the meantime, most of the adsorbents that shown strong fluoride uptake under batch conditions in the laboratory may have failed under field conditions. Therefore, researchers should focus more on fluoride removal in actual environments.

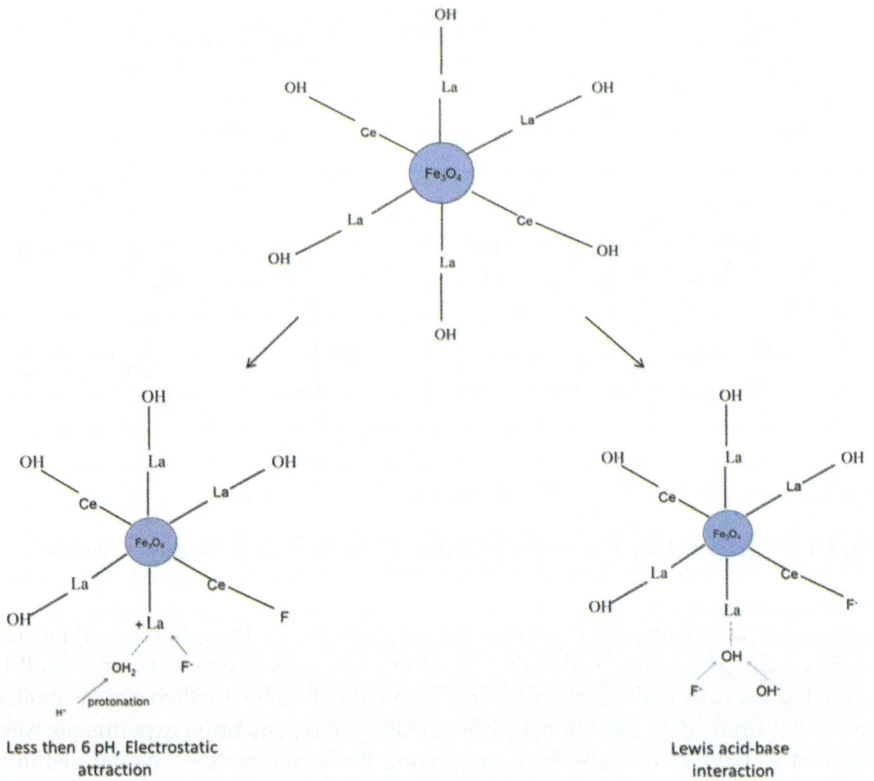

Fig. 4.5 Possible methods of fluoride removal from magnetic core–shell Fe_3O_4@LaCe at various initial solution pH

Acknowledgements The authors thank our prestigious institute and other people who are involved in this study for their direct and indirect contributions.

References

Abo Markeb, A., Alonso, A., Sánchez, A., & Font, X. (2017). Adsorption process of fluoride from drinking water with magnetic core-shell Ce–Ti@Fe_3O_4 and Ce–Ti oxide nanoparticles. *Science of the Total Environment, 598*, 949–958. https://doi.org/10.1016/j.scitotenv.2017.04.191

Ahmad, S., Singh, R., Arfin, T., & Neeti, K. (2022). Fluoride contamination, consequences and removal techniques in water: A review. *Environmental Science: Advances.*

Alhassan, S. I., He, Y., Huang, L., Wu, B., Yan, L., Deng, H., & Wang, H. (2020). A review on fluoride adsorption using modified bauxite: Surface modification and sorption mechanisms perspectives. *Journal of Environmental Chemical Engineering, 8*, 104532. https://doi.org/10.1016/j.jece.2020.104532

Ali, S., Fakhri, Y., Golbini, M., Thakur, S. K., Alinejad, A., Parseh, I., Shekhar, S., & Bhattacharya, P. (2019). Concentration of fluoride in groundwater of India: A systematic review, meta-analysis and risk assessment. *Groundwater for Sustainable Development, 9*, 100224. https://doi.org/10.1016/j.gsd.2019.100224

Ali, S., Thakur, S. K., Sarkar, A., & Shekhar, S. (2016). Worldwide contamination of water by fluoride. *Environmental Chemistry Letters, 14*, 291–315. https://doi.org/10.1007/s10311-016-0563-5

Ammavasi, N., & Mariappan, R. (2018). Enhanced removal of hazardous fluoride from drinking water by using a smart material: Magnetic iron oxide fabricated layered double hydroxide/cellulose composite. *Journal of Environmental Chemical Engineering, 6*, 5645–5654. https://doi.org/10.1016/j.jece.2018.08.071

Arfin, T., & Waghmare, S. (2015). Fluoride removal from water by various techniques: Review. *International Journal of Innovative Science, Engineering & Technology, 2*, 560–571.

Baig, N., Kammakakam, I., & Falath, W. (2021). Nanomaterials: A review of synthesis methods, properties, recent progress, and challenges. *Materials Advances, 2*, 1821–1871. https://doi.org/10.1039/D0MA00807A

Bhaumik, M., Leswifi, T. Y., Maity, A., Srinivasu, V. V., & Onyango, M. S. (2011). Removal of fluoride from aqueous solution by polypyrrole/Fe_3O_4 magnetic nanocomposite. *Journal of Hazardous Materials, 186*, 150–159. https://doi.org/10.1016/j.jhazmat.2010.10.098

Bombuwala Dewage, N., Liyanage, A. S., Pittman, C. U., Mohan, D., & Mlsna, T. (2018). Fast nitrate and fluoride adsorption and magnetic separation from water on α-Fe_2O_3 and Fe_3O_4 dispersed on Douglas fir biochar. *Bioresource Technology, 263*, 258–265. https://doi.org/10.1016/j.biortech.2018.05.001

Bonyadi, Z., Kumar, P. S., Foroutan, R., Kafaei, R., Arfaeinia, H., Farjadfard, S., & Ramavandi, B. (2019). Ultrasonic-assisted synthesis of Populus alba activated carbon for water defluorination: Application for real wastewater. *Korean Journal of Chemical Engineering, 36*, 1595–1603. https://doi.org/10.1007/s11814-019-0373-0

Cai, H. M., Chen, G. J., Peng, C. Y., Zhang, Z. Z., Dong, Y. Y., Shang, G. Z., Zhu, X. H., Gao, H. J., & Wan, X. C. (2015). Removal of fluoride from drinking water using tea waste loaded with al/fe oxides: A novel, safe and efficient biosorbent. *Applied Surface Science, 328*, 34.

Chen, P., Wang, T., Xiao, Y. W., Tian, E. X., Wang, W. L., Zhao, Y. L., Tian, L., Jiang, H. L., & Luo, X. B. (2018). Efficient fluoride removal from aqueous solution by synthetic Fe–Mg–La tri-metal nanocomposite and the analysis of its adsorption mechanism. *Journal of Alloys and Compound, 738*, 118.

Chen, P., Zhang, W., Li, M., Ai, P., Tian, L., & Jiang, H. (2016). Facile synthesis of magnetic La–Zr composite as high effective adsorbent for fluoride removal. *RSC Advances, 6*, 35859–35867. https://doi.org/10.1039/C5RA27929A

Chufa, B. M., Murthy, H. C. A., Gonfa, B. A., & Anshebo, T. Y. (2021). Carbon nanotubes: A review on green synthesis, growth mechanism and application as a membrane filter for fluoride remediation. *Green Chemistry Letters and Reviews, 14*, 647–664. https://doi.org/10.1080/17518253.2021.1991484

da Silva, T. L., da Costa, T. B., Neves, H. S. D. C., da Silva, M. G. C., & Vieira, M. G. A. (2021). Methods of synthesis of magnetic adsorbents. In L. Meili & G. L. Dotto (Eds.), *Advanced magnetic adsorbents for water treatment: Fundamentals and new perspectives* (pp. 25–58). Springer International Publishing.

Dehghani, M. H., Gholami, S., Karri, R. R., Lima, E. C., Mahvi, A. H., Nazmara, S., & Fazlzadeh, M. (2021). Process modeling, characterization, optimization, and mechanisms of fluoride adsorption using magnetic agro-based adsorbent. *Journal of Environmental Management, 286*, 112173. https://doi.org/10.1016/j.jenvman.2021.112173

Dehghani, M. H., Karri, R. R., Lima, E. C., Mahvi, A. H., Nazmara, S., Ghaedi, A. M., Fazlzadeh, M., & Gholami, S. (2020). Regression and mathematical modeling of fluoride ion adsorption from contaminated water using a magnetic versatile biomaterial and chelating agent: Insight on production and experimental approaches, mechanism and effects of potential interferers. *Journal of Molecular Liquids, 315*, 113653. https://doi.org/10.1016/j.molliq.2020.113653

Deletic, A., & Wang, H. (2019). Water pollution control for sustainable development. *Engineering, 5*, 839–840. https://doi.org/10.1016/j.eng.2019.07.013

Dong, S., & Wang, Y. (2016). Characterization and adsorption properties of a lanthanum-loaded magnetic cationic hydrogel composite for fluoride removal. *Water Research, 88*, 852–860. https://doi.org/10.1016/j.watres.2015.11.013

Fu, H., He, H., Usman, M., Chen, Q., Laipan, M., Yang, Y., Zhu, R., & Cai, L. (2020). Facile synthesis of Al/Fe bimetallic (oxyhydr)oxide-coated magnetite for efficient removal of fluoride from water. *Environmental Technology, 41*, 2625–2636. https://doi.org/10.1080/09593330.2019.1575919

Gai, W.-Z., Zhang, S.-H., Yang, Y., Zhang, X., & Deng, Z.-Y. (2021). Separation of excess fluoride from water using amorphous and crystalline AlOOH adsorbents. *ACS Omega, 6*, 16488–16497. https://doi.org/10.1021/acsomega.1c01620

Gao, Y., Li, M., Ru, Y., & Fu, J. (2021). Fluoride removal from water by using micron zirconia/zeolite molecular sieve: Characterization and mechanism. *Groundwater for Sustainable Development, 13*, 100567. https://doi.org/10.1016/j.gsd.2021.100567

Ghosh, S., & Ghosh, D. (2019). Impact of fluoride toxicity on freshwater fishes: A mini-review. *International Journal of Advance Research and Innovation, 6*, 13–18.

Goutam, S., Saxena, G., Roy, D., Yadav, A., & Bharagava, R. (2019). Green synthesis of nanoparticles and their applications in water and wastewater treatment. (pp. 349–379).

Hagnauer, G. L. (1986). Polymers and polymer precursor characterization. In J. W. McCauley & V. Weiss (Eds.), *Materials characterization for systems performance and reliability* (pp. 189–243). Springer US.

Han, M., Zhang, J., Hu, Y., & Han, R. (2019). Preparation of novel magnetic microspheres with the La and Ce-bimetal oxide shell for excellent adsorption of fluoride and phosphate from solution. *Journal of Chemical & Engineering Data, 64*, 3641–3651. https://doi.org/10.1021/acs.jced.9b00434

Hao, Y., Gao, Y., Gao, L., He, Y., Niu, Y., Hussain, S., Gao, R., Pfefferle, L. D., Shahid, M., & Wang, S. (2021). Amphiphilic core–shell magnetic adsorbents for efficient removal and detection of phthalate esters. *Chemical Engineering Journal, 423*, 129817. https://doi.org/10.1016/j.cej.2021.129817

He, J., Yang, Y., Wu, Z., Xie, C., Zhang, K., Kong, L., & Liu, J. (2020). Review of fluoride removal from water environment by adsorption. *Journal of Environmental Chemical Engineering, 8*, 104516. https://doi.org/10.1016/j.jece.2020.104516

Hegde, R. M., Rego, R. M., Potla, K. M., Kurkuri, M. D., & Kigga, M. (2020). Bio-inspired materials for defluoridation of water: A review. *Chemosphere, 253*, 126657. https://doi.org/10.1016/j.chemosphere.2020.126657

Hu, H., Yang, L., Lin, Z., Xiang, X., Jiang, X., & Hou, L. (2018). Preparation and characterization of novel magnetic Fe_3O_4/chitosan/AL(OH)3 beads and its adsorption for fluoride. *International Journal of Biological Macromolecules, 114*, 256–262. https://doi.org/10.1016/j.ijbiomac.2018.03.094

Huang, L., Luo, Z., Huang, X., Wang, Y., Yan, J., Liu, W., Guo, Y., Babu Arulmani, S. R., Shao, M., & Zhang, H. (2022). Applications of biomass-based materials to remove fluoride from wastewater: A review. *Chemosphere, 301*, 134679. https://doi.org/10.1016/j.chemosphere.2022.134679

Huang, L., Yang, Z., He, Y., Chai, L., Yang, W., Deng, H., Wang, H., Chen, Y., & Crittenden, J. (2020). Adsorption mechanism for removing different species of fluoride by designing of core-shell boehmite. *Journal of Hazardous Materials, 394*, 122555. https://doi.org/10.1016/j.jhazmat.2020.122555

Jian, S., Shi, F., Hu, R., Liu, Y., Chen, Y., Jiang, W., Yuan, X., Hu, J., Zhang, K., Jiang, S., Yang, W., & Duan, G. (2022). Electrospun magnetic La_2O_3–CeO_2–Fe_3O_4 composite nanofibers for removal of fluoride from aqueous solution. *Composites Communications, 33*, 101194. https://doi.org/10.1016/j.coco.2022.101194

Jing, F., Hui, B., Yudan, X., Rui, Z., Puli, Z., Duo, B., Zeng, D., Wei, L., & Xuebin, L. (2020). Recycling of iron and aluminum from drinking water treatment sludge for synthesis of a magnetic composite material (ALCS–Fe–Al) to remove fluoride from drinking water. *Groundwater for Sustainable Development, 11*, 100456. https://doi.org/10.1016/j.gsd.2020.100456

Khan, F. S. A., Mubarak, N. M., Khalid, M., Walvekar, R., Abdullah, E. C., Mazari, S. A., Nizamuddin, S., & Karri, R. R. (2020). Magnetic nanoadsorbents' potential route for heavy metals removal—A review. *Environmental Science and Pollution Research, 27*, 24342–24356. https://doi.org/10.1007/s11356-020-08711-6

Kharissova, O. V., Dias, H. V. R., & Kharisov, B. I. (2015). Magnetic adsorbents based on micro- and nano-structured materials. *RSC Advances, 5*, 6695–6719. https://doi.org/10.1039/C4RA11423J

Kyzas, G. Z., Tolkou, A. K., Al Musawi, T. J., Mengelizadeh, N., Mohebi, S., & Balarak, D. (2022). Fluoride removal from water by using green magnetic activated carbon derived from canola stalks. *Water, Air, & Soil Pollution, 233*, 424. https://doi.org/10.1007/s11270-022-05900-6

Lacson, C. F. Z., Lu, M.-C., & Huang, Y.-H. (2021). Chemical precipitation at extreme fluoride concentration and potential recovery of CaF_2 particles by fluidized-bed homogenous crystallization process. *Chemical Engineering Journal, 415*, 128917. https://doi.org/10.1016/j.cej.2021.128917

Liang, P., An, R., Li, R., & Wang, D. (2018). Comparison of La^{3+} and mixed rare earths-loaded magnetic chitosan beads for fluoride adsorption. *International Journal of Biological Macromolecules, 111*, 255–263. https://doi.org/10.1016/j.ijbiomac.2017.12.151

Liu, L., Cui, Z., Ma, Q., Cui, W., & Zhang, X. (2016). One-step synthesis of magnetic iron–aluminum oxide/graphene oxide nanoparticles as a selective adsorbent for fluoride removal from aqueous solution. *RSC Advances, 6*, 10783–10791. https://doi.org/10.1039/C5RA23676B

Liu, S., Ma, C., Ma, M.-G., & Xu, F. (2019). 12—Magnetic nanocomposite adsorbents. In G. Z. Kyzas & A. C. Mitropoulos (Eds.), *Composite nanoadsorbents* (pp. 295–316). Elsevier.

Liu, X., Shan, Y., Zhang, S., Kong, Q., & Pang, H. (2022). Application of metal organic framework in wastewater treatment. *Green Energy & Environment*. https://doi.org/10.1016/j.gee.2022.03.005

Ma, W., Zhao, N., Yang, G., Tian, L., & Wang, R. (2011). Removal of fluoride ions from aqueous solution by the calcination product of Mg–Al–Fe hydrotalcite-like compound. *Desalination, 268*, 20–26. https://doi.org/10.1016/j.desal.2010.09.045

Massoudinejad, M., Ghaderpoori, M., Shahsavani, A., & Amini, M. M. (2016). Adsorption of fluoride over a metal organic framework Uio-66 functionalized with amine groups and optimization with response surface methodology. *Journal of Molecular Liquids, 221*, 279–286. https://doi.org/10.1016/j.molliq.2016.05.087

Mehta, D., Mazumdar, S., & Singh, S. K. (2015). Magnetic adsorbents for the treatment of water/wastewater—A review. *Journal of Water Process Engineering, 7*, 244–265. https://doi.org/10.1016/j.jwpe.2015.07.001

Mohammadi, E., Daraei, H., Ghanbari, R., Athar, S. D., Zandsalimi, Y., Ziaee, A., Maleki, A., & Yetilmezsoy, K. (2019). Synthesis of carboxylated chitosan modified with ferromagnetic nanoparticles for adsorptive removal of fluoride, nitrate, and phosphate anions from aqueous solutions. *Journal of Molecular Liquids, 273*, 116–124.

Mohseni-Bandpi, A., Kakavandi, B., Kalantary, R. R., Azari, A., & Keramati, A. (2015). Development of a novel magnetite–chitosan composite for the removal of fluoride from drinking water: Adsorption modeling and optimization. *RSC Advances, 5*, 73279–73289. https://doi.org/10.1039/C5RA11294J

Mukherjee, I., & Singh, U. K. (2018). Groundwater fluoride contamination, probable release, and containment mechanisms: A review on Indian context. *Environmental Geochemistry and Health, 40*, 2259–2301. https://doi.org/10.1007/s10653-018-0096-x

Mukherjee, S., & Halder, G. (2018). A review on the sorptive elimination of fluoride from contaminated wastewater. *Journal of Environmental Chemical Engineering, 6*.https://doi.org/10.1016/j.jece.2018.01.046

Nehra, S., Raghav, S., & Kumar, D. (2020). Biomaterial functionalized cerium nanocomposite for removal of fluoride using central composite design optimization study. *Environmental Pollution (Barking, Essex: 1987), 258*, 113773. https://doi.org/10.1016/j.envpol.2019.113773

Ni, C., Liu, C., Xie, Y., Xie, W., He, Z., & Zhong, H. (2022). A critical review on adsorption and recovery of fluoride from wastewater by metal-based adsorbents. *Environmental Science and Pollution Research International, 29*, 82740–82761. https://doi.org/10.1007/s11356-022-23416-8

Pandi, K., Periyasamy, S., & Viswanathan, N. (2017). Remediation of fluoride from drinking water using magnetic iron oxide coated hydrotalcite/chitosan composite. *International Journal of Biological Macromolecules, 104*, 1569–1577. https://doi.org/10.1016/j.ijbiomac.2017.02.037

Pandi, K., Viswanathan, N., & Meenakshi, S. (2019). Hydrothermal synthesis of magnetic iron oxide encrusted hydrocalumite-chitosan composite for defluoridation studies. *International Journal of Biological Macromolecules, 132*, 600–605. https://doi.org/10.1016/j.ijbiomac.2019.03.115

Patel, D., Pamidimukkala, P., Chakraborty, D., & Yadav, A. (2022). Bharuch district, Gujarat, India: Factor analysis and geographical distribution of water quality characteristics. *Environmental Nanotechnology, Monitoring & Management, 18*, 100732. https://doi.org/10.1016/j.enmm.2022.100732

Phouthavong, V., Yan, R., Nijpanich, S., Hagio, T., Ichino, R., Kong, L., & Li, L. (2022). Magnetic adsorbents for wastewater treatment: Advancements in their synthesis methods. *Materials, 15*, 1053. https://doi.org/10.3390/ma15031053

Pillai, P., Lakhtaria, Y., Dharaskar, S., & Khalid, M. (2020). Synthesis, characterization, and application of iron oxyhydroxide coated with rice husk for fluoride removal from aqueous media. *Environmental Science and Pollution Research, 27*, 20606.

Prasanna, S. R. V. S., Balaji, K., Pandey, S., & Rana, S. (2019). Chapter 4—Metal oxide based nanomaterials and their polymer nanocomposites. In N. Karak (Ed.), *Nanomaterials and polymer nanocomposites* (pp. 123–144). Elsevier.

Raghav, S., Nehra, S., & Kumar, D. (2019). Biopolymer scaffold of pectin and alginate for the application of health hazardous fluoride removal studies by equilibrium adsorption, kinetics and thermodynamics. *Journal of Molecular Liquids, 284*, 203–214. https://doi.org/10.1016/j.molliq.2019.03.155

Roto, R. (2018). Surface modification of Fe_3O_4 as magnetic adsorbents for recovery of precious metals. In C. Mohammad Asaduzzaman (Ed.), *Advanced surface engineering research*. IntechOpen.

Roy, A. (2021). Microwave-assisted synthesis and characterization of γ-Al_2O_3/γ-Fe_2O_3 composite and evaluating its efficiency in fluoride removal. *Colloids and Surfaces A: Physicochemical and Engineering Aspects, 608*, 125574. https://doi.org/10.1016/j.colsurfa.2020.125574

Sadhu, M., Bhattacharya, P., Vithanage, M., & Padmaja Sudhakar, P. (2021). Adsorptive removal of fluoride using biochar—A potential application in drinking water treatment. *Separation and Purification Technology, 278*, 119106. https://doi.org/10.1016/j.seppur.2021.119106

Sahoo, S. K., & Hota, G. (2018). Surface functionalization of GO with MgO/$MgFe_2O_4$ binary oxides: A novel magnetic nanoadsorbent for removal of fluoride ions. *Journal of Environmental Chemical Engineering, 6*, 2918–2931. https://doi.org/10.1016/j.jece.2018.04.054

Santhosh, C., Daneshvar, E., Tripathi, K. M., Baltrėnas, P., Kim, T., Baltrėnaitė, E., & Bhatnagar, A. (2020). Synthesis and characterization of magnetic biochar adsorbents for the removal of Cr(VI) and acid orange 7 dye from aqueous solution. *Environmental Science and Pollution Research, 27*, 32874–32887. https://doi.org/10.1007/s11356-020-09275-1

Senapati, K. K., Sharma, M., Kalita, P., & Garg, A. (2018). Magnetic nanoparticles as an effective adsorbent for removal of fluoride–A review. *MOJ Ecology & Environmental Sciences*.

Sharma, A., Mangla, D., Shehnaz, Chaudhry, S. A. (2022). Recent advances in magnetic composites as adsorbents for wastewater remediation. *Journal of Environmental Management, 306*, 114483. https://doi.org/10.1016/j.jenvman.2022.114483

Shirin, S., Jamal, A., Emmanouil, C., Singh, V. P., & Yadav, A. K. (2022). Assessment and characterization of waste material used as backfilling in an abandoned mine. *International Journal of Coal Preparation and Utilization*, 1–9. https://doi.org/10.1080/19392699.2022.2118259

Shirin, S., Jamal, A., Emmanouil, C., & Yadav, A. K. (2021). Assessment of characteristics of acid mine drainage treated with fly ash. *Applied Sciences, 11*, 3910. https://doi.org/10.3390/app11093910

Shirin, S., Jamal, A., Ranjan, P., & Yadav, A. K. (2019). Study on assessment of slope stability and mixed disposal of overburden in voids of Singrauli coalfield. *Environmental Quality Management, 28*, 131–139. https://doi.org/10.1002/tqem.21616

Shirin, S., & Yadav, A. K. (2014). Physico chemical analysis of municipal wastewater discharge in ganga river, Haridwar district of Uttarakhand, India. *Current World Environment, 9*, 536–543. https://doi.org/10.12944/CWE.9.2.39

Shirin, S., & Yadav, A. K. (2021). Assessment and suitability for the dumping of flyash in an abandoned mine. *Environmental Challenges, 4*, 100111. https://doi.org/10.1016/j.envc.2021.100111

Srivastava, S. K., Senapati, S., Singh, S. B., & Raul, P. K. (2016). Magnetic Ni/PPy nanocomposite as effective reusable adsorbent for removal of arsenite and fluoride from contaminated water. *RSC Advances, 6*, 113424–113431. https://doi.org/10.1039/C6RA24531E

Takmil, F., Esmaeili, H., Mousavi, S. M., & Hashemi, S. A. (2020). Nano-magnetically modified activated carbon prepared by oak shell for treatment of wastewater containing fluoride ion. *Advanced Powder Technology, 31*, 3236.

Tan, T. L., Krusnamurthy, P. A. P., Nakajima, H., & Rashid, S. A. (2020). Adsorptive, kinetics and regeneration studies of fluoride removal from water using zirconium-based metal organic frameworks. *RSC Advances, 10*, 18740–18752. https://doi.org/10.1039/D0RA01268H

Tee, G. T., Gok, X. Y., & Yong, W. F. (2022). Adsorption of pollutants in wastewater via biosorbents, nanoparticles and magnetic biosorbents: A review. *Environmental Research, 212*, 113248. https://doi.org/10.1016/j.envres.2022.113248

Wan, Z., Chen, W., Liu, C., Liu, Y., & Dong, C. (2015). Preparation and characterization of γ-alooh@ cs magnetic nanoparticle as a novel adsorbent for removing fluoride from drinking water. *Journal of Colloid and Interface Science, 443*, 115–124.

Wang, J., Han, Q., Wang, K., Li, S., Luo, W., Liang, Q., Zhong, J., & Ding, M. (2023). Recent advances in development of functional magnetic adsorbents for selective separation of proteins/peptides. *Talanta, 253*, 123919. https://doi.org/10.1016/j.talanta.2022.123919

Wang, M., & You, X.-Y. (2021). Critical review of magnetic polysaccharide-based adsorbents for water treatment: Synthesis, application and regeneration. *Journal of Cleaner Production, 323*, 129118. https://doi.org/10.1016/j.jclepro.2021.129118

Wang, M., Yu, X., Yang, C., Yang, X., Lin, M., Guan, L., & Ge, M. (2017). Removal of fluoride from aqueous solution by Mg–Al–Zr triple-metal composite. *Chemical Engineering Journal, 322*, 246–253. https://doi.org/10.1016/j.cej.2017.03.155

Wei, Y., Wang, L., Li, H., Yan, W., & Feng, J. (2022). Synergistic fluoride adsorption by composite adsorbents synthesized from different types of materials—A review. *Frontiers in Chemistry, 10*.https://doi.org/10.3389/fchem.2022.900660

WHO. (2011). *Guideline for drinking water quality*. Switzerland World Health Organization.

Yadav, A. K., Sahoo, S. K., Mahapatra, S., Kumar, A. V., Pandey, G., Lenka, P., & Tripathi, R. M. (2014). Concentrations of uranium in drinking water and cumulative, age-dependent radiation doses in four districts of Uttar Pradesh, India. *Toxicological and Environmental Chemistry, 96*, 192–200. https://doi.org/10.1080/02772248.2014.934247

Yadav, A. K., Shirin, S., Hopke, P. K., Pal, D., & Jamal, A. (2022). Process to reduce particulate matter in ambient air using bubbles of sodium palmitate. *Chemical Engineering & Technology, 45*, 1497–1500. https://doi.org/10.1002/ceat.202100454

Yadav, K., & Jagadevan, S. (2021). Chapter 4—Adsorbents for removal of fluoride from water. In M. Hadi Dehghani, R. Karri, & E. Lima (Eds.), *Green technologies for the defluoridation of water* (pp. 89–118).

Yang, W., Dou, X., Li, Y., Mohan, D., Pittman, C. U., & Ok, Y. S. (2016). Performance and mass transfer of aqueous fluoride removal by a magnetic alumina aerogel. *RSC Advances, 6*, 112988–112999. https://doi.org/10.1039/C6RA23532H

Zhang, C., Li, Y., Wang, T.-J., Jiang, Y., & Fok, J. (2017). Synthesis and properties of a high-capacity iron oxide adsorbent for fluoride removal from drinking water. *Applied Surface Science, 425*, 272–281. https://doi.org/10.1016/j.apsusc.2017.06.159

Zhang, C., Li, Y., Wang, T.-J., Jiang, Y., & Wang, H. (2016). Adsorption of drinking water fluoride on a micron-sized magnetic Fe_3O_4@Fe-Ti composite adsorbent. *Applied Surface Science, 363*, 507–515. https://doi.org/10.1016/j.apsusc.2015.12.071

Zhang, L.-T., Guo, J., Li, M., Tshinkobo, O., Wang, C., Changguo, X., Liu, Y., & Li, J.-J. (2020). Low-cost magnetic adsorbent for efficient Cu(II) removal from water. *Materials Research Express, 7*, 105503. https://doi.org/10.1088/2053-1591/abbe3f

Zhao, X., Wang, J., Wu, F., Wang, T., Cai, Y., Shi, Y., & Jiang, G. (2010). Removal of fluoride from aqueous media by Fe_3O_4@Al(OH)3 magnetic nanoparticles. *Journal of Hazardous Materials, 173*, 102–109. https://doi.org/10.1016/j.jhazmat.2009.08.054

Zhou, J., Liu, Y., Han, Y., Jing, F., & Chen, J. (2019). Bone-derived biochar and magnetic biochar for effective removal of fluoride in groundwater: Effects of synthesis method and coexisting chromium. *Water Environment Research, 91*, 588–597.

Zuo, H., Chen, L., Kong, M., Qiu, L., Lü, P., Wu, P., Yang, Y., & Chen, K. (2018). Toxic effects of fluoride on organisms. *Life Sciences, 198*, 18–24. https://doi.org/10.1016/j.lfs.2018.02.001

Chapter 5
Efficient and Cost Effective Groundwater De-fluoridation Adsorbents with Focus on Rural Hilly India: A Comprehensive Review

Rahul Singh Thakur and Ankit Modi

Abstract From ancient times, underground water resources such as wells and bawaris have been used for drinking and other daily activities. Half of the world's population still relies on groundwater to satisfy their drinkable water needs. Consequently, its quality is degraded due to industrialization and human interference. The most prevalent concern is groundwater contamination from sewage, industrial effluents, pesticides, and other pollutants. As a result, one of the most critical and complex environmental concerns confronting all life forms on Earth is supplying adequate clean drinking water to every human being to ensure survival. Groundwater is estimated to be the source of domestic water for 80% of the rural and 50% of its urban areas. Fluoride-rich groundwater exposure produced everything from dental fluorosis to devastating skeletal fluorosis in humans and animals. Although numerous defluoridation methods are available, including coagulation, reverse osmosis, and nanofiltration, these technologies have proven ineffective in rural areas, particularly in rural hilly areas, due to high costs and a lack of skilled operators. The present review aims to corroborate the uses of organic products, mainly from agriculture, agroforestry, and forest waste. In this chapter, we discuss the preparation of various adsorbents from these products, the efficiency of fluoride removal, the cost–benefit analysis, and market economy. The chapter provides insight into some cost-effective mitigating

R. S. Thakur (✉)
Centre for Energy and Environmental Engineering, National Institute of Technology Hamirpur, Hamirpur 177005, India
e-mail: rsthakur.0145@gmail.com

A. Modi
Environmental Engineering and Management Lab, Department of Civil Engineering, Indian Institute of Technology Kanpur, Kanpur 208016, India

Discipline of Civil Engineering, Indian Institute of Technology Gandhinagar, Gujarat 382424, India

A. Modi
e-mail: ankit.mbm.civil@gmail.com

A. K. Yadav et al. (eds.), *Advanced Treatment Technologies for Fluoride Removal in Water*, Water Science and Technology Library 125,
https://doi.org/10.1007/978-3-031-38845-3_5

approaches for ground water defluoridation at the household and community level in rural hilly areas of India.

Keywords Fluoride · Rural · Hilly · Defluoridation · India · Mitigation · Groundwater

5.1 Introduction

Fluoride is a naturally occurring mineral found in rocks, soil, and water. It occurs naturally in many water sources in an ionic form. Fluoride is part of the human life cycle and is helpful for human health when consumed within a specific range. Fluoride is added to municipal water systems as a public health intervention to prevent adverse health effects when fluoride levels are low. In contrast, excess fluoride is eliminated from water in cases of excessive fluoride consumption, which can be harmful to human well-being. Excess fluoride causes skeletal and dental fluorosis. Skeletal fluorosis is a more dangerous disorder that affects bones and joints, while dental fluorosis is a cosmetic condition that damages tooth enamel. In addition to teeth and bones, serious harm has also been reported to critical organs such as the brain, liver, and kidney.

Defluoridation is the process of removing excess fluoride from water or other sources. Defluoridation procedures include reverse osmosis, electro-dialysis, adsorption, nanofiltration, ion exchange, electrocoagulation, precipitation, and distillation. Defluoridation is necessary in locations where fluoride levels in potable water exceed the acceptable limit of 1.5 mg/L. It is also used in companies that employ fluoride as a raw material or a by-product. Defluoridation is a critical issue in India because more than 20 states have high levels of fluoride in groundwater, up to 48 ppm. According to the World Health Organization (WHO), the acceptable limit of fluoride in potable water is 1.5 mg/L, although fluoride levels in several Indian districts exceed 2.5 mg/L. Defluoridation is a serious issue in India, and government and non-governmental organizations are working together to provide clean drinking water to affected populations. Proper defluoridation methods must go a long way toward providing clean drinking water to millions of Indians.

Adsorption is a commonly used process for defluoridation in India. Activated alumina is the most widely used adsorbent in defluoridation, following activated carbon and bone char. The Nalgonda technique, also known as the Permutt process, is a widely used method for the defluoridation of drinking water in India that involves the use of activated alumina as an adsorbent to remove fluoride ions from water. Activated alumina is a porous material that effectively removes fluoride ions from water. Some states where activated alumina is used for defluoridation include Kerala, Telangana, Gujarat, Andhra Pradesh, Karnataka, Tamil Nadu, Rajasthan, Maharashtra, Madhya Pradesh, and Uttar Pradesh. In some states, high fluoride levels are found in groundwater and activated alumina is used to remove fluoride from water. Generally, the cost of the activated alumina defluoridation method can range from Rs. 3 to 6 per

1000 L of treated water. This cost includes the capital cost of the treatment plant, the cost of activated alumina media, electricity and labor, and maintenance costs.

Reverse osmosis is a popular water treatment method using a semi-permeable membrane to remove fluoride and other contaminants. Reverse osmosis (RO) technology is used for defluoridation in several states of India. Some states in which RO-based defluoridation plants have been established include Karnataka, Tamil Nadu, Andhra Pradesh, Kerala, Rajasthan, and Gujarat. The cost of reverse osmosis (RO) for defluoridation in India can differ depending on numerous factors, such as the capacity of the RO plant, the quality of the membranes used, the cost of electricity and the level of automation of the system. A small-scale RO plant producing 1000 L of treated water per hour can cost around 5–6 lakh INR.

Distillation is a process that involves boiling water to produce steam, which is then condensed to produce purified water. This process removes fluoride and other impurities from the water. The state of Tamil Nadu in India uses distillation as one of the methods for defluoridation of drinking water. In particular, the Nilgiris district in Tamil Nadu has implemented a solar distillation-based defluoridation plant that uses solar energy to evaporate water, leaving behind fluoride-free water for drinking purposes. This plant has successfully provided clean drinking water to local communities in the region. Generally, distillation is considered to be a relatively expensive method of defluoridation compared to other techniques, such as adsorption or precipitation. Typically, the cost of distillation in defluoridation can range from Rs. 2 to 5 per liter of treated water. The price of defluoridation distillation in India can differ depending on numerous factors, including the size of the treatment plant, the type of equipment used, the amount of fluoride in the type of water, the fuel type and the location of the plant. This cost may be higher in remote areas with high transportation costs or water scarcity. In India, the cost effectiveness of distillation as a defluoridation method may depend on factors such as the availability of alternative methods and the severity of fluoride contamination in the area.

Electrocoagulation is a water treatment technique that can be used to remove fluoride from water and it involves using an electric current to coagulate and precipitate impurities in water. In the electrocoagulation process, an electric current passes through a solution containing metal electrodes, which causes the release of metal ions into the solution. These metal ions react with the fluoride ions in the water to form insoluble metal fluoride compounds, which can then be removed by sedimentation or filtration. The advantages of electrocoagulation for defluoridation include its low cost, simplicity, and ease of operation. It does not require any chemicals or special equipment, and it can easily be scaled up or down depending on the volume of water to be treated. The uninterrupted electricity supply is a significant limitation of the electrocoagulation technique for defluoridation. In addition, the quality of the metal electrodes and the pH of the water can also influence the process efficiency. Some states where electrocoagulation is used for defluoridation include Karnataka, Andhra Pradesh, Telangana, Rajasthan, and Gujarat. The cost of EC technology can range from INR 2 to INR 12 per 1000 L of treated water (MoJS, 2023).

Electro-dialysis is a process in which an electric field separates ions from a solution. This process involves placing a membrane between two electrodes and passing an electric current through the solution. The membrane selectively allows specific ions to pass through while preventing others from passing through. This process is used to remove salt and other ionic contaminants from water. Electro-dialysis is not widely used for fluoride removal from water in India. However, some studies and pilot projects have been conducted in different states of India to test the feasibility of electro-dialysis for fluoride removal. One such study was conducted in Tamil Nadu, where electrodialysis removed fluoride from groundwater (Karthikeyan et al., 2010). Another study was conducted in Uttar Pradesh, where electrodialysis treated fluoride-contaminated water in a rural village. Electro-dialysis can cost between INR 3 and INR 5 per liter of treated water in India.

Nanofiltration is a type of water filtration technology that uses a semi-permeable membrane with very small pore sizes to remove dissolved ions and other particles from water. It is commonly used in water treatment for desalination and to remove specific contaminants, such as fluoride. Nanofiltration is effective in removing fluoride from water. The semipermeable membrane used in nanofiltration has a pore size of about 1 nm, which is small enough to filter out fluoride ions (which have a diameter of about 0.13 nm) and other dissolved salts and impurities. Several states in India have implemented this technology to provide safe drinking water to their citizens. In 2015, the Karnataka government launched a project to remove fluoride from groundwater using nanofiltration technology. The project was implemented in the fluoride affected regions of the state, including Belagavi, Bagalkot, and Gadag. Andhra Pradesh has also implemented nanofiltration technology to remove fluoride from water. The project was launched in 2019 and is being implemented in the fluoride-affected areas of the Nalgonda district. The Telangana government has set up several nanofiltration plants to remove fluoride from water in areas affected by fluoride. These plants are located in Nalgonda, Mahabubnagar, and Ranga Reddy districts. The Rajasthan government has also implemented nanofiltration technology to remove fluoride from water. The technology is being used in fluoride affected areas of the Barmer, Jaisalmer, and Bikaner districts.

The precipitation method is commonly used to remove fluoride from water in India. The process involves adding a chemical coagulant to the water, which causes the fluoride ions to precipitate out of the water and form solid particles that can be removed by filtration. One of the most commonly used chemical coagulants in India for fluoride removal is alum (aluminum sulfate). Alum is added to the water and the pH of the water is adjusted to around 6.5–7.0 using lime or another alkaline material. The alum reacts with the fluoride ions in the water, forming aluminum fluoride particles that can be easily filtered out of the water. The effectiveness of the precipitation method depends on several factors, including the concentration of fluoride in the water, the pH of the water, and the type and amount of coagulant used. In general, the precipitation method effectively reduces fluoride levels in the water but may not be suitable for treating large volumes of water or water with very high fluoride concentrations.

In general, these defluoridation techniques have successfully reduced fluoride levels in drinking water in India. However, the effectiveness of each method depends on several factors, including the type and concentration of fluoride in the water, the amount of water to be treated, and the availability of resources and infrastructure. Table 5.1 further shows that most of the concerns with these solutions are related to cost, operation, and maintenance. Standard defluoridation techniques used in urban India are activated alumina, reverse osmosis, distillation, ion exchange, and electrocoagulation. However, in rural areas, the installation of these technologies has rarely been proven to be sustainable (MoJS, 2023). This failure can be attributed to many issues, including lack of technological support, insufficient material supply, insufficient O&M, and irregular monitoring.

5.2 Current Defluoridation Practices in Hilly Regions of Rural India

In rural India, where groundwater is often the main source of potable water, excess fluoride in the water is a significant problem. Adsorption is a highly suitable method for fluoride removal in hilly rural India due to its cost-effectiveness, ease of implementation, high efficiency, low maintenance, and safety. Adsorbents, which are materials that remove excess fluoride from drinking water, are crucial in this process. As a result, this process can be improved by using locally available materials, for example, activated alumina, bone char, activated carbon, or clay, rather than employing expensive commercial adsorbents.

One common defluoridation adsorbent used in the hilly regions of rural India is activated alumina. Activated alumina is a porous material made of aluminum oxide that effectively removes fluoride ions from water. Activated alumina is a variety of aluminum oxides with a high surface area and is highly porous. It works by adsorbing fluoride ions from water onto its surface. However, it also has many limitations in hilly regions, such as the need for frequent regeneration. It can only be produced by aluminum ore and transporting it to remote locations is costly and time-consuming.

Another popular defluoridation adsorbent used in hilly rural India is bone charcoal. Bone charcoal, or bone char, is activated carbon that is heated to high temperatures without air. This process creates a porous black material that can absorb impurities from water. When bone charcoal is used for defluoridation, it adsorbs fluoride ions onto its surface. The process depends on several factors, such as fluoride concentration in the water, contact time between the water and bone charcoal, and the pH of the water. Bone charcoal for defluoridation is typically packed into a container or filter, passing water through it. Bone charcoal will absorb fluoride ions, reducing the fluoride concentration in water. However, it is essential to note that bone charcoal can also adsorb other impurities in the water, so it is crucial to properly maintain and replace the filter to ensure consistent and effective defluoridation. In general, bone charcoal is an effective and low-cost method of defluoridation in hilly regions of rural India. However, bone charcoal is prepared using animal bones, which may be unacceptable to some communities in rural areas.

Table 5.1 Fluoride removal methods in India and their limitations

Methodology	Limitations for rural parts of India
Electro-dialysis	• High cost: Electro-dialysis defluoridation requires significant capital investment in equipment and infrastructure, making it too expensive for rural communities to implement and maintain • High energy consumption: The process requires considerable electricity, which can be challenging in rural areas where access to reliable electricity may be limited • Skilled operation and maintenance: Electro-dialysis defluoridation requires skilled operators and regular maintenance to ensure effective operation, which may be challenging in rural areas where trained personnel may be scarce • Low capacity: The process may not be able to handle large volumes of water, making it impractical for use in areas with high water demand • Chemical waste disposal: The process generates significant chemical waste, which must be safely disposed of to prevent environmental contamination. It can be challenging in rural areas where adequate disposal facilities are not available
Distillation	• High cost: Distillation defluoridation requires specialized equipment, which can be expensive and difficult to maintain. This can be difficult to implement in rural areas, where resources are often limited • High-energy requirements: The distillation process requires significant energy to boil the water, which can be a difficulty in regions with limited access to electricity or fuel • Low efficiency: The distillation process is not very efficient in removing fluoride and may not be able to reduce fluoride levels to the safe recommended levels • Labour-intensive: The distillation process requires constant attention and monitoring, which can be time-consuming and labour-intensive • Limited capacity: The amount of water that can be processed by distillation is limited by the size of the equipment, which may not be sufficient to meet the needs of rural communities
Reverse osmosis	• High cost: Reverse osmosis systems can be costly to install and maintain, making them unaffordable for many rural communities in India • High energy consumption: Reverse osmosis systems require a significant amount of energy to function, which can be an issue in locations with intermittent or limited access to electricity • Low water recovery rate: Reverse osmosis systems have a low water recovery rate, which implies that a significant amount of water is wasted during treatment. It can be a major issue when water is scarce or hard to find • Dependence on technology: Reverse osmosis systems require particular technical expertise to operate and maintain. This can be a significant challenge in rural areas where such expertise may be lacking

(continued)

Table 5.1 (continued)

Methodology	Limitations for rural parts of India
Electro-coagulation	• Energy consumption: Electrocoagulation uses a large amount of energy to function, which might be difficult in locations with unstable or restricted electricity supply. It can increase the operating costs of the treatment process • Equipment cost: The equipment needed for electrocoagulation can be expensive, making it difficult for small-scale water treatment systems to implement • Maintenance requirements: Electrocoagulation systems require regular maintenance to ensure that they work properly. It can be challenging in areas with limited resources or expertise • Limited effectiveness: Electrocoagulation may not be effective in removing all types of contaminants from water and may not be as effective as other treatment technologies in removing fluoride • Sludge disposal: Electrocoagulation produces sludge as a by-product, which can be difficult and expensive to dispose of safely • Chemical requirements: Electrocoagulation may require chemicals to enhance the coagulation process, which can increase costs and environmental impact
Nano-filtration	• Cost: Nanofiltration is expensive, and its installation and maintenance costs can be high. It may make it less accessible to communities in rural or low-income areas where fluoride contamination is a significant issue • Energy consumption: Nanofiltration requires energy, which can be challenging in areas where electricity is unreliable or readily unavailable. It may limit nanofiltration use for fluoride removal in remote or off-grid areas • Membrane fouling: Membrane fouling is a common issue in nanofiltration systems, which can reduce their efficiency and lifespan. Fouling can occur in the removal of fluoride due to a high concentration of dissolved materials in the water, which can clog the membrane and impair its ability to remove fluoride • Limited capacity: Nanofiltration systems have limited ability and may not be suitable for large-scale fluoride removal projects. It could be challenging to address the widespread problem of fluoride contamination in India, where millions of people are affected • Disposal of concentrate: Nanofiltration produces a concentrated stream that contains the removed contaminants and must be disposed of properly. Disposal of concentrate can be a challenge, especially in areas with limited infrastructure for waste management

(continued)

Table 5.1 (continued)

Methodology	Limitations for rural parts of India
Precipitation	• Cost: The precipitation method requires chemicals such as aluminum sulfate or calcium hydroxide, which can be expensive and may not be affordable for small-scale water treatment facilities in rural areas • Wastewater Treatment: The chemicals used in the precipitation method can create sludge and other waste products that require proper disposal. In many areas of India, wastewater treatment facilities are not readily available, making this an additional challenge • pH control: The precipitation method requires careful pH control to ensure that the fluoride ions are effectively removed from the water. However, pH control can be difficult when water quality is variable and may require additional testing and monitoring • Effectiveness: The precipitation method may not be effective in removing other contaminants from water, such as arsenic and heavy metals. The occurrence of other ions in water, such as sulfate, chloride, and bicarbonate, can also alter the effectiveness of the process • Sustainability: The long-term sustainability of the precipitation method is uncertain. The continued use of chemicals for treatment can lead to the depletion of natural resources and environmental degradation. Alternatives to chemical-based treatment methods, such as adsorption or ion exchange, might be more sustainable in the long run
Adsorption	• Cost: Adsorption depends on the cost of adsorbed material, and the installation and maintenance of defluoridation units can be prohibitive for rural communities with limited financial resources • Maintenance: The defluoridation units require regular maintenance and replacement of the adsorbed material, which can be challenging to manage in rural areas with limited access to skilled technicians and spare parts • Water availability: The defluoridation process requires a continuous flow of water to maintain the system's efficiency. However, many rural areas in India face water scarcity and diverting large volumes of water for defluoridation may not be feasible • Cultural acceptability: The defluoridation process can alter the taste and color of water, which may be unacceptable for some communities in rural areas • Sustainability: The long-term sustainability of the activated alumina defluoridation method is questionable in rural areas, as it requires a regular supply of activated alumina, which may not be available locally

Activated carbon is effective in removing fluoride from water. Activated carbon is a porous material with a large surface area that can absorb impurities from water. When activated carbon is added to water, it adsorbs fluoride ions onto its surface, effectively removing them from the water. Adjusting the pH of water and the contact period between activated carbon and water can improve the adsorption ability of activated carbon. One potential issue with the use of activated carbon for defluoridation is the cost. Therefore, low-cost methods can be used to produce activated carbon using locally available materials, such as coconut shells or wood. These low-cost methods can make activated carbon more accessible and affordable for communities in rural India. In general, activated carbon is a promising technology for defluoridation in rural areas of India. In mountainous regions, agriculture is limited due to the adverse conditions for growth. Therefore, forest waste products can be used in the development of low-cost adsorbents. Forest waste products are affordable and readily available in the hilly regions of rural India. Some common agroforestry waste products are wood, coir pith, nut shells, paddy husk, Aramark, tea waste, cashew nut shell, bamboo, pine cone, corn cobs, and sawdust (Table 5.2).

5.3 State-of-the-Art Adsorbents for Hilly Regions in Rural India

State-of-the-art research in the field of defluoridation adsorbents has led to the development of several effective materials due to their benefits over other methods such as

(i) Simple operation process and easy to handle,
(ii) Low operating and maintenance costs,
(iii) A wide range of adsorbents are available,
(iv) Gravity-based and does not require electricity, and
(v) Most adsorbents/polymeric resins can be regenerated using simple chemicals that enhance the media life and for routine O&M without requiring highly skilled human resources.

Agricultural waste (biomaterials), plant materials, and industrial waste are abundant in nature. Biomass from vegetation, crop residues, and industrial waste can be used for effective fluoride assimilation that is inexpensive and environmentally beneficial. As a result, researchers are concentrating on developing low-cost adsorbents that may be easily manufactured and applied in hilly rural areas. The following subsections describe the waste products that will be explored to make them a feasible choice in defluoridation.

5.3.1 Agroforestry-Based Adsorbents

Researchers have used crop residues such as rice husks, coconut shells, and sugarcane bagasse to make activated carbon for fluoride removal. Similarly, forest waste materials such as neem bark, sawdust, bamboo, pine cone, and coir pith have been analyzed to make adsorbents. These low-cost adsorbents have shown promising results and could be a viable option for rural communities (Table 5.3).

Table 5.2 Adsorbents for fluoride removal in hilly rural India and their limitations

Methodology	Limitations for rural parts of India
Activated alumina	• High cost: Activated alumina is expensive compared to other defluoridation techniques, such as bone char, making it less affordable for low-income communities • Limited lifespan: Activated alumina has a limited lifespan and requires frequent replacement, which can be a logistical challenge in remote areas with limited access to replacement materials • Adsorption capacity: The adsorption capacity of activated alumina decreases as the pH of water reduces, which can limit its effectiveness in acidic water sources • Pre-treatment requirements: Activated alumina requires pre-treatment to remove turbidity and other impurities, which can be challenging in remote areas where water sources may be highly contaminated • Waste disposal: The spent activated alumina must be properly disposed of, which can be challenging in areas with limited waste management infrastructure • Health risks: The production of activated alumina involves strong acids and high temperatures, which can pose health risks to workers and the environment if not properly managed
Bone charcoal and alkali-treated bones	• Limited lifespan: Bone char has a limited lifespan and needs to be replaced frequently, which can be costly for rural communities • Inconsistent performance: The efficiency of bone char for fluoride removal can vary according to bone quality and the manufacturing process. It can result in inconsistent performance, making it difficult to rely on bone char for consistent fluoride removal • Health risks: The production of bone char involves burning animal bones at high temperatures, which can release harmful gases and particles into the environment. It can pose health risks to those involved in the production process and the community • Maintenance issues: Bone char filters require regular maintenance to ensure proper functioning, including cleaning and cleaning. In rural areas with limited resources and infrastructure, this can be challenging • Limited scalability: Bone-char filters are not easily scalable, meaning that they may not be a practical solution for larger communities with higher fluoride levels in their drinking water
Activated carbon	• Regeneration: Activated carbon needs to be regenerated periodically to maintain its effectiveness. This process requires specialized equipment and can be costly • Limited capacity: The defluoridation capacity of activated carbon is limited and may not be suitable for treating large volumes of water. It could be a significant issue in rural areas, where water scarcity is a common problem • Filter clogging: Activated carbon filters can become clogged over time, reducing efficiency and requiring frequent maintenance • Disposal: The disposal of spent activated carbon can be a significant issue. It is considered hazardous waste and requires special handling and disposal methods • Efficiency: Activated carbon is not always effective in removing all types of fluoride compounds, especially those with high concentrations • Environmental impact: The manufacture and disposal of activated carbon can have significant environmental consequences, such as greenhouse gas emissions and water pollution

Table 5.3 Agroforestry waste in hilly rural India suitable for adsorbents for fluoride removal

Methodology	Specification
Bamboo	• Bamboo is a fast-growing grass that is commonly found in hilly areas. It can produce activated carbon, which is effective in defluoridation • Bamboo is increasingly being researched as a potential source of activated carbon (Wendimu et al., 2017) • Bamboo-based activated carbon can be produced using various techniques, including pyrolysis and physical and chemical activation. Each approach has benefits as well as drawbacks, and the method used is determined by the individual application and the required activated carbon qualities • Bamboo is found in various hilly areas of India, including the north-eastern region of the country, the western Ghats and the eastern Ghats
Corn cobs	• Corn cobs are a by-product of the corn industry and can be used to produce activated carbon • Corn cob-based activated carbon has been reported to be effective in defluoridation water (Parmar et al., 2006) • The mountainous regions of Jammu and Kashmir, Himachal Pradesh, Uttarakhand, and the northern provinces of Meghalaya and Nagaland are where maize is grown regularly in India
Sawdust	• Sawdust is a by-product of the timber industry and can be used to produce activated carbon • Sawdust-based activated carbon has been reported to have a high fluoride adsorption capacity (Joshi et al., 2022) • Several states, such as Maharashtra, Karnataka, Tamil Nadu, and Kerala, have hilly areas that generate sawdust. The mountain range in the eastern part of India is home to the most biodiverse forests. Arunachal Pradesh, Sikkim, and West Bengal have hilly regions with significant forestry industries that produce sawdust. Similarly, in central India, the Vindhya range has states Madhya Pradesh, Chhattisgarh, and Uttar Pradesh have hilly areas that generate sawdust
Rice Husk	• Rice husk is an abundant agricultural waste that can be used to make activated carbon. Activated carbon has a high surface area and can effectively adsorb fluoride from water (Pratha & Prabakar, 2020) • North East states (Assam, Arunachal Pradesh, Meghalaya, Nagaland, and Tripura), Eastern Himalayan region (Sikkim, Darjeeling district of West Bengal), Western Ghats (Parts of Kerala, Tamil Nadu, Karnataka and Maharashtra), and Central India (Chhattisgarh, Madhya Pradesh, and Odisha) have many hilly regions of India where rice is grown and processed • The fluoride adsorption capacity was 15.08 mg/g in the batch investigation, while it was 9.5 mg/g in the column study. The highest removal of fluoride was reached at pH 5.0, demonstrating that the initial fluoride content did not affect adsorption. The adsorption capacity of the materials utilized in the filter was determined to be 9 mg/g (Ganvir & Das, 2011)

(continued)

Table 5.3 (continued)

Methodology	Specification
Biochar from wood waste	• Biochar from wood waste, such as activated carbon based on Eucalyptus bark, has been reported to be effective in defluoridation due to its high surface area and adsorption capacity (Khoshnamvand et al., 2018) • Many other trees in the hilly regions can be used to produce activated carbon, such as Acacia, Teak, Mahogany, Oak, and Willow • Pine needle-based activated carbon has been reported to have a high fluoride adsorption capacity (Thakur et al., 2020) • Pine trees are commonly found in the hilly regions of the Himalayas in India, particularly in the states of Uttarakhand, Himachal Pradesh, and Jammu & Kashmir
Biomass—based fly ash	• Biomass fly ash is the ash produced from biomass combustion, including wood, crop residues, animal manure, and municipal waste. It is similar to coal fly ash in terms of composition, but it has a lower content of heavy metals and other toxic compounds typically found in coal fly ash • In recent years, there has been growing interest in using biomass fly ash for the defluoridation of water (Manna et al., 2018)
Tea waste	• Tea waste is a by-product of tea processing and is commonly available in hilly areas of Bengal, Assam, Tamil Nadu, Kerala, Tripura, Karnataka, Himachal Pradesh, Uttarakhand, Arunachal Pradesh, and Manipur • Tea waste was reported to have a high fluoride adsorption capacity (Roy & Das, 2016). Mondal et al. (2012) investigated water defluoridation utilizing activated tea ash (AcTAP) created by the AcTAP tea residue had a Langmuir adsorption capacity of 8.55 mg/g and exhibited pseudo second order kinetics. The ideal pH for the most intense removal of fluoride is 6.0
Cashew nut and Wall nutshell	• Cashew nuts and wall nut shell are a by-product of the cashew processing industry in Jammu and Kashmir, Maharashtra, Karnataka, Kerala, Tamil Nadu, West Bengal, and Odisha • Cashew nutshell has been reported to have high fluoride adsorption capability (Alagumuthu & Rajan, 2010). Sivabalan et al. (2003) found that the maximum percentage of fluoride removal for the 8000 ppm adsorbent dose of activated carbon from the cashew nut sheath was 87.6% at pH 6.9 • Rajan and Alagumuthu (2012) investigated the fluoride removal capacity of zirconium-treated walnut shell carbon (ZIWSC). At pH 3, ZIWSC removed 94% of the fluoride

(continued)

Table 5.3 (continued)

Methodology	Specification
Tamarind gel and seed	• In India, tamarind is commonly grown in states of the western Ghats, such as Maharashtra, Kerala, and Karnataka • Tamarind gel and seed were reported to have a high fluoride adsorption capacity (Walia et al., 2021) • The concentration of fluoride from a solution of sodium fluoride of 10 mg/L can be reduced to 2 mg/L by adding tamarind gel alone and to 0.05 mg/L by adding a small amount of chloride with the tamarind gel (MoJS, 2023)
Activated carbons obtained from the fruit peel	• Fruit crops grown in hilly areas of India include apples, pears, peaches, apricots, cherries, plums, strawberries, kiwis, oranges, and lemons. These high-value crops can provide peel-based activated carbon based on peels from waste material generated in the peeling process of fruits • Dwivedi et al. (2014) studied the batch reactor defluoridation capacity of Mosambi peel powder. The adsorbent removed 82.5% of the fluoride when the original fluoride content in the water was 5 mg/L

5.3.2 Soil-Based Adsorbents

In hilly areas of India, limestone and clay are common soils. Limestone is a sedimentary rock composed mainly of calcium carbonate. It is often found in hilly areas of India, such as the Shivalik range, the Aravalli range, and the western Ghats. Special soils (red soils, black soils, and laterite soils) are also found in hilly areas of India. These soils are often rich in minerals and nutrients and are suitable for agriculture. Red soils are found in hilly regions of eastern and southern India and are rich in iron oxide. Black soils, also known as regur soils, are found in the Deccan plateau region and are known for their high fertility. Laterite soils in the western Ghats are known for their high clay content.

Limestone effectively removes fluoride from drinking water (Mohan & Dutta, 2020). The removal process of fluoride from water using limestone is known as adsorption. When water containing fluoride is passed through a limestone bed, the fluoride ions adsorb onto the surface of the limestone particles, effectively removing them from the water. The use of limestone for fluoride removal has been successfully implemented in several areas of India, including hilly regions. The process is relatively simple and inexpensive, making it an attractive solution for communities with limited resources.

Fluoride removal by adsorption on low-cost minerals such as kaolinite, bentonite, and lithium was also explored (Das et al., 2022; Uddin et al., 2019). Clay mineral scan be added directly to water sources, such as wells or ponds, for fluoride removal. The clay will settle to the bottom of the water source and the fluoride will be removed as the water passes through the clay. The treated water can then be collected from the surface of the water source. Arecent study on the fluoride removal efficiency of clay minerals, published in the journal Environmental Technology & Innovation, evaluated the potential of montmorillonite clay minerals to remove fluoride from water (Guiza et al., 2021). The results showed that montmorillonite showed good removal efficiency, up to 40 mg/g. The study also noted that water pH and fluoride concentration were critical factors that affected the efficiency of clay minerals for fluoride removal. The researchers concluded that clay minerals have the potential to be an effective and affordable solution for fluoride removal, where the problem of fluoride contamination is widespread. However, it should be noted that further research is needed to determine the feasibility of using clay minerals for large-scale fluoride removal, as well as to investigate any potential negative effects of long-term use of these minerals on water quality and soil fertility.

In addition to developing effective adsorbents, researchers have also explored various methods to optimize their performance. An approach is to modify the surface of the adsorbent material to enhance its fluoride removal capacity. Another way is to use a combination of adsorbents to improve efficiency. In general, research in defluoridation adsorbents has made significant progress in recent years, offering hope for hilly rural communities in India affected by fluoride contamination in drinking water. However, more research and development is needed to ensure that these technologies can be scaled up and implemented effectively in resource-limited settings.

5.4 Viability of State-of-the-Art Adsorbents

Several organizations, start-ups, social firms, research institutes, and NGOs in India are working independently or in collaboration with international agencies to establish fluoride removal systems using agroforestry waste, soil, and other means to make it feasible and viable for the country's remote areas. The following subsections summarized few critical technologies that have been established and can be utilized directly or in conjunction with the replacement of agroforestry waste appropriate to the locality in hilly rural parts of India, as follows.

5.4.1 CMERI Fluoride and Iron Removal Technology

The Central Mechanical Engineering Research Institute (CMERI) in India has developed a fluoride and iron removal technology for drinking water. The technology pre-treatment is carried out through a sand filter to remove suspended particles and impurities. The water is passed through a specially designed adsorption filter containing sand, gravel, and adsorbent materials such as activated alumina, activated carbon, and bone char. These materials adsorb fluoride and other contaminants from the water. In post-treatment, the water is finally passed through a carbon filter to remove any residual impurities and improve the taste and odor. This technology has several advantages over other fluoride and iron removal technologies, such as low cost, low maintenance, and high efficiency. It is particularly suitable for rural areas where access to safe drinking water is a significant problem.

5.4.2 Eawag Technology

Eawag is a Swiss water research institute that has developed a method for fluoride removal using bone char. This process can be used by substituting local agroforestry waste for bone char.

5.4.3 The Inter-country Center for Oral Health (ICOH) Defluoridator

The Inter-Country Center for Oral Health (ICOH) of the University of Chiang Mai (Thailand) developed a domestic defluoridation that removes excess fluoride from drinking water using crushed charcoal and bone char. The technology is simple to build and use, making it an economical and practical alternative for populations with excessive levels of fluoride in drinking water.

5.4.4 Fluoridation Using Fired Bricks

In Sri Lanka, a device was tested to remove fluoride from drinking water. The unit has a bed of freshly burnt brick fragments topped with burnt coconut shells and pebbles. In an up-flow mode, water is passed through the unit. According to reports, the efficiency is determined by the quality of freshly burned bricks. The technology can be upgraded for India's hilly regions using locally available soils, stones, and agroforestry waste.

5.4.5 Solar-Operated Treatment Plant (Aqua Sphere Green Tech)

The system uses solar energy and is suitable for treating arsenic, fluoride, and TDS. Fluoride is removed with re-generable or disposable granular media. Microfiltration and ultrafiltration are used to treat surface water. The compact system has a capacity of 5.7 KLD, and the mini WS System has a capacity of 40–50 KLD.

Each technique has strengths and weaknesses in terms of efficiency, effectiveness, and economy. The choice of technology/adsorbent for defluoridation will depend on various aspects, including the application's specific requirements, the availability of the adsorbent material, and the project budget constraints.

The ability of an adsorbent to remove fluoride from water is termed its efficiency. The efficiency of an adsorbent is determined by several aspects, including its surface area, pore size distribution, chemical characteristics, and surface properties. Adsorbents with a large surface area, a wide pore size distribution, and a specific chemical makeup that promotes fluoride adsorption are more effective in removing fluoride from water.

Effectiveness refers to the ability of an adsorbent to maintain its efficiency over time. Some adsorbents may lose effectiveness over time due to fouling, clogging, or saturation. Adsorbents resistant to fouling and clogging tend to be more effective in maintaining their efficiency over time.

The economy refers to the cost-effectiveness of using an adsorbent for defluoridation. The economics of an adsorbent depends on various factors, including the cost of the adsorbent material, the availability of the adsorbent material, and the operating costs associated with the use of the adsorbent. Adsorbents that are readily available and inexpensive and have low operating prices tend to be more economically viable for defluoridation.

In summary, the choice of defluoridation adsorbent for rural hilly areas of India depends on factors such as cost, availability, fluoride removal capacity, and ease of regeneration. Activated alumina and bone char are the most widely used adsorbents in India as a result of their low cost and high fluoride removal capacity. However, activated carbon and synthetic resins can also be used if price is not a limiting factor. Finally, the adsorbent should be chosen on the basis of the particular demands and accessibility of the community resources.

5.5 Conclusion

In general, defluoridation adsorbents are an effective and inexpensive solution to decrease fluoride levels in drinking water. However, the sustainability of these practices depends on the availability and affordability of the defluoridation units and the willingness to maintain and operate them. On rural hillsides, defluoridation units should be set up at the village or community level, where water can be treated prior to distribution. These adsorbents can be prepared using locally available agroforestry waste in the hilly regions of rural India. The village community or a local organization should typically operate and maintain these units. The treated water is then distributed to households for consumption. Ongoing research and development is needed to improve the effectiveness and cost-effectiveness of these adsorbents and ensure that they are accessible to all communities in need.

Acknowledgements The authors thank our prestigious institute and other people who are involved in this study for their direct and indirect contributions.

References

Alagumuthu, G., & Rajan, M. (2010). Equilibrium and kinetics of adsorption of fluoride onto zirconium impregnated cashew nut shell carbon. *Chemical Engineering Journal, 158*, 451–457. https://doi.org/10.1016/j.cej.2010.01.017

Das, S., Pramanik, A., Das, R., & Chatterjee, A. (2022). An evolving perspective on the fluoride mitigation techniques. *International Journal of Environmental Science and Technology*, 1–32.

Dwivedi, S., Mondal, P., & Balomajumder, C. (2014). Removal of fluoride using Citrus limetta in batch reactor: Kinetics and equilibrium studies. *Research Journal of Chemical Sciences, 4*, 50–58. 2231-606X

Ganvir, V., & Das, K. (2011). Removal of fluoride from drinking water using aluminum hydroxide coated rice husk ash. *Journal of Hazardous Materials, 185*, 1287–1294. https://doi.org/10.1016/j.jhazmat.2010.10.044

Guiza, S., Brouers, F., & Bagane, M. (2021). Fluoride removal from aqueous solution by montmorillonite clay: Kinetics and equilibrium modeling using new generalized fractal equation. *Environmental Technology and Innovation, 21*, 101187.

Joshi, S., Garg, M., & Jana, S. (2022). Thermal activated adsorbent from D. sissoo sawdust for fluoride removal: Batch study. *Journal of the Institution of Engineers India: Series E, 103*, 323–337.

Karthikeyan, K., Nanthakumar, K., Velmurugan, P., Tamilarasi, S., & Lakshmanaperumalsamy, P. (2010). Prevalence of certain inorganic constituents in groundwater samples of Erode district, Tamilnadu, India, with special emphasis on fluoride, fluorosis and its remedial measures. *Environmental Monitoring and Assessment, 160*, 141–155. https://doi.org/10.1007/s10661-008-0664-0

Khoshnamvand, N., Bazrafshan, E., & Kamarei, B. (2018). Fluoride removal from aqueous solutions by NaOH-modified eucalyptus leaves. *Ssu-Jehsd, 3*, 481–487.

Manna, S., Roy, D., Adhikari, B., Thomas, S., & Das, P. (2018). Biomass for water defluoridation and current understanding on biosorption mechanisms: A review. *Environmental Progress & Sustainable Energy, 37*, 1560–1572.

Mohan, R., & Dutta, R. K. (2020). A study of suitability of limestone for fluoride removal by phosphoric acid-crushed limestone treatment. *Journal of Environmental Chemical Engineering, 8*, 104410.

MoJS. (2023). Handbook on drinking water treatment technologies.

Mondal, N. K., Bhaumik, R., Baur, T., Das, B., Roy, P., & Datta, J. K. (2012). Studies on defluoridation of water by tea ash: An unconventional biosorbent. *Chemical Science Transactions, 1*, 239–256.

Parmar, H. S., Patel, J. B., Sudhakar, P., & Koshy, V. (2006). Removal of fluoride from water with powdered corn cobs. *Journal of Environmental Science & Engineering, 48*, 135–138.

Pratha, A. A., & Prabakar, J. (2020). Defluoridation potential of rice husk, groundnut shell as a conventional alternative for fluoride removal—A Review. *Journal of Pharmaceutical Research International, 32*, 124–131.

Rajan, M., & Alagumuthu, G. (2012). Study of fluoride affinity by zirconium impregnated walnut shell carbon in aqueous phase: Kinetic and isotherm evaluation. *Journal of Chemistry, 2013*, 235048. https://doi.org/10.1155/2013/235048

Roy, S., & Das, P. (2016). Assessment on the defluoridation using novel activated carbon synthesized from tea waste: Batch, statistical optimization and mathematical modeling. *Journal of Industrial Pollution Control, 32*.

Sivabalan, R., Rengaraj, S., Arabindoo, B., & Murugesan, V. (2003). Cashewnut sheath carbon: A new sorbent for defluoridation of water. *Indian Journal of Chemical Technology, 10*, 217–222.

Thakur, R. S., Katoch, S. S., & Modi, A. (2020). Assessment of pine cone derived activated carbon as an adsorbent in defluoridation. *SN Applied Sciences, 2*, 1407. https://doi.org/10.1007/s42452-020-03207-x

Uddin, M. K., Ahmed, S. S., & Naushad, M. (2019). A mini update on fluoride adsorption from aqueous medium using clay materials. *Desalination and Water Treatment, 145*, 232–248.

Walia, R., Chauhan, A., & Kumar, S. (2021). Literature review on the elimination of fluoride ions from industrial wastewater utilizing tamarind. *International Journal of Advanced Engineering Management, 3*, 110–120. https://doi.org/10.35629/5252-0307110120

Wendimu, G., Zewge, F., & Mulugeta, E. (2017). Aluminium-iron-amended activated bamboo charcoal (AIAABC) for fluoride removal from aqueous solutions. *Journal of Water Process Engineering, 16*, 123–131.

Chapter 6
Experimental Evaluation of Remediation of Fluoride-Contaminated Water Using Limestone Powder

Ravindra Budania, Prashant Bhadula, and Sanyam Dangayach

Abstract The presence of fluoride in Groundwater is a significant issue. The fluoride in the water used for human consumption does not exceed 1.5 mg/l, according to the WHO; however, many countries already exceed that limit. The human health risks of fluoride contamination are well documented, prompting widespread efforts to find effective defluoridation strategies. This study aims to investigate the efficacy of inexpensive limestone powder as an adsorbent for treating fluoride-contaminated Groundwater. The efficacy of limestone powder was investigated through batch studies with different physical and chemical parameters, including pH, contact duration, adsorbent dosage, and fluoride concentration. The removal of fluoride using limestone was most effective at a pH of 7.22, with a sorption capacity of 2.57 mg/g. It took 90 min to reach equilibrium. Groundwater samples were also used in the sorption investigation; the study revealed that limestone powder could decrease fluoride from 3.48 to 0.878 ppm. SEM and FTIR spectroscopy were used to ascertain the surface's functional group and morphology.

Keywords Limestone powder · Fluoride contamination · Adsorption

R. Budania · P. Bhadula · S. Dangayach (✉)
Department of Civil Engineering, Malaviya National Institute of Technology, Jaipur, Rajasthan 302017, India
e-mail: sanyamd.ce@mnit.ac.in

R. Budania
e-mail: 2019rce9058@mnit.ac.in

P. Bhadula
e-mail: 2021rce9514@mnit.ac.in

A. K. Yadav et al. (eds.), *Advanced Treatment Technologies for Fluoride Removal in Water*, Water Science and Technology Library 125,
https://doi.org/10.1007/978-3-031-38845-3_6

6.1 Introduction

Groundwater is a crucial drinking water resource for numerous communities and is recharged by precipitation that infiltrates the ground and stores it in the underground water system. Groundwater quality can suffer due to various activities that may arise from natural or human-induced sources, such as excessive use of fertilizers in agriculture, waste discharge from industrial or residential areas, and changes made to the hydrology of land surfaces. Groundwater can also be contaminated with various substances, including chemicals, pathogens, and naturally occurring minerals such as fluoride, nitrate, and arsenic (Li et al., 2021; Yadav et al., 2014).

Fluoride occurs naturally, and its concentration in the world's aquifers varies (Zhou et al., 2018). Although fluoride is essential for good health, excessive exposure can cause health problems, especially in children (Solanki et al., 2022). An excessive amount of fluoride present in water intake can affect the teeth and skeletal system (Budania & Dangayach, 2023). Fluorosis can cause tooth discoloration and mottling and can cause bones to become fragile and prone to fractures (Lawrinenko et al., 2023). Almost 200 million people drink water that exceeds the WHO 1.5 ppm fluoride guideline (Kut et al., 2016). Groundwater fluoride contamination can be mitigated by consistently monitoring and assessing groundwater quality and implementing preventive measures.

Fluoride contamination of Groundwater can occur due to natural processes, such as volcanic activity and weathering, or human activities, such as the disposal of fluoride-containing waste. There are several techniques available that can be used to eliminate water fluoride contamination, including electrocoagulation (Vasudevan & Oturan, 2014), precipitation (Budyanto et al., 2015), ion exchange (Jia et al., 2015), membrane processes (Shen et al., 2015), and adsorption (Chai et al., 2013). The most appropriate method depends on the specific site conditions and the fluoride concentration. Electrocoagulation and membrane technologies are adequate but not ideal for underprivileged areas due to cost and availability. Adsorption and precipitation have been proven to be effective and environmentally safe methods of removing excess fluoride from drinking water. They work well, are easy to use, and are economical.

Numerous studies have been conducted to develop an affordable, efficient and easily accessible adsorbent to remove fluoride from Groundwater. Several researchers have examined the efficacy of various adsorbents in the removal of fluoride, including activated alumina (Ahamad et al., 2018), activated carbon (Fito et al., 2019), calcite (Wang et al., 2022), activated sawdust (Dhanasekaran & Sahu, 2021), active coconut shell carbon (Bhamare et al., 2022), bone charcoal (Cruz-Briano et al., 2021), soil adsorbent (Wambu et al., 2016), and others.

Among these several approaches, fluoride adsorption using limestone powder appeared exciting and potentially practical. Limestone powder is the fine powder that is produced by crushing limestone rocks. It mainly comprises calcium carbonate ($CaCO_3$) and may contain other minerals and trace elements (Nath & Dutta, 2010). It finds common usage in diverse industrial applications, including, but not limited to, construction, agriculture, manufacturing, and water treatment.

Limestone is widely used in water treatment applications because it neutralizes acidic water and removes contaminants such as metallic substances, fluoride, and organic compounds (Silva et al., 2010). When limestone is added to water contaminated with fluoride, the following reaction occurs:

$$CaCO_3 + 2F^- \rightarrow CaF_2 + CO_2 \tag{6.1}$$

Calcium carbonate ($CaCO_3$) in limestone combines with fluoride ions (F^-) in water to form insoluble calcium fluoride (CaF_2) and carbon dioxide (CO_2). Insoluble calcium fluoride will precipitate from the solution and easily remove (Tanvir Arfin, 2015).

Therefore, it is crucial to find easy and sustainable ways to remove fluoride. In this context, this research investigates the viability of using limestone powder as a remediation material for fluoride-contaminated water. In this study, laboratory experiments will be conducted to assess the efficacy of limestone powder in removing fluoride under different conditions, including contact time, pH, and dosage.

6.2 Materials and Method

6.2.1 Material

Limestone powder with a particle of less than 150 μm was acquired from Pinakin minerals, Jaipur, Rajasthan, India. Deionized water was utilized as the solution solvent because it is free of dissolved minerals and ions that might interfere with the experiment. Stock and standard fluoride solutions were prepared using analytical reagent grade NaF (sodium fluoride). The experiments were conducted in polypropylene flasks to avoid contamination from outside sources.

6.2.2 Characterization

FTIR and SEM/EDS are analytical techniques commonly used to characterize materials and their properties. FTIR can detect functional and chemical species on a material's surface by revealing chemical bonds. Generally, the range of wavelengths used for FTIR analysis is between 4000 and 400 cm^{-1}, with different regions providing information about different aspects of the material's structure and composition.

SEM produces high-resolution images of the surface of the adsorbent, which can reveal important details about its morphology and texture. The material's elemental composition can be identified through EDS, which analyzes the X-ray emissions generated by bombarding the material with a high-energy electron beam.

6.2.3 Batch Studies

A study was conducted using a batch experiment to assess how varying environmental and chemical factors affect the fluoride removal efficiency of a solution. The study aimed to govern the optimal parameters for the removal of fluoride from the solution by altering these factors.

At room temperature, a series of studies were carried out in 250 ml Teflon containers (Tarsons brand) placing a known adsorbent dose in a 100 ml fluoride solution of known concentration. To regulate the pH of the fluoride-polluted solution in this investigation, 0.1 N HCl or 0.1 N NaOH was employed. The entire batch study was carried out at room temperature. Groundwater was also studied after batch tests were performed using synthetic water.

6.3 Result and Discussion

6.3.1 Fourier Transform Infrared Spectroscopy (FTIR)

The spectrum displayed distinct absorption bands related to the molecular vibrations of its diverse chemical constituents. As shown in Fig. 6.1, the 706, 875 and 1427 cm^{-1} peaks in the spectrum indicate the occurrence of carbonate (CO_3) stretching and bending vibrations before adsorption. Limestone mainly comprises calcium carbonate ($CaCO_3$), which contains carbonate groups in its crystal structure.

The spectrum peaks indicate the existence of calcite at 1823 and 2512 cm^{-1}. These findings show that calcite constitutes the bulk of the limestone studied. Furthermore, the spectrum at 3448 cm^{-1} is related to the crystalline structure's hydroxyl (–OH)

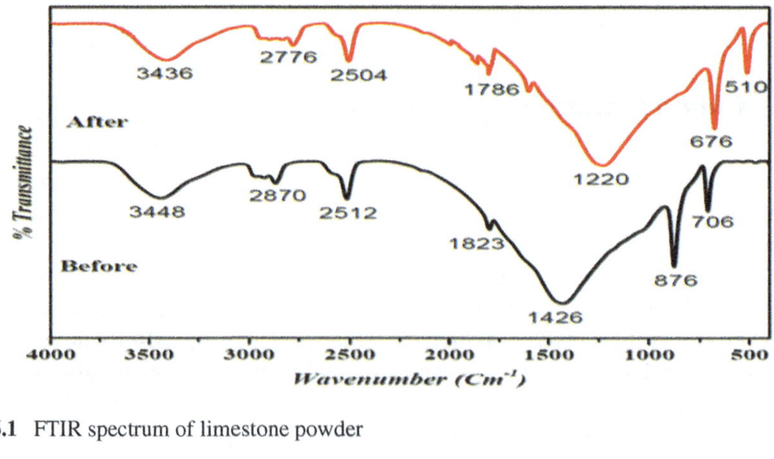

Fig. 6.1 FTIR spectrum of limestone powder

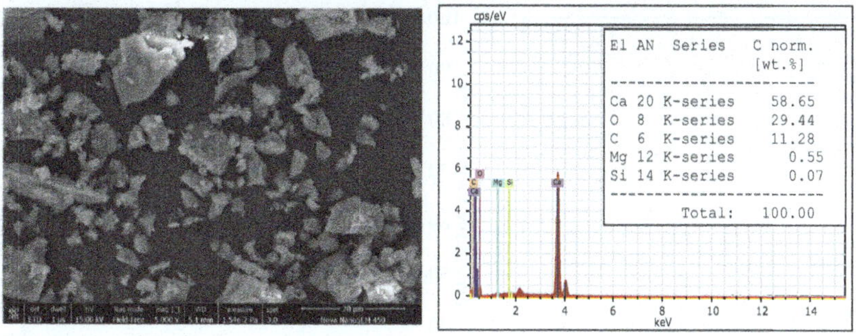

Fig. 6.2 SEM–EDS micrographs of limestone powder

group. This vibration is associated with the O–H bond and indicates the presence of water molecules or hydroxyl ions within the lattice of the limestone mineral.

The Ca–F stretching vibration of CaF_2 peaked in the spectra at 676 cm^{-1} after adsorption. Similar observations were drawn by Hwidi et al. (2018) for limestone. The bending of the adsorbent showed higher transmittance before adsorption than after adsorption. Since fluoride molecules filled all active sites after adsorption, transmittance was reduced.

6.3.2 SEM–EDS

Figure 6.2 displays SEM images of limestone powder magnified 5000 times, revealing LS's surface morphology to be angular and rough, with visible pores and cracks. Dust particles were also observed on the surface. Moreover, the morphology of the samples displays angular quartz grains of specific sizes coated with fine particles. Calcite grains are also easily discernible.

Identification of calcium (Ca), oxygen (O) and carbon (C) as vital elements in the sample indicates the significant role of carbonate phases, particularly the $CaCO_3$ compound, in the sample (Zhang et al., 2017). $CaCO_3$ has important implications for their physical and chemical properties, including its potential use as an adsorbent material, as $CaCO_3$ has been shown to possess adsorption capabilities for various pollutants (Aziz et al., 2020). The SEM–EDS analysis of LS showed no presence of toxic elements, indicating that it is safe for further study as an adsorbent material.

6.3.3 Impact of Contact Time

Figure 6.3 shows the investigation results for contact times ranging from 30 to 180 min, focusing on the percentages of adsorption and removal. The results show that the remediation initially but soon achieved saturation as the contact duration

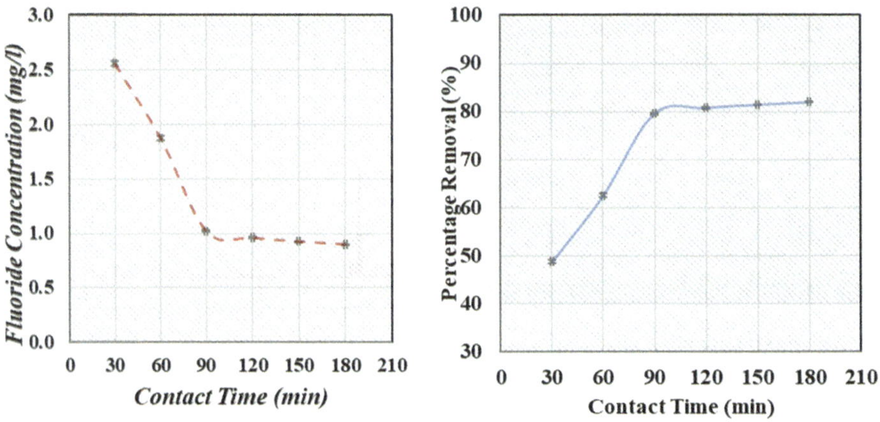

Fig. 6.3 Impact of contact duration on removal efficiency

extended. Specifically, we found that the removal percentage was approximately 80% at a concentration of 5 ppm after 90 min, and this contact time was deemed optimal for further investigation.

The high adsorbate concentration and significant free adsorbent surface may trigger the initial adsorption. After equilibrium, fluoride absorption slowed due to the inadequate movement of the adsorbate molecule from the solution to the material surface. These results are significant in a prior investigation on the treatment of fluoride using industrial adsorbents such as hydrated cement and marble powder (Suneetha et al., 2015).

Limestone has virtuous adsorption characteristics and could remove fluoride and other contaminants from contaminated water. The 90-min contact duration is optimal for further study to enhance adsorption.

6.3.4 Impact of pH

Figure 6.4 illustrates the results of a study investigating the Impact of pH on fluoride treatment using limestone. According to the study, the most significant fluoride remediation was achieved at a pH of 6.0–8.0. However, the efficiency decreased when the pH was below 6.0 or above 8.0.

The efficacy of fluoride adsorption is reduced in acidic solutions because of the presence of faintly ionized hydrofluoric acid, which decreases the concentration of unbound ions accessible for reaction. As shown in Fig. 6.4b, pHzc was 7.22. Similarly, as pH > pHzc, decreases the positively charged surface, leads to a lower concentration of adsorbed anions. As a result, the elevated concentration of

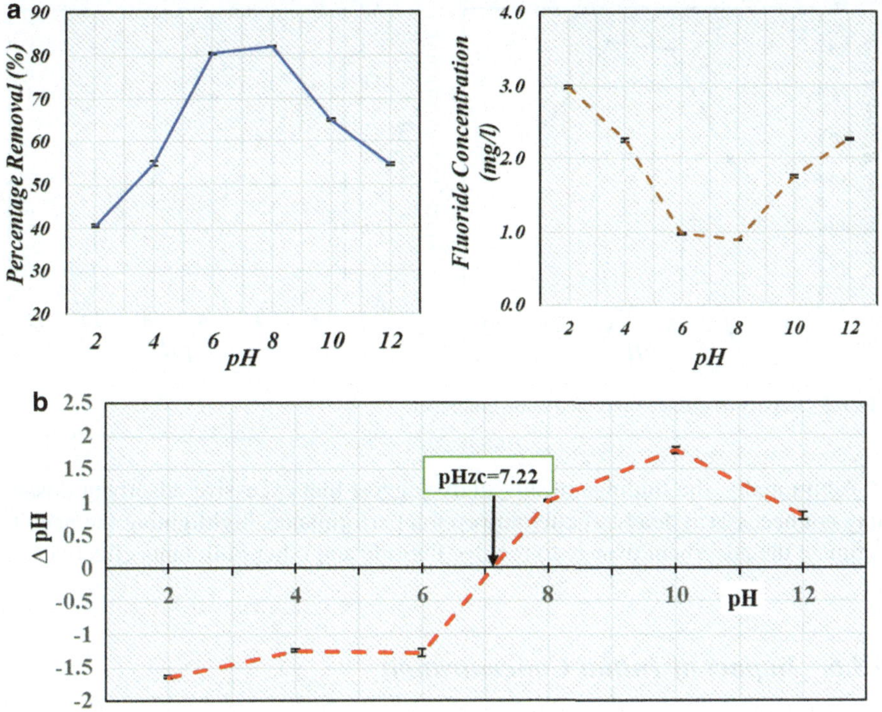

Fig. 6.4 **a** Impact of pH on remediation efficiency. **b** Point of zero charge (pHzc)

OH– anions in the solution affects the interaction between fluoride and hydroxide (OH–) ions at the binding site. Other studies have reported similar findings on optimal fluoride treatment using rare earth oxides and lateritic soil as adsorbents (Gebrewold et al., 2019).

These findings suggest that a pH of 6.0–8.0 is optimal for fluoride adsorption using limestone samples. Therefore, further studies can optimize the pH conditions for fluoride removal. Limestone could potentially be used to remove fluoride and other contaminants as a cost-effective adsorbent.

6.3.5 *Impact of Absorbent Dose*

After identifying the optimal pH and duration of contact, the Impact of dosage on fluoride remediation was explored using limestone. Figure 6.5 illustrates the results of the study, which demonstrate that the optimal dosage to remove fluoride is 3 mg/l. Beyond this dose, there was no significant improvement in fluoride removal due to particle aggregation, active site overlaps at higher doses, or a reduction in effective surface area (Lamayi et al., 2018).

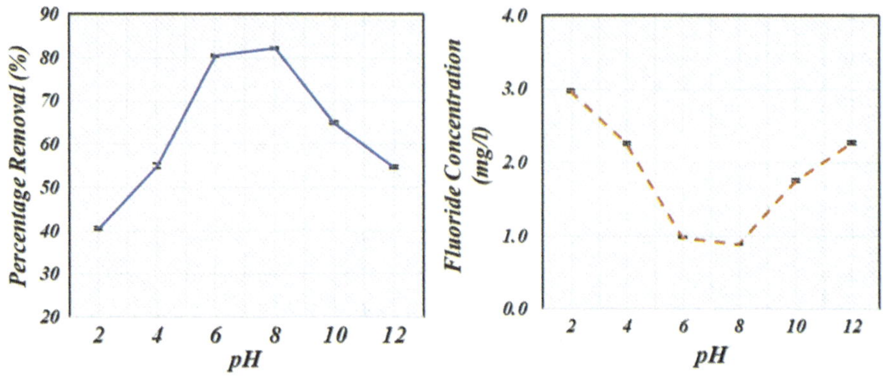

Fig. 6.5 Impact of dosage on remediation efficiency

Additionally, the findings of this study suggest that excessive adsorbent dosage may not necessarily lead to better adsorption performance, highlighting the need to optimize the adsorbent dosage to remove fluoride and other pollutants effectively.

6.3.6 Impact of Initial Concentration

Limestone was used as an adsorbate to remove fluoride concentration. These experiments were carried out under optimal pH conditions, adsorbent dosage, and contact duration. Figure 6.6 showed that as the fluoride content in a solution increased, the limestone powder became less efficient in removing fluoride from the water (Kamala et al., 2005).

As the ions in the solution increase, the graph depicting the efficacy of fluoride removal becomes flatter, indicating a decrease in remediation competence. The adsorption rate is influenced by the nature of the material and the sites available for adsorption, as indicated by the findings. When the fluoride concentration is low, fluoride ions can access the limestone surface site more, resulting in greater removal efficiency. However, the ratio is reduced for higher concentrations, leading to a small percentage of removal (Kamala et al., 2005).

Figure 6.6b shows that limestone powder has an adsorbent capacity of 2.57 mg/g. Limestone has potential for fluoride removal, but its effectiveness may be limited by the initial concentration of fluoride in water. Activated granular carbon and other adsorbent materials have also shown similar results in the treatment of fluoride (Kamala et al., 2005). Therefore, further investigations can explore the mechanisms underlying the concentration-dependent effect on fluoride removal using limestone and identify the mechanism for enhancing limestone adsorption capacity for higher fluoride concentrations in contaminated water.

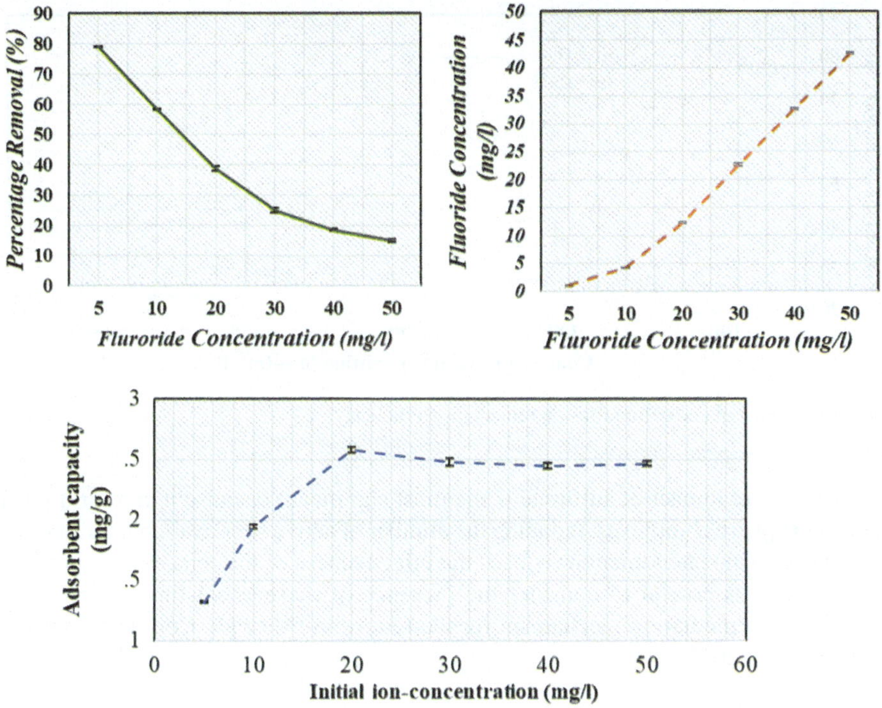

Fig. 6.6 Impact of concentration on remediation efficiency

6.3.7 Impact of Coexisting Ions

Groundwater is limited not only to fluoride, but it may also comprise dissolved ions, such as chloride, sulfate, nitrate, bicarbonate, and calcium. To explore the Impact of other anions on the removal of fluoride using limestone, this study added varying concentrations (100–500 mg/l) of nitrate, chloride, and sulfate anions to water (5 mg/l). Nitrate has a weaker affinity for the active sorption sites and interacts through outer-sphere complexation. As a result, its Impact on the effectiveness of remediation is relatively lower, as shown in Fig. 6.7 (Mohan & Dutta, 2020).

At the same time, chloride can have positive and negative effects depending on the concentration. At low levels, chloride ions can increase the effectiveness of fluoride removal by creating a protective coating on the limestone surface, inhibiting the erosion of limestone particles. However, chloride ions can inhibit fluoride removal at high concentrations by competing with fluoride ions for adsorption (Eskandarpour et al., 2008).

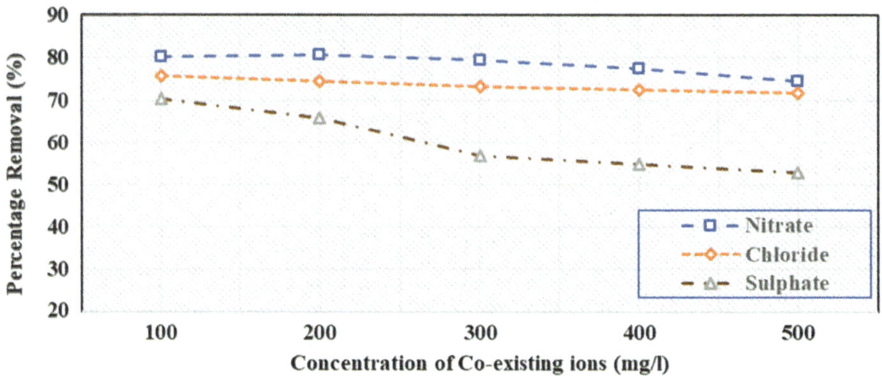

Fig. 6.7 Impact of coexisting ions on remediation efficiency

Instead, the Impact of sulfurate is the most significant because it can form inner and outer sphere complexes, reducing the number of active sites available for fluoride adsorption. This phenomenon reduces the effectiveness of the remediation process. Therefore, it is crucial to consider the presence of these coexisting anions when designing a remediation mechanism for the removal of fluoride from Groundwater (Zhang et al., 2015).

6.3.8 Remediation of Groundwater

This experiment aimed to evaluate the effectiveness of limestone treatment in removing fluoride from Groundwater obtained from the Churu district in Rajasthan, India. The water characteristics of the groundwater sample collected, including its initial fluoride content of 3.48 mg/l, are shown in Table 6.1.

After batch studies, the groundwater sample was treated with limestone at optimal parameters. As shown in Table 6.1, fluoride concentrations were reduced using limestone from 3.48 to 0.878 mg/l.

The experiment conducted on water samples from the Churu area of Rajasthan inferred that limestone is an efficient fluoride remover from Groundwater. The water parameter did not affect the efficacy of the removal, and the approach could address the problem of high fluoride concentration at various locations.

Table 6.1 Groundwater characteristics before and after limestone treatment

Parameters (mg/l except for pH)	WHO standard	Initial	After treatment
pH	6.5–8.5	7.8	7.41
Dissolved solids	600	280	232
Hardness	200	130	180
Alkalinity	200	132	145
Chloride	250	32	19.3
Nitrate	50	1.34	1.32
Sulphate	500	54	48
Sodium	200	87	82
Iron	0.3	0.071	Less than 0.001
Fluoride	1.5	3.48	0.878
Potassium	NS (Not Specified)	3.98	3.43

6.4 Conclusion

In this study, the equilibrium characteristics of fluoride adsorption in limestone powder were examined. Controlled conditions were used for batch experiments. The results showed 90 min to reach a stable equilibrium state. As the pH of the solution increased from 4 to 8, the absorption of fluoride was also increased, and the initial adsorption rates were found to increase with higher pH levels. However, at high alkaline pH, fluoride efficacy decreased as a result of the interaction between hydroxide and fluoride ions. The equilibrium studies were conducted at pH 7.5, similar to natural water. There was an adsorbent capacity of 2.57 mg/g for limestone powder, which could be enhanced using different defluoridation strategies. The efficiency of limestone fluoride removal can vary significantly depending on its composition and groundwater conditions. Therefore, it is crucial to carefully assess the quality and composition of the limestone powder before using it for the removal of fluoride. Limestone is an attractive alternative for fluoride treatment because it is widely available and inexpensive.

Acknowledgements The authors thank the Materials Research Center (MRC) at MNIT, Jaipur, for generously providing access to their sample characterization facility.

References

Ahamad, K. U., et al. (2018). Equilibrium and kinetics modeling of fluoride adsorption onto activated alumina, alum and brick powder. *Groundwater for Sustainable Development, 7*(April), 452–458. https://doi.org/10.1016/j.gsd.2018.06.005

Aziz, H. A., et al. (2020). Iron and manganese removal from groundwater using limestone filter with iron-oxidized bacteria. *International Journal of Environmental Science and Technology, 17*(5), 2667–2680. https://doi.org/10.1007/s13762-020-02681-5

Bhamare, P., et al. (2022). Coconut shell derived coal as low-cost adsorbent for the fluoride removal from drinking water. *IOP Conference Series: Earth and Environmental Science, 1086*(1), 012038. https://doi.org/10.1088/1755-1315/1086/1/012038

Budania, R., & Dangayach, S. (2023). A comprehensive review on permeable reactive barrier for the remediation of groundwater contamination. *Journal of Environmental Management, 332*(February), 117343. https://doi.org/10.1016/j.jenvman.2023.117343

Budyanto, S., Kuo, Y. L., & Liu, J. C. (2015). Adsorption and precipitation of fluoride on calcite nanoparticles: A spectroscopic study. *Separation and Purification Technology, 150*, 325–331. https://doi.org/10.1016/j.seppur.2015.07.016

Carrard, N., Foster, T., & Willetts, J. (2019). Groundwater as a source of drinking water in southeast Asia and the Pacific: A multi-country review of current reliance and resource concerns. *Water (Switzerland), 11*(8). https://doi.org/10.3390/w11081605

Chai, L., et al. (2013). Sulfate-doped Fe_3O_4/Al_2O_3 nanoparticles as a novel adsorbent for fluoride removal from drinking water. *Water Research, 47*(12), 4040–4049. https://doi.org/10.1016/j.watres.2013.02.057

Cruz-Briano, S. A., et al. (2021). Bone char from an invasive aquatic specie as a green adsorbent for fluoride removal in drinking water. *Water, Air, and Soil Pollution, 232*(9). https://doi.org/10.1007/s11270-021-05286-x

Dhanasekaran, P., & Sahu, O. (2021). Arsenate and fluoride removal from groundwater by sawdust impregnated ferric hydroxide and activated alumina (SFAA). *Groundwater for Sustainable Development, 12*, 100490. https://doi.org/10.1016/j.gsd.2020.100490

Eskandarpour, A., et al. (2008). Removal of fluoride ions from aqueous solution at low pH using schwertmannite. *Journal of Hazardous Materials, 152*(2), 571–579. https://doi.org/10.1016/j.jhazmat.2007.07.020

Fito, J., et al. (2019). Fluoride removal from aqueous solution onto activated carbon of Catha edulis through the adsorption treatment technology. *Environmental Systems Research, 8*(1), 1–10. https://doi.org/10.1186/s40068-019-0153-1

Gebrewold, B. D., et al. (2019). Fluoride removal from groundwater using chemically modified rice husk and corn cob activated carbon. *Environmental Technology (United Kingdom), 40*(22), 2913–2927. https://doi.org/10.1080/09593330.2018.1459871

Hwidi, R. S., Tengku Izhar, T. N., & Mohd Saad, F. N. (2018). Characterization of limestone as raw material to hydrated lime. *E3S Web of Conferences, 34*(January). https://doi.org/10.1051/e3sconf/20183402042

Jia, Y., et al. (2015). Porous 2-line ferrihydrite/bayerite composites (LFBC): Fluoride removal performance and mechanism. *Chemical Engineering Journal, 268*, 325–336. https://doi.org/10.1016/j.cej.2015.01.080

Kamala, C. T., et al. (2005). Removal of arsenic(III) from aqueous solutions using fresh and immobilized plant biomass. *Water Research, 39*(13), 2815–2826. https://doi.org/10.1016/j.watres.2005.04.059

Kut, K. M. K., et al. (2016). A review of fluoride in African groundwater and local remediation methods. *Groundwater for Sustainable Development, 2–3*, 190–212. https://doi.org/10.1016/j.gsd.2016.09.001

Lamayi, D. W., et al. (2018). Aqueous phase removal of fluoride as fluorosis agent using montmorillonite clay as a natural nanoadsorbent. *Nanochemistry Research, 3*(2), 219–226. https://doi.org/10.22036/ncr.2018.02.012

Lawrinenko, M., Kurwadkar, S., & Wilkin, R. T. (2023). Long-term performance evaluation of zero-valent iron amended permeable reactive barriers for groundwater remediation—A mechanistic approach. *Geoscience Frontiers, 14*(2), 101494. https://doi.org/10.1016/j.gsf.2022.101494

Li, P., et al. (2021). Sources and consequences of groundwater contamination. *Archives of Environmental Contamination and Toxicology, 80*(1), 1–10. https://doi.org/10.1007/s00244-020-00805-z

Mohan, R., & Dutta, R. K. (2020). A study of suitability of limestone for fluoride removal by phosphoric acid-crushed limestone treatment. *Journal of Environmental Chemical Engineering, 8*(6), 104410. https://doi.org/10.1016/j.jece.2020.104410

Nath, S. K., & Dutta, R. K. (2010). Fluoride removal from water using crushed limestone. *Indian Journal of Chemical Technology, 17*(2), 120–125.

Shen, J., et al. (2015). Renewable energy powered membrane technology: Fluoride removal in a rural community in northern Tanzania. *Separation and Purification Technology, 149*, 349–361. https://doi.org/10.1016/j.seppur.2015.05.027

Silva, A. M., et al. (2010). Manganese and limestone interactions during mine water treatment. *Journal of Hazardous Materials, 181*(1–3), 514–520. https://doi.org/10.1016/j.jhazmat.2010.05.044

Solanki, Y. S., et al. (2022). Fluoride occurrences, health problems, detection, and remediation methods for drinking water: A comprehensive review. *Science of the Total Environment, 807*, 150601. https://doi.org/10.1016/j.scitotenv.2021.150601

Suneetha, M., Sundar, B. S., & Ravindhranath, K. (2015). Removal of fluoride from polluted waters using active carbon derived from barks of Vitex negundo plant. *Journal of Analytical Science and Technology, 6*(1), 1–19. https://doi.org/10.1186/s40543-014-0042-1

Tanvir Arfin, S. S. W. (2015) Fluoride removal from water by calcium materials: A state-of-the-art review. *International Journal of Innovative Research in Science, Engineering and Technology, 4*(9), 8090–8102.https://doi.org/10.15680/ijirset.2015.0409013

Vasudevan, S., & Oturan, M. A. (2014). Electrochemistry: As cause and cure in water pollution-an overview. *Environmental Chemistry Letters, 12*(1), 97–108. https://doi.org/10.1007/s10311-013-0434-2

Velis, M., Conti, K. I., & Biermann, F. (2017). Groundwater and human development: Synergies and trade-offs within the context of the sustainable development goals. *Sustainability Science, 12*(6), 1007–1017. https://doi.org/10.1007/s11625-017-0490-9

Wambu, E. W., et al. (2016). Review of fluoride removal from water by adsorption using soil adsorbents—An evaluation of the status. *Journal of Water Reuse and Desalination, 6*(1), 1–29. https://doi.org/10.2166/wrd.2015.073

Wang, J., et al. (2022). Mechanisms of fluoride uptake by surface-modified calcite: A 19F solid-state NMR and TEM study. *Chemosphere, 294*, 133729. https://doi.org/10.1016/j.chemosphere.2022.133729

Yadav, A. K., et al. (2014). Concentrations of uranium in drinking water and cumulative, age-dependent radiation doses in four districts of uttar pradesh, india. *Toxicological and Environmental Chemistry, 96*, 192–200. https://doi.org/10.1080/02772248.2014.934247

Zhang, Y. G., et al. (2015). Ultralong hydroxyapatite nanowires synthesized by solvothermal treatment using a series of phosphate sodium salts. *Materials Letters, 144*, 135–137. https://doi.org/10.1016/j.matlet.2015.01.031

Zhang, Y., Sun, Q., & Geng, J. (2017). Microstructural characterization of limestone exposed to heat with XRD, SEM and TG-DSC. *Materials Characterization, 134*(March), 285–295. https://doi.org/10.1016/j.matchar.2017.11.007

Zhou, J., et al. (2018). Arsenic and fluoride removal from contaminated drinking water with Haix-Fe–Zr and Haix-Zr resin beads. *Journal of Environmental Management, 9*(3), 100248. https://doi.org/10.1016/j.jclepro.2017.03.061

Chapter 7
Utilization of Inexpensive Bio-sorbents for Water Defluoridation

Veera Brahmam Mukkanti, A. R. Tembhurkar, and Rajesh Gupta

Abstract Groundwater contamination with excess fluoride concentration is a serious concern for various nations around the world. Consumption of drinking water with excess fluoride poses a serious threat to human health through dental/ skeletal fluorosis. Therefore, defluoridation of water has attracted global attention and different techniques have been developed to ensure its safety for human consumption. However, the higher operating costs and low efficiency of these techniques make them difficult to use for field applications. Among the various techniques for water defluoridation. Adsorption is widely adopted as a result of its low cost, ease of availability, easy utilization, high efficiency, and simple physical technique. Although activated carbon has been widely used for the removal of various contaminants, including fluoride, it is sometimes restricted due to its higher cost. Recently, efforts to develop low-cost adsorbents from waste materials have led to the development of bio-sorbents from plants and animal waste that have shown considerable potential for water defluoridation. The main emphasis of this chapter is to gather in-depth knowledge concerning the utilization of inexpensive adsorbents for water defluoridation.

Keywords Defluoridation · Adsorption · Bio-sorbents · Biomass · Mariculture waste

V. B. Mukkanti · A. R. Tembhurkar · R. Gupta (✉)
Civil Engineering Department, Visvesvaraya National Institute of Technology, Nagpur, Maharashtra 440010, India
e-mail: drrajeshgupta123@hotmail.com

V. B. Mukkanti
e-mail: brahmi518@gmail.com

A. R. Tembhurkar
e-mail: artembhurkar@gmail.com

A. K. Yadav et al. (eds.), *Advanced Treatment Technologies for Fluoride Removal in Water*, Water Science and Technology Library 125,
https://doi.org/10.1007/978-3-031-38845-3_7

7.1 Introduction

Supply of safe drinking water, especially for rural regions, is a major issue in the world due to various emerging natural and anthropogenic contaminants (Narsimha & Rajitha, 2018). Groundwater is the major drinking water source for humans in many parts of the world. However, extensive pollution with various contaminants, especially fluoride, poses serious health risks to humans around the globe (Naik et al., 2021). Consumption of groundwater that contains excess fluoride concentrations (> 1.5 mg/L) may cause dental/skeletal fluorosis. It has been estimated that about 200 million people in 25 nations are facing the problem of groundwater with an excess fluoride concentration (Chen et al., 2016). The excess concentration of fluoride in groundwater may be caused by the dissolution of fluoride-containing rocks such as biotite, topaz, and fluorite or the release of fluoride from the fluorochemical industries, including the steel, aluminum, and glass industries (Annan et al., 2021).

Among various defluoridation techniques, adsorption is widely used for water defluoridation because of its simple design, economic viability, and ease of operation. Various adsorbents including nanomaterials, biomass, carbon materials, chitosan, MOFs, bone char, and plant materials have been successfully applied for water defluoridation. Recently, increasing attention has been paid to the development of biosorbents from waste materials for water defluoridation due to their easy availability and low cost (Mukkanti & Tembhurkar, 2023). The present chapter presents the feasibility and practical applicability of various novel biosorbents developed from the waste materials for water defluoridation. Furthermore, this chapter also elaborately explained the merits and demerits of the different defluoridation techniques.

7.2 Adsorption for Water Defluoridation

Basically, water defluoridation can be a household defluoridation that is done by household individuals, and a community defluoridation which is done at suburban, town, and village levels. Many treatment techniques have been implemented and are currently being used for water defluoridation, such as coagulation, ion exchange, electrocoagulation, membrane separation, and adsorption processes. However, each of the aforementioned techniques has its own advantages and disadvantages, which make them difficult to implement for large-scale applications.

Compared to other defluoridation techniques, adsorption is the most prominent to use as a result of its simplicity, cost effectiveness, and accessibility of a wide range of sorbents. Moreover, adsorption is easy to implement for practical applications, especially for small groups. Adsorption also has the benefit of being effective in a wide pH range with less residue than precipitation (Alhassan et al., 2020). Adsorption includes the formation of a bond between the sorbent surface and molecules in the bulk solution by physical or chemical forces. When the fluoride solution passes through a sorbent bed, the fluoride is eliminated by surface chemical reaction or ion exchange (Waghmare & Arfin, 2015). Adsorption is frequently applied for defluoridation of water, especially when water contains low F^- concentrations but above permissible limits. As a result, numerous investigations have been reported on the application of various sorbents for the defluoridation of water defluoridation have been reported (Mukherjee et al., 2021; Mukkanti & Tembhurkar, 2021; Saghi et al., 2020). The studies reported in the literature have mainly focused on the development of alternative sorbents that are superior in performance, easily available, and economical. Although synthetic sorbents have shown good defluoridation capacity but are costlier for large-scale applications, the availability of large amounts of natural materials or wastes from industries or agricultural processes may be a potential low-cost alternative material for water defluoridation. Therefore, low-cost biosorbents developed for water defluoridation over the past decade, especially from waste materials, are discussed below.

7.3 Development of Novel Biosorbents

Biosorbents are generally developed from natural materials or chemically modified natural materials for F^- removal from water. Waste materials generated from the domestic, agricultural and industrial sectors have been used to develop biosorbents for fluoride removal due to their low cost and easy availability. The commercial use of biosorbents has been limited as a result of the use of synthetic sorbents. However, various unconventional biosorbents that are low-cost and equally efficient as commercial defluoridation materials have been tested to replace conventional sorbents (Hegde et al., 2020). As such, researchers have shown great interest in developing biosorbents from waste materials such as biomass materials and mariculture waste materials.

7.3.1 Biomass-Based Adsorbents

In recent decades, biomass materials have shown great potential for water treatment because of their high efficiency, renewable power, low cost, and abundant nature. In addition, they contain hydroxyl, carboxyl and other functional groups along with loose and porous surface, structures, making them effective and promising sorbents (Sahu et al., 2019). Therefore, biomass sorbents such as agricultural wastes, chitosan, bone char, etc. have been applied for water defluoridation. Recently, in search of low-cost, environmentally friendly and locally available materials, researchers have shown great interest developing biosorbents from agricultural waste materials. Various methods have been applied to convert these waste materials into potential biosorbents. For example, Khound and Bharali (2018) developed a biosorbent from sandalwood leaf powder using a simple grinding technique to biosorb fluoride. The increase in pH and the contact time to equilibrium increased the defluoridation performance, while a decrease in temperature increased the defluoridation performance. The sorbent exhibited a 75% defluoridation efficiency and F^- uptake capacity of 4.66 mg/g for a contact time of 120 min, pH 6.8, and a temperature of 303 K. Furthermore, researchers have applied acid treatment using various acids with different concentrations for plant leaves to enhance its surface area and thereby its defluoridation performance. As such, acid treatment with nitic acid (1 N HNO_3) has been applied as a pre-treatment prior to thermal treatment to develop a biosorbent from Ficus benghalensis leaf powder (George & Tembhurkar, 2018). The highly acidic and alkali conditions have shown a lower defluoridation efficiency compared to that under near-neutral pH conditions, and the increase in the adsorbate concentration had a negative impact as a result of the decrease in the number of active sites. The sorbent showed a significant defluoridation efficiency (92%) and a defluoridation capacity of 2.242 mg/g for 5 mg/L of F^- concentration and 90 min contact period at pH 7. Langmuir isotherm and second-order kinetics were best fitted, which indicates the monolayer chemisorption mechanism. In another study, phosphoric acid (0.01 N H_3PO_4) was applied as a acid pre-treatment step for Manihot esculenta leaves, and obtained a F^- uptake capacity of 1.568 mg/g for 20 g/L of adsorbent dose (Pongener et al., 2018). All of these biosorption studies on plant leaves showed a similar tendency for the influence of process parameters and produced a maximum defluoridation efficiency at near neutral pH and lower adsorbate concentration. Among these, acid pre-treatment given for surface area enhancement proven to be a good method to adopt as it showed a higher surface area and defluoridation efficiency for the benghalensis leaf powder (302.42 m^2/g) and the Manihot esculenta leaves powder (812 m^2/g) over the sandal leaf powder from simple grinding (23.85 m^2/g).

Recently, coconut tree plant roots were also used to develop the biosorbent by nitric acid (1 M HNO_3) pre-treatment followed by thermal treatment (George & Tembhurkar, 2019). The increase in biosorbent dose resulted in an increase in defluoridation efficiency and a decrease in defluoridation capacity (1.918–0.376 mg/g) which may be due to decrease in F^-/binding sites ratio. While the increase in defluoridation capacity (0.2305–1.765 mg/g) was observed with increase in adsorbate concentration which might be resulted from an increase in F^-/binding sites ratio. The suitability of the endothermic sorption mechanism from a thermodynamic study indicates its potential application at higher temperatures. Furthermore, the biosorbent showed a significant defluoridation performance for real water samples with a defluoridation efficiency greater than 90% for all the samples for which the F^- concentration ranging from 1.71 to 3.12 mg/L.

Some different types of fruit seeds obtained from food wastes were also used to develop biosorbents for water defluoridation. Araga et al. (2017) synthesized a activated carbon-based biosorbent from *Syzygium cumini* seeds by acid (KOH) activation followed by thermal treatment (pyrolyzed at 900 °C), and compared its defluoridation performance with the inactivated sorbent. Compared to inactivated/pyrolyzed biosorbent, acid-activated biosorbent (AJS: 3.65 mg/g) showed a higher defluoridation capacity over inactivated/pyrolyzed biosorbent (PJS: 0.8 mg/g). The increase in the defluoridation capacity for AJS could be attributed to the tremendous increase in the surface area and porous nature of the sorbent. It was observed that the surface area increased from 13.7 m^2/g (PJS) to 747.45 m^2/g (AJS). The gradual decrease in the adsorption rate with contact time was observed until equilibrium attainment and thereafter no change in F^- sorption with time. The saturation of active sites with time resulted in a decrease in sorption rate with time and the left over active sites after equilibrium are becomes difficult to adsorb F^- ions due to the repulsion between F^- ions on sorbent surface and in the liquid. Furthermore, acidic conditions (pH 3) provided a higher defluoridation efficiency compared to alkaline and neutral conditions due to the surface positive charge of the sorbent present in acidic conditions, which eventually became negative from acidic to alkaline conditions.

The avocado food plant native to south-central Mexico is widely distributed and a fast-growing flowering plant. But avocado seeds are having no other use which usually discarded as waste material. Some researchers have shown interest in the conversion of these avocado waste materials into potential biosorbents for the removal of pollutants from water and wastewater such as organic compounds, cadmium, phenol, and methyl blue. Therefore, the activated carbon-based biosorbents from avocado seed waste eventually emerged as a promising and potential biosorbents for the treatment of water and wastewater. Tefera et al. (2020) used the combination of thermal activation (600 °C) followed by chemical activation (0.1 M $CaCl_2 \cdot 2H_2O$) for developing an avocado seed based biosorbent (ACAS) for water defluoridation. The composition of ACAS includes oxides of iron (11.1%), aluminium (14.2%), potassium (18%), and silicon (47.1%). As previous studies proved, the hydroxides and oxides of metals are the major responsible for the defluoridation which are present in much less composition for the ACAS. Therefore, ACAS was required a huge amount of sorbent dose (19 g/L) to obtain a defluoridation efficiency of 86% and a defluoridation capacity of 1.2 mg/g.

Carrot grass biomass is another agricultural waste that poses a skin allergy which often reduces crop production. These are often discarded along agricultural fields, railway tracks, and roadsides. Therefore, the researchers focused on use of these waste materials to develop useful adsorbents for the removal of pollutants. A comparative study of defluoridation performance was carried out in biosorbents developed from *Parthenium hysterophorus*/carrot grass biomass (PHAC) and Aegle marmelos/bael shells (BAC) by steam activation (Mukherjee et al., 2018). BAC with less sorbent dose (6 g/L) was observed to exhibit a higher defluoridation capacity (16.85 mg/g) than PHAC which exhibited a lower defluoridation capacity (6.22 mg/g) even for a higher sorbent dose (10 g/L). This can be explained from the fixed carbon determined from the proximate analysis being 51.65% and 78.35% for PHAC and BAC, respectively. As BAC contains more amount of fixed carbon, which provides more surface area (1625 m^2/g), it helped to obtain higher defluoridation capacity for less sorbent dose than PHAC (308 m^2/g). The cost analysis revealed that the preparation cost for 1 kg of BAC (1.122 USD) is slightly more expensive than that for PHAC (1.0615 USD). Although the BAC is slightly costlier, the higher defluoridation performance of BAC makes it more efficient for water defluoridation over PHAC.

Recently, the authors have also focused on using biological materials such as algae (Padina sp. Alga), which contain carboxyl functional groups that help in biosorption of various pollutants and especially for F^- sorption. Mohamed et al. (2020) investigated the feasibility of biosorbent powder developed from Padina sp. Alga by simple grinding. The authors adopted both the conventional approach (OFAT) and the response surface methodology (RSM) for the optimization of the process parameters to obtain a maximum defluoridation efficiency. Comparatively, the optimum parameters obtained from RSM exhibited a higher defluoridation efficiency compared to those from OFAT parameters. The defluoridation efficiency of 94.57% and 85.95% was obtained for optimum parameters from RSM and OFAT, respectively. This might be attributed to the longer contact time and slightly acidic conditions from RSM optimization might have resulted in a higher defluoridation efficiency. Mondal et al. (2021) also explored the feasibility of a red algae as a biosorbent for water defluoridation by optimization of the process parameters using RSM. The optimum process variables obtained from RSM, such as contact period: 23.17 min, temperature: 302.63 K, initial adsorbate: 9.77 mg/L, and red algae dose: 10.8 g/L showed a defluoridation capacity of 5.484 mg/g. Ion exchange is the main mechanism involved in the bioaccumulation of F^- onto algae which may be due to the presence of functional groups of amine, carbonyl, carboxyl, and carboxylic acid on the cell walls of algae. Regeneration of exhausted sorbent was performed with different desorbing agents to obtain maximum desorption efficiency. Among the applied desorbing agents such as 0.1 N NaOH, distilled water, 0.05 N EDTA, and 0.1 N HCl, NaOH shown a greater desorption efficiency, making it most promising desorbing agent to be applied for reuse of the sorbent. Overall, biological materials such as Padina sp. Algae and red algae have shown significant defluoridation performance. However, the higher surface area and more favorable functional groups present in the red algae resulted in a greater defluoridation performance making it a more potent and effective biosorbent over Padina sp. Alga biosorbent.

7.3.2 Adsorbents Derived from Mariculture Waste

The biocompatibility with the human body, the high affinity of calcium for F^- removal, and low cost of the use of adsorbents developed from calcium enriched materials for F^- removal (Mondal & George, 2015). Recently, the calcium-based adsorbents from mariculture waste have attracted the researchers as a result of their rich calcium content and abundant availability. From the literature, it was found that waste from mariculture materials such as fish, snail shells, oysters, and clamshells have been studied for water defluoridation.

7.3.2.1 Fish

Bone char is extensively used in the sugar industry to extract the colour from the sugar solutions and has found its applications in water treatment, especially for water defluoridation. Bone char usually contains calcium in terms of hydroxyapatite ($Ca_5(PO_4)_3(OH)$) which has a high affinity for defluoridation. Several studies have reported the use of different types of bone chars such as cattle bones, pig bones, camel bones, goat bones and chicken bones for F^- removal. In addition, some researchers also checked the potential feasibility of bone char prepared from the mariculture waste i.e., fish bones for water defluoridation. Brunson and Sabatini (2009) have investigated the effect of the charring temperature (300–600 °C) on bone char prepared from fish bones for water defluoridation. It was observed that FBC-500 °C which is the fish bone char at 500 °C, exhibited a higher surface area and defluoridation efficiency over FBC at lower pyrolysis temperatures (FBC-300 °C). The lower surface area obtained for FBC-300 °C (one fourth of FBC-500 °C) due to the incomplete removal of organic matter could have resulted in a lower defluoridation efficiency for lower pyrolysis temperatures and FBC-300 °C. Furthermore, the effluent obtained after sorption treatment with FBC-300 °C and FBC-500 °C showed dark yellow and colour free water, respectively. So, the authors reported that the FBC-500 °C was the optimum charring temperature for water defluoridation from both an aesthetic and a performance point of view, which showed a defluoridation capacity of 4.42 mg/g. Cruz-Briano et al. (2021) used devilfish bones to investigate its defluoridation performance with charring temperature. They reported an increase in defluoridation performance with charring temperature up to 500 °C and a decrease in performance with further increase in charring temperature. The presence of more base sites on the surface of the sorbent and the damage to the apatite structure for the combustion temperature above 600 °C which plays important role in F^- sorption might have resulted in a lower defluoridation performance for the combustion temperature above 500 °C. So, they also reported 500 °C as the optimal combustion temperature to obtain the maximum defluoridation capacity, which is 7.12 mg/g for devilfish bones. Therefore, providing a longer detention period and a larger surface area of the devilfish bones exhibited a higher defluoridation performance over the FBC-500 °C.

Some authors have also used biosorbents developed from fish scales as potential biosorbents for water defluoridation, as they are abundantly available as waste materials because of the bulk quantity of daily fish consumption. Bhaumik et al. (2017) used *Catla catla* fish scales powder for water defluoridation by optimizing the process parameters using the response surface methodology. The optimal variables obtained from RSM such as pH: 9.93, adsorbate concentration: 8.49 mg/L, contact period: 179.72 min, and adsorbent dose: 22.6 g/L produced a defluoridation capacity of 4.89 mg/g. FTIR spectral analysis suggest the F^- might be interacted with the functional groups such as OH^- and metal oxides present on the fish scale sorbent. Furthermore, the exhausted sorbent was also successfully regenerated approximately 94.3% using 0.5 N NaOH for a 120 min contact period. Nkansah et al. (2022) also investigated fluoride biosorption of fluoride using $Al(OH)_3$ modified fish scales of

Sciaenops ocellatus and Tilapia Sp. Both sorbents have shown a significant fluoride removal for which the defluoridation efficiency of 76% and 70% and a defluoridation capacity of 0.065 mg/g and 0.045 mg/g were obtained for Tilapia Sp. and Sciaenops ocellatus fish scales, respectively. It is clear from both studies that the catla fish scales powder has a higher defluoridation capacity, but it requires a longer contact period and a large amount of adsorbent dose.

7.3.2.2 Snail Shells

The snail shells which contain 89–99% of $CaCO_3$ have find various applications in the preparation of ornaments, food supplements, wound healing, house decoration etc. Therefore, the authors attempted to utilize the $CaCO_3$ of snail shells to develop biosorbents for water defluoridation, which will also help in biowaste management. As such, a nanocrystalline hydroxyapatite was synthesized using a second precipitation method from snail shells (Limacine artica shells), and its feasibility for water was investigated (Nayak et al., 2017). The adsorbent was found to be quite effective for water defluoridation with a defluoridation efficiency of 96% and 28.57 mg/g of sorption capacity. The H bonding, ligand exchange, and electrostatic attraction mechanisms were involved in the F^- sorption. Kinetic and isotherm analysis indicates the F^- sorption is a monolayer chemisorption onto a homogeneous surface. Furthermore, the increase in temperature had a positive impact on the defluoridation efficiency, which indicates endothermic sorption. In another study, a modified chemical precipitation method was used to develop a hydroxyapatite-based biosorbent from Achatina achatina snail shells (Asimeng et al., 2018). In which, the stirring time was varied to control the trace elements and size of the crystallite. Various Hap sorbents such as HAp 1, HAp 3, HAp 6, HAp 9, and HAp 12 for stirring time of 1, 3, 6, 9 and 12 h. An increase in stirring time was observed to reduce the crystallite size producing a smaller crystallite for higher stirring time, i.e., the crystallite size of HAp1 is higher than of HAp 12. However, trace element concentrations such as Na and K increased with stirring time. However, it was also observed that the crystallite size of HAp does not affect its defluoridation performance and a maximum defluoridation efficiency was obtained for HAp1 which successfully reduced the F^- from 20 to 1.59 mg/L. This may be attributed to the HAp with more surface ions, that is, HAp1 resulted in a higher defluoridation efficiency.

7.3.2.3 Oysters

The oyster farm is increasing rapidly due to its large consumption around the world. It was estimated that more than 30 thousand tons of oysters were produced annually, especially in Taiwan. Therefore, the use of these waste oyster shells, which is a calcium-rich material, could help in oyster farming, making it more sustainable. However, a limited literature on oyster shells as sorbents for water defluoridation. Chang et al. (2019) used $Al(OH)_3$ coated oyster shells for water defluoridation and obtained a maximum F^- uptake capacity of 5.8 mg/g at pH 8.07 and 1.5 mg/g at pH 8.64 for $Al(OH)_3$ modified and raw oyster shells, respectively. It was observed that the Al coated shells have produced a higher defluoridation capacity over the raw shells, which may be due to the enhancement of the surface area of the oyster shells by the coated Al ions and also the high affinity of Al ions towards F^- ions. The interaction between Al ions on the oyster shells surface and F^- ions in the solution is further confirmed by the peaks related to AlF_3 along with CaF_2 peaks from the XPS analysis. Moreover, the F^- peaks shown by EDS spectra at higher concentrations of adsorbate indicate its greater defluoridation performance for higher F^- concentrations. However, more in-depth studies are required for sorbent regeneration and the influence of anions before it is recommended for field applications.

7.3.2.4 Clamshells

Clams are one of the most edible shellfish, which are mass-produced and consumed. Apart from ornamental usage, the shells of clams have no other use, which eventually end up being disposed of as waste products. But it is composed of plenty of calcium that could be used to develop sorbents for the removal especially defluoridation of water, as the high affinity of calcium materials for F^- removal was reported in the previous literature. The effect of pyrolysis temperature on defluoridation efficiency was investigated using biosorbent developed at various pyrolysis temperatures (100–900 °C) of *Mytilus coruscus* shells (MCS) (Lee et al., 2021). It was observed that the pyrolysis temperature up to 500 °C has shown a negligible defluoridation capacity, while a higher defluoridation capacity was observed for a temperature above 700 °C and obtained a maximum defluoridation capacity at 800 °C (82.93 mg/g). The sorbent with a highly porous surface, a larger surface area, and a reduction in zeta potential formed at higher pyrolysis temperatures might have resulted in a higher defluoridation performance. Choi et al. (2022) also investigated the influence of pyrolysis temperature on defluoridation efficiency using *Venerupis philippinarum* shells (VPS). They have also confirmed that 800 °C and above are the optimum pyrolysis temperature to obtain the maximum defluoridation performance. This could be attribute to the presence of more amount of CaO and $Ca(OH)_2$ for sorbent at 800 °C and above, which plays a significant role in F^- sorption. Furthermore, the *Venerupis philippinarum* shells have shown a defluoridation efficiency of about 90% for all acidic and alkaline conditions, including its suitability irrespective of the initial pH for water defluoridation.

The biosorbent powder was developed by simply crushing from bivalve shells (Corbula trigona), which has produced a defluoridation efficiency of 74% and 68% for an initial F^-concentration of 4 mg/L and 2.2 mg/L, respectively (Yapo et al., 2021). Acidic conditions (pH 2.5) were favorable to obtain maximum defluoridation efficiency (80%). This may be due to the conversion of $CaCO_3$ into CO_3^{2-} and Ca^{2+} at acidic pH, which might have favoured the protonation and precipitation of CO_3^{2-} to HCO_3, and F^- to CaF_2. The adsorbent also shown a good reuse performance with 0.5 M NaOH. However, the defluoridation efficiency was reduced with each regeneration cycle due to the incomplete desorption of previously attached F^- ions on the sorbent surface, which caused a decrease in active sites on the sorbent surface for the next cycle. Yapo et al. (2022) also used the same Corbula trigona shells to synthesize a biosorbent based on hydroxyapatite using the hydrothermal method using phosphoric acid. In which the authors obtained a significant defluoridation efficiency (89%) and sorption capacity (4.517 mg/g) for pH 7.5, 175 min contact time and 5 g/L of adsorbent dose.

Mtavangu et al. (2022) have also synthesized a biosorbent based on hydroxyapatite using chemical precipitation method from Anadara granosa shells (97.4% of $CaCO_3$), and obtained a F^- uptake capacity of 15.374 mg/. The study on the impact of anions suggests the coexistence of CO_3^{2-} and HCO_3^- have more negative impact on defluoridation due to an increase in pH by releasing of OH^- ions from the hydrolysis of HCO_3^- and CO_3^{2-}. Moreover, a very low defluoridation efficiency was observed when the presence of CO_3^{2-} than CO_3^-, which might be attributed to the release of more OH^- ions by CO_3^{2-} due to its higher dissociation constant. The authors reported that electrostatic attraction, ligand exchange, and H-bonding mechanisms were involved in the attachment of F^- onto hydroxyapatite. Furthermore, HAp has also shown a good reusability, as only a slight decrease in defluoridation efficiency was observed after the fourth cycle of regeneration. These results demonstrate that HAp from clam shells could be an effective and efficient biosorbent over biosorbents developed from calm shells by other forms of treatment.

7.4 Summary and Conclusion

This chapter presents the necessity of water defluoridation and a special emphasis on various novel biosorbents developed from waste materials of biomass and calcium-rich materials for water defluoridation. The reviews indicated that the use of biosorbents developed from biomass materials and modified biomass materials could provide a green approach for water defluoridation. However, a lower defluoridation capacity, a high sorbent dose and presence of high volatile matter are the drawbacks of these biosorbents. On the other hand, the increasing number of research publications, in which biosorbents developed from calcium-rich materials, especially from the mariculture waste materials, signifies the potentiality of these adsorbents as emerging biosorbents for water defluoridation. The literature reviewed in this chapter also indicates the higher F^- removal capability for mariculture waste materials over biomass adsorbents. However, many of these sorption studies on both biomass and mariculture waste adsorbents were limited to the lab scale and defluoridation from synthetic samples. Many factors affect the defluoridation efficiency of an adsorbent, so laboratory experiments may vary from the field applications. Therefore, it is recommended to check the practical applicability and the reuse and regeneration performance of these biosorbents using field water samples. Moreover, all of the biosorbents reviewed were tested for water defluoridation under batch experiments, so it is necessary to conduct a column feasibility of these biosorbents to implement them for large-scale applications. Furthermore, there are only limited reports on hybrid defluoridation techniques. Therefore, future research should also focus on combining two or more defluoridation techniques to mitigate the demerits of individual techniques.

Acknowledgements The authors thank our prestigious institute and other people who are involved in this study for their direct and indirect contributions.

References

Alhassan, S. I., He, Y., Huang, L., Wu, B., Yan, L., Deng, H., & Wang, H. (2020). A review on fluoride adsorption using modified bauxite: Surface modification and sorption mechanisms perspectives. *Journal of Environmental Chemical Engineering, 8*, 104532. https://doi.org/10.1016/j.jece.2020.104532

Annan, E., Nyankson, E., Agyei-Tuffour, B., Armah, S. K., Nkrumah-Buandoh, G., Hodasi, J. A. M., & Oteng-Peprah, M. (2021). Synthesis and characterization of modified kaolin-bentonite composites for enhanced fluoride removal from drinking water. *Advances in Materials Science and Engineering, 2021*.https://doi.org/10.1155/2021/6679422

Araga, R., Soni, S., & Sharma, C. S. (2017). Fluoride adsorption from aqueous solution using activated carbon obtained from KOH-treated jamun (Syzygium cumini) seed. *Journal of Environmental Chemical Engineering, 5*, 5608–5616. https://doi.org/10.1016/j.jece.2017.10.023

Asimeng, B. O., Fianko, J. R., Kaufmann, E. E., Tiburu, E. K., Hayford, C. F., Anani, P. A., & Dzikunu, O. K. (2018). Preparation and characterization of hydroxyapatite from *Achatina achatina* snail shells: Effect of carbonate substitution and trace elements on defluoridation of water. *Journal of Asian Ceramic Societies, 6*, 205–212. https://doi.org/10.1080/21870764.2018.1488570

Bhaumik, R., Mondal, N. K., & Chattoraj, S. (2017). An optimization study for defluoridation from synthetic fluoride solution using scale of Indian major carp Catla (Catla catla): An unconventional biosorbent. *Journal of Fluorine Chemistry, 195*, 57–69. https://doi.org/10.1016/j.jfluchem.2017.01.015

Brunson, L. R., & Sabatini, D. A. (2009). An evaluation of fish bone char as an appropriate arsenic and fluoride removal technology for emerging regions. *26*

Chang, H. Y. H., Kuo, Y., & Liu, J. C. (2019). Science of the total environment fluoride at waste oyster shell surfaces—Role of magnesium. *Science of the Total Environment, 652*, 1331–1338. https://doi.org/10.1016/j.scitotenv.2018.10.238

Chen, L., Zhang, K., He, J., Cai, X. G., Xu, W., & Liu, J. H. (2016). Performance and mechanism of hierarchically porous Ce–Zr oxide nanospheres encapsulated calcium alginate beads for fluoride removal from water. *RSC Advances, 6*, 36296–36306. https://doi.org/10.1039/c6ra01337f

Choi, M. Y., Lee, C. G., & Park, S. J. (2022). Conversion of organic waste to novel adsorbent for fluoride removal: Efficacy and mechanism of fluoride adsorption by calcined venerupis philippinarum shells. *Water, Air, & Soil Pollution, 233*.https://doi.org/10.1007/s11270-022-05757-9

Cruz-Briano, S. A., Medellín-Castillo, N. A., Torres-Dosal, A., Leyva-Ramos, R., Moreno-Piraján, J. C., Giraldo-Gutiérrez, L., Díaz-Flores, P. E., Reyes-López, S. Y., & Ocampo-Pérez, R. (2021). Bone char from an invasive aquatic specie as a green adsorbent for fluoride removal in drinking water. *Water, Air, & Soil Pollution, 232*.https://doi.org/10.1007/s11270-021-05286-x

George, A. M., & Tembhurkar, A. R. (2018). Biosorptive removal of fluoride from aqueous solution onto newly developed biosorbent from Ficus benghalensis leaf: Evaluation of equilibrium, kinetics, and thermodynamics. *Sustainable Chemistry and Pharmacy, 10*, 125–133. https://doi.org/10.1016/j.scp.2018.10.007

George, A. M., & Tembhurkar, A. R. (2019). Analysis of equilibrium, kinetic, and thermodynamic parameters for biosorption of fluoride from water onto coconut (Cocos nucifera Linn.) root developed adsorbent. *Chinese Journal of Chemical Engineering, 27*, 92–99. https://doi.org/10.1016/j.cjche.2018.03.004

Hegde, R. M., Rego, R. M., Potla, K. M., Kurkuri, M. D., & Kigga, M. (2020). Bio-inspired materials for defluoridation of water: A review. *Chemosphere, 253*, 126657. https://doi.org/10.1016/j.chemosphere.2020.126657

Khound, N. J., & Bharali, R. K. (2018). Biosorption of fluoride from aqueous medium by Indian sandalwood (Santalum Album) leaf powder. *Journal of Environmental Chemical Engineering, 6*, 1726–1735. https://doi.org/10.1016/j.jece.2018.02.010

Lee, J. I., Kang, J. K., Hong, S. H., Lee, C. G., Jeong, S., & Park, S. J. (2021). Thermally treated Mytilus coruscus shells for fluoride removal and their adsorption mechanism. *Chemosphere, 263*, 128328. https://doi.org/10.1016/j.chemosphere.2020.128328

Mohamed, W. S. E. D., Hamad, M. T. M. H., & Kamel, M. Z. (2020). Application of statistical response surface methodology for optimization of fluoride removal efficiency by Padina sp. alga. *Water Environment Research, 92*, 1080–1088.https://doi.org/10.1002/wer.1305

Mondal, N. K., Bhaumik, R., Sen, K., & Debnath, P. (2021). Adsorption of fluoride in aqueous solutions using saline water algae (Rhodophyta sp.): An insight into isotherm, kinetics, thermodynamics and optimization studies. *Modeling Earth Systems and Environment*. https://doi.org/10.1007/s40808-021-01320-3

Mondal, P., & George, S. (2015). A review on adsorbents used for defluoridation of drinking water. *Reviews in Environmental Science & Biotechnology, 14*, 195–210. https://doi.org/10.1007/s11157-014-9356-0

Mtavangu, S. G., Mahene, W., Machunda, R. L., van der Bruggen, B., & Njau, K. N. (2022). Cockle (Anadara granosa) shells-based hydroxyapatite and its potential for defluoridation of drinking water. *Results in Engineering, 13*.https://doi.org/10.1016/j.rineng.2022.100379

Mukherjee, S., Dutta, S., Ray, S., & Halder, G. (2018). A comparative study on defluoridation capabilities of biosorbents: Isotherm, kinetics, thermodynamics, cost estimation, and ecotoxicological study. *Environmental Science and Pollution Research, 25*, 17473–17489. https://doi.org/10.1007/s11356-018-1931-4

Mukherjee, S., Kamila, B., Paul, S., Hazra, B., Chowdhury, S., & Halder, G. (2021). Optimizing fluoride uptake influencing parameters of paper industry waste derived activated carbon. *Microchemical Journal, 160*, 105643. https://doi.org/10.1016/j.microc.2020.105643

Mukkanti, V. B., & Tembhurkar, A. R. (2021). Defluoridation of water using adsorbent derived from the Labeo rohita (rohu) fish scales waste: Optimization, isotherms, kinetics, and thermodynamic study. *Sustainable Chemistry and Pharmacy, 23*, 100520. https://doi.org/10.1016/j.scp.2021.100520

Mukkanti, V. B., & Tembhurkar, A. R. (2023). Taguchi optimization for water defluoridation by thermally treated biosorbent developed from the waste snail shells developed from the waste snail shells. *Journal of Dispersion Science and Technology*, 1–10.https://doi.org/10.1080/01932691.2023.2194383

Naik, M. R., Barik, M., Jha, V., Sahoo, S. K., & Sahoo, N. K. (2021). Spatial distribution and probabilistic health risk assessment of fluoride in groundwater of Angul district, Odisha, India. *Groundwater for Sustainable Development, 14*, 100604. https://doi.org/10.1016/j.gsd.2021.100604

Narsimha, A., & Rajitha, S. (2018). Spatial distribution and seasonal variation in fluoride enrichment in groundwater and its associated human health risk assessment in Telangana State, South India. *Human and Ecological Risk Assessment, 24*, 2119–2132. https://doi.org/10.1080/10807039.2018.1438176

Nayak, B., Samant, A., Patel, R., & Misra, P. K. (2017). Comprehensive understanding of the kinetics and mechanism of fluoride removal over a potent nanocrystalline hydroxyapatite surface. *ACS Omega, 2*, 8118–8128. https://doi.org/10.1021/acsomega.7b00370

Nkansah, M. A., Dua, A. B., Aryee, G. A., & Adusei-Gyamfi, J. (2022). Evaluation of scales of Tilapia Sp. and Sciaenops ocellatus as low cost and green adsorbent for fluoride removal from water. *Frontiers in Chemistry, 10*, 1–11. https://doi.org/10.3389/fchem.2022.813533

Pongener, C., Bhomick, P. C., Supong, A., Baruah, M., Sinha, U. B., & Sinha, D. (2018). Adsorption of fluoride onto activated carbon synthesized from Manihot esculenta biomass—Equilibrium, kinetic and thermodynamic studies. *Journal of Environmental Chemical Engineering, 6*, 2382–2389. https://doi.org/10.1016/j.jece.2018.02.045

Saghi, M. H., Qasemi, M., Mohammadi, A. A., Kowsari, M. H., & Shams, M. (2020). Fluoride removal from aqueous solution by municipal solid waste compost ash: Kinetics, and isotherms studies. *International Journal of Environmental Analytical Chemistry, 00*, 1–13. https://doi.org/10.1080/03067319.2020.1761349

Sahu, U. K., Sahu, S., Mahapatra, S. S., & Patel, R. K. (2019). Synthesis and characterization of magnetic bio-adsorbent developed from Aegle marmelos leaves for removal of As(V) from aqueous solutions. *Environmental Science and Pollution Research, 26*, 946–958. https://doi.org/10.1007/s11356-018-3643-1

Tefera, N., Mulualem, Y., & Fito, J. (2020). Adsorption of fluoride from aqueous solution and groundwater onto activated carbon of avocado seeds. *Water Conservation Science and Engineering, 5*, 187–197. https://doi.org/10.1007/s41101-020-00093-7

Waghmare, S. S., & Arfin, T. (2015). Fluoride removal from water by various techniques. *International Journal of Innovative Science, Engineering and Technology, 2*, 560–571.

Yapo, N. S., Aw, S., Briton, B. G. H., Drogui, P., Yao, K. B., & Adouby, K. (2022). Removal of fluoride in groundwater by adsorption using hydroxyapatite modified Corbula trigona shell powder. *Chemical Engineering Journal Advances, 12*.https://doi.org/10.1016/j.ceja.2022.100386

Yapo, N. S., Briton, B. G. H., Aw, S., Reinert, L., Drogui, P., & Adouby, K. (2021). Bivalve shells (Corbula trigona) as a new adsorbent for the defluoridation of groundwater by adsorption-precipitation. *Journal of Environmental Science and Health—Part A Toxic/Hazardous Substances and Environmental Engineering, 56*, 694–704. https://doi.org/10.1080/10934529.2021.1917937

Chapter 8
Adsorptive Removal of Fluoride from Water Using Iron Oxide-Hydrogen Nanoparticles

Archana Kushwaha, Zeenat Arif, and Bineeta Singh

Abstract Fluoride, a naturally occurring element, is released from rocks into the land, water, and air. The WHO recommends 1 mg/L of fluoride in drinking water as the ideal or recommended amount. An abundance of fluoride ions can cause dental/skeletal fluorosis, muscular and bone damage, chronicle issues, inhibit the photosynthesis process, and enzymatic and metabolic activities in aquatic organisms. Membrane filtration processes, reverse osmosis, electrodialysis are widely used treatment processes for the removal of fluoride; however, they are expensive and complex. The adsorption method is considered to be an effective technique because of the low operating costs, the capacity to hold metal ions at low concentrations, and the availability of a variety of adsorbents for treatment applications, which makes it an attractive option. If there are opposing anions present including chloride, iodide, and sulfate, then the iron oxide-hydroxide nanoparticles demonstrate their effectiveness as a repeatable and efficient adsorbent medium for defluoridating water. This chapter deals with adsorptive removal of fluoride removal using iron-based adsorption techniques and highlights different parameters affecting the performance of adsorption.

Keywords Fluoride · Reverse osmosis · Adsorption · Optimization · Wastewater

A. Kushwaha · Z. Arif (✉) · B. Singh
Chemical Engineering Department, Harcourt Butler Technical University, Kanpur 208002, India
e-mail: dariflt@hbtu.ac.in

A. Kushwaha
e-mail: archanakushwaha6r6@gmail.com

B. Singh
e-mail: bsinghch@hbtu.ac.in

A. K. Yadav et al. (eds.), *Advanced Treatment Technologies for Fluoride Removal in Water*, Water Science and Technology Library 125,
https://doi.org/10.1007/978-3-031-38845-3_8

8.1 Introduction

Development and change in consumption pattern create an exponential increase in water demand, simultaneously generates and degrades the quality of discharge water, commonly referred to as wastewater (Ashraf & Mohd Hanafia, 2019; Reddy et al., 2012; Shirin & Yadav, 2014). Organic and inorganic pollutants, heavy metals, suspended and dissolved particles, agricultural pollutants, and thermal and radioactive pollutants are some of the different types of pollutants found in wastewater. Through sewage and industrial effluents, mixed pollutants are discharged into water bodies, adversely affect living organisms and the ecosystem. Among the different pollutant, the removal of fluoride from wastewater is gaining attention among researchers, as they are considered a highly toxic element for living organisms if consumed in excess amount.

Fluoride is an inorganic monoatomic anion of fluorine. Natural sources of fluoride originate from the dissolution of a variety of rocks and minerals containing fluorine (fluorspar, cryolite, and fluorapatite). The main causes and sources for the generation of a significant quantity of fluoride are industrialization and urbanization. Metalworking companies including those involved in the production of steel and iron, electrolysis process of electrolysis, zinc and lead smelting, rare earth separation, and lithium-ion batteries are the principal producers of fluoride compounds (Ahmadi-jokani et al., 2021; Shirin et al., 2022; Wan et al., 2021; Yadav et al., 2022). Fluoride concentrations can lead to fluorosis of the teeth, the skeleton, etc. Coagulation is a typical technique for the elimination of fluoride from water (Waghmare & Arfin, 2015), and membrane procedures are sophisticated filtering techniques that make use of the separation capabilities of the finely porous polymeric or inorganic film adsorption, ion exchange (Grzegorzek et al., 2020; Obotey Ezugbe & Rathilal, 2020), electrodialysis (Sharma et al., 2018). The advantages and drawbacks of various fluoride removal procedures are shown in Table 8.1 (Chong et al., 2019; Zhang et al., 2017).

8.2 Elimination of Fluoride by the Adsorption Technique

8.2.1 Adsorption

Adsorption is a phenomenon on the surface whereby molecules, ions, or atoms of a fluid are drawn to a solid surface (Tolkou et al., 2021). Recent years have seen a lot of research on the removal of fluoride using adsorption techniques. Fluoride normally absorbs into a solid absorbent through three steps (mechanism).

- The term "external mass transfer" describes how fluoride ions diffuse or move beyond the boundary of the layer around the absorbent particle to reach the external surface.

Table 8.1 Techniques for fluoride removal: advantage and disadvantage

Methods	Advantages	Disadvantages
Coagulation and flocculation	• Simple and inexpensive • Limited investment	• The sludge production is high • Disposal problems • Operational costs
Membrane processes	• Fast with low space requirement • High separation efficiency	• Limited lifetime • High cost • Easy fouling
Ion exchange	• High regeneration and separation efficiency	• Expensive technique
Electrodialysis	• High recovery	• Limited to the removal of ions • Expensive technique
Adsorption	• Low cost	• Poor selectivity

- Fluoride ion adsorption on particle surfaces.
- Adsorbed fluoride ions may interchange with structural components, or they may diffuse within porous materials, depending on the solid chemistry (Habuda et al., 2014).

8.2.2 *Types of Adsorption*

- *Physical adsorption*: The most fundamental technique of immobilization, known as physical adsorption or physisorption, involves bonding atoms or molecules to a surface through hydrophobic interactions, hydrogen bonds, or weak connections such as van der Waals forces.
- *Chemical adsorption*: When particles adsorb on solid materials, a process known as chemical adsorption or chemisorption, the surfaces of the adsorbent and adsorbate share a significant amount of electrons, resulting in the development of a covalent or ionic connection. Chemisorption is demonstrated by the adsorption of gases such as hydrogen, nitrogen, and other gases at high temperatures on the surface of adsorbents such as ferrous catalysts (Khan & Ganai, 2020). Figure 8.1: Illustrates the chemical and Physical adsorption mechanisms

8.2.3 *Adsorbent*

Adsorbents are solid materials that are used to draw solute molecules out of a liquid or gas.

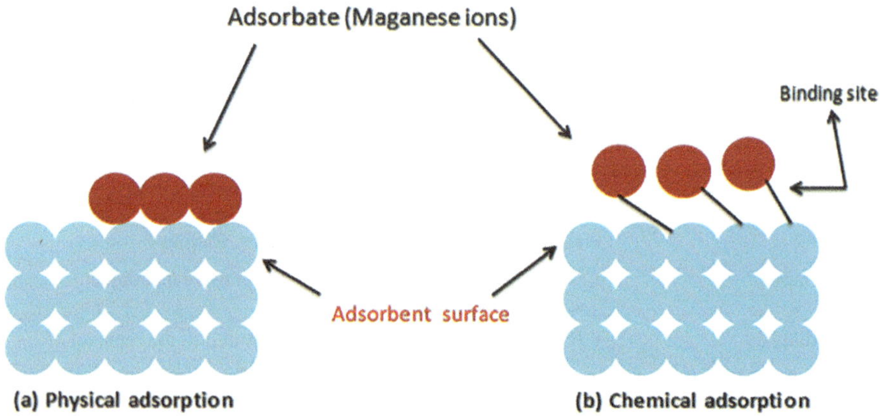

Fig. 8.1 Types of adsorption **a** physical **b** chemical adsorption

8.2.3.1 Types of Adsorbents

Broadly, adsorbents are classified into two categories, as mentioned below.

1. Natural adsorbents are adsorbents that are naturally occurring. For example, extracts from plants such as herbal drugs and natural pigments, fermented antibiotics such as penicillins and cephalosporin, peptides such as insulin, proteins, vitamins, nutraceuticals such as polyphenols, etc. (Wadidi et al., 2022).
2. Synthetic adsorbents: are adsorbents that are not naturally occurring and are made by humans. For example: Carbon materials (including activated carbon and carbon fibers), silica gel, activated alumina, zeolites, mesoporous silicas, metal–organic frameworks, and metal oxides are examples of natural adsorbents.

8.2.4 Fluoride Removal Using Various Adsorbents

Fluoride-related problems are now widespread on a global scale. Fluorosis has been an issue in many developing nations. Groundwater becomes contaminated with fluorides as a result of anthropogenic and natural activity. Both benefits and drawbacks to human health exist with fluoride in drinking water. Fluorides are necessary for all living things at minimal levels to enhance bone and dental enamel; nevertheless, skeletal and dental fluorosis can result at doses greater than 1.5 mg/L (Aryal et al., 2019). Figure 8.2 shows different sources of fluoride.

Different types of adsorbents include synthetic polymers, activated carbon (AC), bone char, and biochar, carbon nanotubes, graphene- and graphene-based materials, silica-based adsorbents, and iron-based adsorbents.

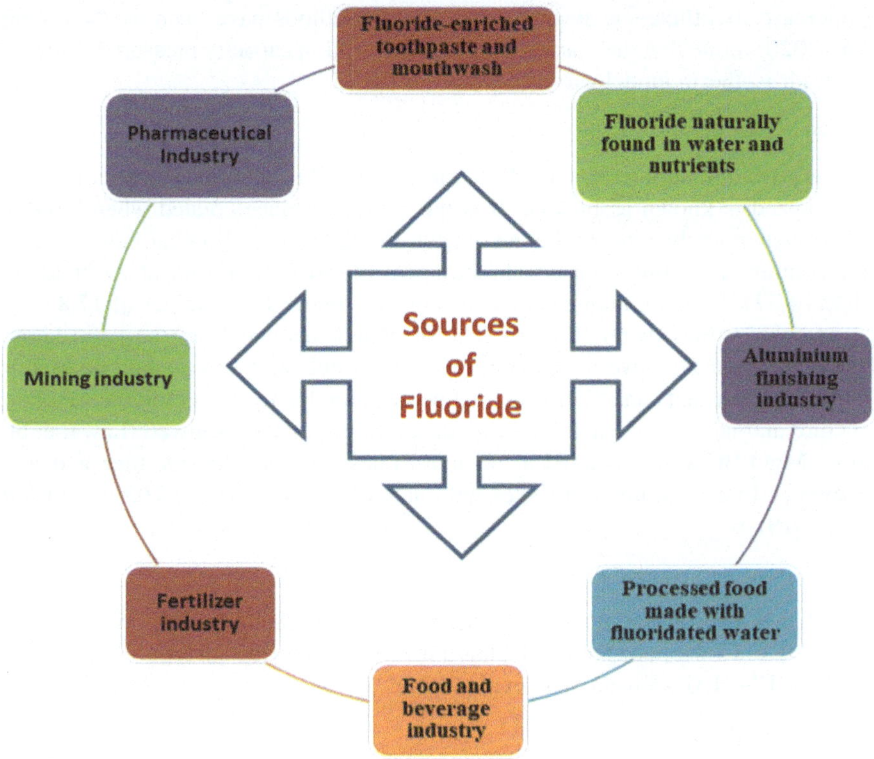

Fig. 8.2 Sources of fluoride

Synthetic polymer:

Humans have made synthetic polymers, which are polymers. Thermoplastics, thermosets, elastomers, and synthetic fibers are four categories into which they are separated. The versatility of synthetic polymers is unmatched. It is great that synthetic polymers have useful characteristics including strength, flexibility, resistivity, and chemical inertness. The coated polyacrylonitrile nanofibrous mat (CPNM), with an average fiber diameter of 189 nm, has an adsorption capacity of 1708 mg/g, according to Lou et al. (2019). The pseudo-second-order kinetics model accurately described how acid blue-113 adsorption on CPNM occurred. Based on the Langmuir isotherm, Raghav and Kumar (2019) reported the adsorption capacity (Qe) for the Pectin-Fe–Al–Ni adsorbent at 285 and Alginic-Fe–Al–Ni as 200 mg/g and concluded that PFAN had a much higher Qe for F- than did AFAN.

Activated carbon:

Activated carbon, because of its low price, significant surface area, and high renewability, has enormous potential in the field of water treatment. Although it has a tiny capacity for adsorption, activated carbon has a weak attraction for fluoride.

To increase its efficiency, several cutting-edge methods have been devised. Pang et al. (2020) found that the maximal fluoride adsorption capacity measured using the Langmuir isotherm model was 28.50 mg/L at 25 °C.

Biochar:

The light-weight, dark residue made of carbon and ashes that remains after biomass is pyrolyzed is known as biochar. It is the solid substance created when biomass is thermochemically converted in an oxygen-limited setting. Biochar manufactured from commercial cellulose has the highest possible adsorption capacity, with iodine values of 371.40 mg/g, tannic acid values of 86.7 mg/L, COD values of 17.89 mg/g, and color values of 60.35 mg/g. The biochar generated from the palm kernel shell is the least effective for adsorption. High cellulosic biomass may be used to produce biochar that can effectively treat wastewater due to changes in the nanopore structure and adsorption capacities that the cellulose content showed (Lawal et al., 2021). MgO-BC was predicted to be able to adsorb more fluoride than activated alumina and magnesium composites with natural carriers, up to 83.05 mg/g (Wan et al., 2019).

Bone char:

Bone char is a porous, granular, and black substance made by burning animal bones. According to Cow bone char (CBC) has a maximum fluoride adsorption capacity of 0.788 mg-F/g-HAP (Sawangjang et al., 2021).

Carbon nanotubes:

Carbon nanotubes, a carbon allotrope, called a carbon nanotube, is a tube-shaped structure comprised of carbon atoms. Due to their exceptional low weight, mechanical, electrical, and thermal capabilities, carbon nanotubes are one of the most investigated nanomaterials. The saturation adsorption capacities of the rhodamine B (RB) and crystal violet (CV) dyes, respectively, were 0.57–0.86 mmol/g and 0.75–0.88 mmol/g at 298–328 K. The adsorption capacity of the hybrid material, which is around 975.4 mg/g, implies that it has the capacity to remove fluoride from a real matrix (Affonso et al., 2020).

Graphene:

The arrangement of a thick sheet of carbon atoms in a hexagonal lattice is called graphene. The essential element of graphite, which is used, among other things, in pencil tips, is graphene. However, graphene is a wonderful material with several excellent features when it is used alone. The Langmuir isotherm and pseudo-second-order kinetic models were well matched by the maximum adsorption capacities for methylene blue (480.76 mg/g) and methyl orange (55.56 mg/g). Due to the beneficial structure of graphite, which has many mesopores, a maximum adsorption capacity of 118.7 mg/g is achieved for fluoride species (Sun et al., 2016). Even after 50 cycles of operation, the fluoride removal capacity of a reduced graphene oxide/hydroxyapatite composite (rGO/HA) of graphite is 0.21 mmol/g and its regeneration efficiency is around 96%. This material is durable and reusable without suffering significant capacity loss (Park et al., 2021).

Silica-based adsorbents:

Because of their enormous surface areas, extensive porosity, and exceptional mechanical and thermal stabilities, silica-based adsorbents are particularly effective, and silica-based adsorbents are employed to remove organic wastes. The oxygen atoms that make up silica particles are suspended and bound to protons. The hydroxyl groups in silica make it polar, and these are utilized for wastewater adsorption. With an adsorption capacity of 12 mg/g, the results showed astonishing effectiveness in removing fluoride (Pillai et al., 2020a).

8.2.5 *Iron Based Adsorbents*

Fluoride species in water can be removed by using iron-based adsorbents. Iron-based adsorbents have attracted interest because of their high efficacy in removing fluoride, environmental friendliness, and accessibility on earth. Examples of adsorbents produced on an industrial scale are zero-valent iron and granular ferric hydroxide (GFH). In order to remove organic and inorganic impurities from water, iron-based nanoparticles are being used more and more. Chemically, iron-based nanoparticles are totally composed of nano zero-valent iron (nZVI), iron oxides (FeO, Fe_2O_3, or Fe_3O_4), or metallic iron cores wrapped in thin layers of iron oxide. Additionally, because of its smaller size than metal iron, nanometallic iron provides more active sites and a larger surface area for the adsorption process to take place. However, in a wet environment, metallic iron is more likely to oxidize at the nanoscale, inducing the formation of ferric hydroxide by releasing Fe^{2+} ions, which are then converted to Fe^{3+} ions by the presence of dissolved oxygen in fluids. To stop oxidation, Fe^{2+} ions in nano zero valent iron (nZVI) particles are routinely alloyed or doped with other metal atoms, such as Ag and Cu (Biftu et al., 2020). In addition to adsorption capacities, the thermodynamics, kinetics, and mass-transfer mechanisms of adsorption processes are also looked at. The detailed investigation of the specific interaction between fluoride and iron involved several spectrum technologies. The Langmuir isotherm model was the most successful, with a maximum fluoride removal of 96% and an adsorption capacity of 67.9 mg/g (Jeyaseelan et al., 2021; Pillai et al., 2020b). Additionally, some iron-based adsorbents have magnetic properties that make it simple to separate saturated compounds from water using an external magnetic field. Table 8.2 lists different pollutant removal, mechanism using iron-based adsorbents.

8.2.5.1 Iron-Based Adsorbent Synthesis Method

Iron/iron oxide nanoparticles may be made in a variety of physical and chemical ways for adsorption. Through the use of appropriate agent capping, such as organic

Table 8.2 Water pollution remediation techniques using iron-based adsorbents

S. No	Mechanism	Adsorbent	Pollutants	Adsorbent capacity (mg/g) or % of removal	References
1	Adsorption	Alpha-Fe_2O_3	Malachite green (MG)	86.13%	Dehbi et al. (2020)
		Alpha-Fe_2O_3	Direct red 28 dye (DR28)	97%	Aragaw (2020)
		Zerovalent iron	Dye (textile)	90%	Mandal and Ghosh (2018)
		$Fe(OH)_3$@cellulose Congo red	Congo red	689.65 mg/g	Zhao et al. (2017)
2	Photocatalytic degradation	Alpha-Fe_2O_3	Reactive blue	95.08%	Bibi et al. (2019)
3	Degradation	Co_3O_4/delta-FeOOH	Lomefloxacin (LMO)	> 82%	Zhang et al. (2019)
4	Reduction/adsorption	Nano zerovalent Fe	DDT insecticide	98%	Altuntas and Debik (2017)
5	Reduction method	Zerovalent Fe NPs	Cr(VI)	97%	Rashmi et al. (2013)
6	Chemical Coprecipitation cum adsorption	γ-Fe_2O_3	As(III)	67.02 mg/g	Lin et al. (2012)
7	Coprecipitation cum adsorption method	Modifed magnetic iron oxide nanoparticles	Hg(II)	0.6 mg/g or 96.2%	Parham et al. (2012)
8	Solgel hydrolysis–precipitation cum adsorption	Magnetic oxide (Fe_3O_4)	As(III)	23.8 mg/g or 82%	Khodabakhshi et al. (2011)
9	Reduction/adsorption	nZVI	4-chloro-3-methyl phenol	63%	Xu and Wang (2011)
10	Co precipitation–hydrolysis cum adsorption	Amino-functionalized magnetic nanoparticles	Cu(II) ions from aqueous	25.77 mg/g or 98%	Hao et al. (2010)

stabilizers and surfactants, and the ferrous or ferric ions using a variety of reducing chemicals, particularly $NaBH_4$, zero-valent iron nanoparticles (nZVI) of desired size can be synthesized. To achieve the appropriate size of the nanoparticles, capping agents are used to stop iron particles from aggregating and to control their growth. The effectiveness of chelating compounds (complexing agents), such as ethylene diamine tetraacetic acid (EDTA), nitrile triacetic acid (NTA), diethylene triamine pentaacetic acid (DTPA), and cyclohexa diamine teraacetic acid (CDTA), can act as stabilizing or capping compounds during the production of nZVI.

Co-precipitation, sol–gel, thermal decomposition, hydrothermal, microemulsion, solvothermal, polyols, electrochemical synthesis, microwave-assisted synthesis, biosynthesis and other methods are the most frequently used techniques for making iron-based nanoparticles. Table 8.3 highlights the synthesis techniques, their pros and cons used to synthesize adsorbents.

8.2.5.2 Use of Iron-Based Adsorbents to Remove Fluoride from Various Industrial Sectors

To meet the growing demand for agricultural goods, more land must be used for crop production, but doing so has negative effects on the environment. This need is brought on by the ongoing increase in the world's population. By using chemical fertilizers, existing croplands have seen an increase in production to meet this demand (Affonso et al., 2020). However, the addition of industrial manufactured chemicals can raise fluoride concentrations to dangerous levels. Many industries, including the production of phosphatic fertilizers, coal cleaning, semiconductors, and metals, use hydrofluoric acid (HF) as a reactant. The rinse water contains more fluoride than is allowed (100–10,000 mg/L), which has a hazardous effect on the environment. Many businesses use sodium fluoride [NaF], including the manufacture of coated papers with pesticides and wood preservatives, the pickling stainless steel, and the remelting of aluminum (Biswas et al., 2018). Untreated effluent was sometimes just discharged into the natural environment. Aluminum fluoride manufacturing in Tunisia makes use of hydrofluoric acid (HF) in large amounts. As a result, the effluents from this procedure have high levels of fluoride. Fluoride levels typically range from 100 to 6500 mg/L (Ezzeddine et al., 2015). According to Bagastyo et al. (2017), wastewater generated by the fertilizer industry can include up to 4540 mg/L of phosphate and 9720 mg/L of fluoride, respectively. The list of different industrial fluoride sources and their removal capacity is listed in Table 8.4

8.2.5.3 Fluoride Concentration and Iron-Based Adsorbent Dosage as a Function of Adsorption Removal Capability

The dosage effects of Fe and its compound as an adsorbent on fluoride were examined using adsorbent doses ranging from 0.025 to 0.25 g per 50 mL against a starting fluoride concentration of 10 mg/L. The defluoridation capacity (DC) of Fe-based

Table 8.3 Chemical preparation procedures for iron oxide-based nanomaterials

Methods	Formation mechanism	Factors	Advantage	Disadvantage
Co-precipitation	The interaction of a water-soluble base with a somewhat oxidized iron salt	✓ Type of salt, ✓ Fe^{2+}/Fe^{3+} ratio, ✓ pH, ✓ Ionic strength, and temperature	Easy and successful	High pH requirement
Sol–gel	The hydroxylation process and condensate of molecules in solutions are then "sol" drying or "gelling," either as a result of a chemical reaction or the solvent being drained away	✓ pH, ✓ The kind and quantity of the salt precursor, ✓ The kinetics, ✓ The properties of the gel, ✓ The temperature	Well-managed aspect ratio, size, and internal structure	Low wear resistance and weak bonding
Thermal decomposition	Thermal breakdown is a chemical process that occurs when a material degrades under heat	✓ Litter quality, temperature, ✓ pH	Easy and economical	The nanocrystal is only soluble in nonpolar solvents and extremely monodisperse Limited size range
Hydrothermal	The aqueous solution as the reaction medium in a unique, closed reaction vessel	✓ The purity, grain size, and form ✓ Temperature, pressure, and time of the reaction	Size and form are easily adjustable	High pressure and temperature are necessary for the reaction
Solvothermal	It involves sealing the precursor in an autoclave, heating it to a temperature above the boiling point of the desired solvent, and then mixing the solvent and precursor together	✓ Temperature and pressure	Ultra-pure nanocrystals	Prolonged reaction times, relatively wide size distribution

(continued)

Table 8.3 (continued)

Methods	Formation mechanism	Factors	Advantage	Disadvantage
Microwave assisted synthesis	Uses the excitation induced by microwave electromagnetic radiations as its main operating mechanism to align material dipoles in an external field	✓ Temperature of the reaction mixture, ✓ Solvent nature, ✓ Particle size, ✓ Power level of microwaves	High purity, small, narrow particle size distribution, energy-saving procedure, reproducibility, and simplicity. Shorter preparation time and greater yield	Since this is an open system that cannot be compressed, superheating to greater temperatures is not feasible
Biosynthesis	A multistep, enzyme-catalyzed procedure in which living things transform their substrates into ever-more complex products	✓ Reaction temperature, ✓ pH, ✓ Reaction stirring time	The best process for creating iron oxide nanoparticles	Expensive equipment, high energy consumption, and low production yield

Table 8.4 Enlist different sources of fluoride generating industries and their adsorption removal capacity using iron adsorbents

S. No	Iron based adsorbent	Industrial source	Adsorption capacity/% removal	Reference
1	Iron ox-hydroxide (FeOOH)	Fertilizer industry	16.70 mg/g	Raul et al. (2012)
2	Iron oxide coated activated sludge granules	Textile industry	91.60%	Hashemifar et al. (2014)
3	Zerovalent-graphene (GO)	Pharmaceutical industry	425.72 mg/g	Li et al. (2016)
4	Feroxyhyte (Delta-FeOOH)	Sugar industry	94.10%	Li et al. (2019)
5	Gamma-Fe_2O_3-functionalized with glycine	Mining industry	625 mg/g	Feitoza et al. (2014)
6	L-cystine-functionalized delta-FeOOH (Cys-delta-FeOOH)	Food and beverage industry	217 mg/g	Maia et al. (2019)
7	Tangerine peel. Fe_3O_4 nanocomposite	Petrochemicals industry	95%	Lingamdinne et al. (2020)
8	Fe_3O_4–gelatin	Slaughter houses	98.884%, 1250 mg/g	Alinejad-Mir et al. (2018)

adsorbents has been observed to increase with dose, reaching saturation at 0.1 g. The DC value for Fe-based compounds decreased from 0.05 to 0.1 mg/g with an adsorbent dose of 0.025–0.25 g due to improved binding sites, which increased DC. Fluoride adsorption of more than 80% was achieved as dosage of Mixed Al/Fe/Ca metallic oxide mixed Al/Fe/Ca dose was increased from 1 to 5 g at neutral pH. Comparing the Fe-Al-Mn@chitosan composite's adsorption capability was done by Chaudhary et al. (2021) and the composite offers minimum resistance to the internal pore diffusion and fluoride adsorption is more in polymer composite due to smaller particle size of NPs available which is not the case with mixed metal oxyhydroxide (Fe–Al–Mn) NP alone. When the dose was varied between 0.25 and 2.5 g/L, an observed increase in the was observed fluoride removal percent from 70.88 to 0.82% using iron-aluminium nanocomposite was produced.

The first ions in the solution are also a significant factor that affects the adsorption. For Fe and Fe composites, it was shown that there is a direct link between the initial ion solution and DC as the first ion solution increases by 2–10 mg/L (Jeyaseelan et al., 2021).

8.3 Adsorption Performances

8.3.1 Adsorption Kinetics

To explore the adsorption kinetics, pseudo-first-order, pseudo-second-order, and intraparticle diffusion models are created.

Pseudo-first-order: To calculate the adsorption capacity in a liquid–solid system, Lagergren defined pseudo-first-order kinetics in Eq. 8.1.

$$da_e/dt = k_1(a_e - a_t) \tag{8.1}$$

Integrating above equation gives solution as

$$\ln(a_e - a_t) = \ln a_e - k_1 t \tag{8.2}$$

where a_e and at: fluoride adsorbed at equilibrium and time t, respectively, k_1 adsorption constant rate can be obtained graphically from plot of ln $(a_e - a_t)$ and time.

Pseudosecond-order kinetics Ho and Mckay in 1988 introduced this model to determine adsorption capacity (Bhan et al., 2021) for the solid-phase system defined by Eq. 8.3

$$da_t/dt = k_2(a_e - a_t)^2 \tag{8.3}$$

whereas k_2 is the equilibrium constant in (g/mg min). Integration and rearranging the above equation give

$$t/a_t = 1/k_2 a_e^2 + t/a_e \tag{8.4}$$

Intraparticle diffusion model: Weber and Morris created the model in 1963. The model aids in the analysis of the rate limiting elements that control the adsorption process. The mathematical expression to define the above model is expressed by Eq. 8.5

$$a_t = k_i t^{1/2} + C \tag{8.5}$$

where k_i (rate constant) and C (constant in relation to thickness of boundary layer. Plot of q_t verses $t^{½}$ is used to estimate k_i and C (Sahu, et al., 2021).

8.3.2 Adsorption Isotherm

Temperature being a sensitive parameter affecting the adsorption process, we therefore consider adsorption capacity. The adsorption isotherms at constant temperature relates equilibrium relation between the adsorbent and the adsorbates.

Langmuir model: In 1932, Irving Langmuir introduced the Langmuir adsorption model with the following presumptions:

- Monolayer deposition means that once the adsorption sites are occupied by adsorbates, no further adsorption takes place and thus binding energy remains constant.
- According to this model, at low adsorption densities, linear adsorption while at larger solute concentrations, maximum surface coverage will occur.

The Langmuir equation has the form:

$$q_e = \frac{q_m K_L C_e}{1 + K_L C_e} \tag{8.6}$$

where the amount of sorption (q_e in mg g^{-1}), Equilibrium fluoride concentration (C_e in mg L^{-1}), q_m (mg g^{-1}) is the saturated adsorption capacity of the materials, K_L (L mg^{-1}) is the Langmuir adsorption constant.

In addition to it, this model is simpler to use and adaptable when using computer simulations.

Freundlich adsorption isotherm model: Herbert Freundlich in 1909 provided *an* empirical equation which describes the nonideality of adsorption behaviours and relates the amount of gas (adsorbate) that is adsorbed by a unit mass of solid adsorbent at constant temperature but varying pressure.

The following curve represents the plot of this equation, which is a straight line.

$$\frac{x}{m} = K_p^{1/n} \tag{8.7}$$

$$\log \frac{x}{m} = \log k + \frac{1}{n} \log P \tag{8.8}$$

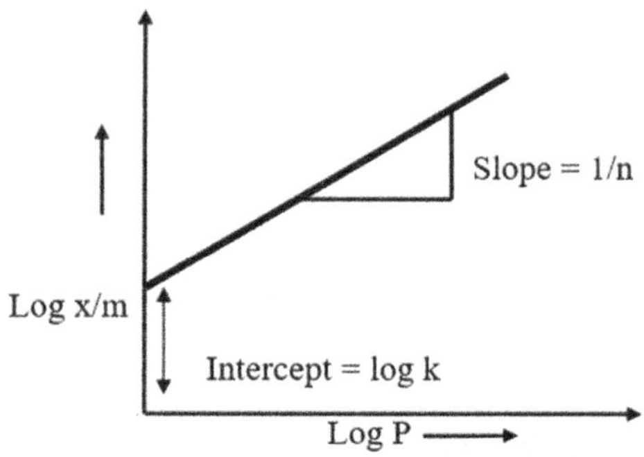

$$q_e = K_F C_e^{1/n} \tag{8.9}$$

where K_F (rate constant in mg g^{-1}), 1/n is the heterogeneity factor (Cadaval et al., 2015) and

$$q_e = \frac{q_m a_S C_e^{1/n}}{1 + a_S C_e^{1/n}} \tag{8.10}$$

where a_S (L mg^{-1}) is the Sips constant (refers to the adsorption energy) (Grover et al., 2012; Wang et al., 2018).

Temkin isotherm

This isotherm emphasizes the uniform distribution of bonding energies to the maximum binding energy. It shows an inverse relationship between the temperature of the adsorption molecules and the temperature of the adsorbate-adsorbent molecules. Mathematically, it is expressed as follows:

$$a_e = B\ln A + B\ln(c_e) \tag{8.11}$$

where a_e and c_e represent the amount and concentration at equilibrium, respectively. B is a constant (reference to heat of adsorption) and A is the Temkin constant.

8.4 Conclusions and Future Prospects

The chapter attempts to include a variety of adsorbents and treatment methods that have been used to remove fluoride from aqueous solutions. Despite numerous methods on fluoride removal methods, the simplicity, environmental friendliness, and economic viability of adsorption emerges out as promising approach for defluoridation. Defluoridation methods have been discussed and are categorized as follows: adsorption, precipitation, membrane, and ion exchange methods. The hydrophobic surface chemistry of iron oxide NPs limits their solubility to nonpolar solvents such as toluene and hexane, which is a drawback. In recent years, much effort has been put towards changing the surface chemistry of iron oxide NP that should exhibit a hydrophilic characteristic and at the same time be biocompatible. Designing magnetic nanoparticles with efficient surface coatings that offer top performance in biological applications both in vitro and in vivo is a significant problem for all of the approaches. On the basis of this chapter, the following observations were made as a conclusion.

- The process of adsorption is simple to those of operate compared to other existing methods. Modified adsorbents show WHO by showing high defluorination capacity.
- The potential of metal oxide nanomaterials reflects the most promising alternative for defluorination from wastewater. At the same time, impacts and risks involved in using nanomaterials on the environment should also be committed to.
- The creation of novel and affordable advanced methodologies that can selectively reflect high performance is still at the forefront of research. Further research scope lies in the use of locally available effective adsorbents, so that many people can afford the price.

Acknowledgements I would like to acknowledge Department of Chemical Engineering, HBTU for providing facility to use web of science for journal access.

References

Affonso, L. N., Marques, J. L., Jr., Lima, V. V., Gonçalves, J. O., Barbosa, S. C., Primel, E. G., Burgo, T. A., Dotto, G. L., Pinto, L. A., & Cadaval, T. R., Jr. (2020). Removal of fluoride from fertilizer industry effluent using carbon nanotubes stabilized in chitosan sponge. *Journal of Hazardous Materials, 388*, 122042.

Ahmadijokani, F., Molavi, H., Rezakazemi, M., Aminabhavi, T. M., & Arjmand, M. (2021). Simultaneous detection and removal of fluoride from water using smart metal-organic framework-based adsorbents. *Coordination Chemistry Reviews, 445*, 214037.

Alinejad-Mir, A., Amooey, A. A., & Ghasemi, S. (2018). Adsorption of direct yellow 12 from aqueous solutions by an iron oxide-gelatin nanoadsorbent; kinetic, isotherm and mechanism analysis. *Journal of Cleaner Production, 170*, 570–580.

Altuntas, K., & Debik, E. (2017). DDT removal by nano zero valent iron: Influence of pH on removal mechanism. *The Eurasia Proceedings of Science Technology Engineering and Mathematics, 1*, 339–346.

Aragaw, T. A. (2020). Recovery of iron hydroxides from electro-coagulated sludge for adsorption removals of dye wastewater: Adsorption capacity and adsorbent characteristics. *Surfaces and Interfaces, 18*, 100439.

Aryal, R. L., Poudel, B. R., Gautam, S. K., Paudyal, H., Ghimire, K. N., & Pokhrel, M. R. (2019). Removal of fluoride from aqueous solution using biomass-based adsorbents: A review. *Journal of Nepal Chemical Society, 40*, 44–51.

Ashraf, M. A., & Mohd Hanafiah, M. (2019). Sustaining life on earth system through clean air, pure water, and fertile soil. *Environmental Science and Pollution Research, 26*(14), 13679–13680.

Bagastyo, A. Y., Anggrainy, A. D., & Nindita, C. S. (2017). Electrodialytic removal of fluoride and calcium ions to recover phosphate from fertilizer industry wastewater. *Sustainable Environment Research, 27*(5), 230–237.

Bhan, C., Singh, J., & Sharma, Y. C. (2021). Development of adsorbent from Mentha plant ash and its application in fluoride adsorption from aqueous solution; A mechanism, isotherm, thermodynamic, and kinetics studies. *International Journal of Phytoremediation, 23*(11), 1113–1123.

Bibi, I., Nazar, N., Ata, S., Sultan, M., Ali, A., Abbas, A., Jilani, K., Kamal, S., Sarim, F. M., Khan, M. I., & Jalal, F. (2019). Green synthesis of iron oxide nanoparticles using pomegranate seeds extract and photocatalytic activity evaluation for the degradation of textile dye. *Journal of Materials Research and Technology, 8*(6), 6115–6124.

Biftu, W. K., Ravindhranath, K., & Ramamoorty, M. (2020). New research trends in the processing and applications of iron-based nanoparticles as adsorbents in water remediation methods. *Nanotechnology for Environmental Engineering, 5*(2), 1–12.

Biswas, G., Thakurta, S. G., Chakrabarty, J., Adhikari, K., & Dutta, S. (2018). Evaluation of fluoride bioremediation and production of biomolecules by living cyanobacteria under fluoride stress condition. *Ecotoxicology and Environmental Safety, 148*, 26–36.

Cadaval, T. R., Dotto, G. L., & Pinto, L. A. (2015). Equilibrium isotherms, thermodynamics, and kinetic studies for the adsorption of food azo dyes onto chitosan films. *Chemical Engineering Communications, 202*(10), 1316–1323.

Chaudhary, M., Jain, N., & Maiti, A. (2021). A comparative adsorption kinetic modeling of fluoride adsorption by nanoparticles and its polymeric nanocomposite. *Journal of Environmental Chemical Engineering, 9*(5), 105595.

Chong, W. C., Choo, Y. L., Koo, C. H., Pang, Y. L., & Lai, S. O. (2019, September). Adsorptive membranes for heavy metal removal—A mini review. In *AIP Conference Proceedings* (Vol. 2157, No. 1, p. 020005). AIP Publishing LLC.

Dehbi, A., Dehmani, Y., Omari, H., Lammini, A., Elazhari, K., & Abdallaoui, A. (2020). Hematite iron oxide nanoparticles (α-Fe_2O_3): Synthesis and modeling adsorption of malachite green. *Journal of Environmental Chemical Engineering, 8*(1), 103394.

Ezzeddine, A., Bedoui, A., Hannachi, A., & Bensalah, N. (2015). Removal of fluoride from aluminum fluoride manufacturing wastewater by precipitation and adsorption processes. *Desalination and Water Treatment, 54*(8), 2280–2292.

Feitoza, N. C., Goncalves, T. D., Mesquita, J. J., Menegucci, J. S., Santos, M. K. M., Chaker, J. A., Cunha, R. B., Medeiros, A. M., Rubim, J. C., & Sousa, M. H. (2014). Fabrication of glycine-functionalized maghemite nanoparticles for magnetic removal of copper from wastewater. *Journal of Hazardous Materials, 264*, 153–160.

Grover, V. A., Hu, J., Engates, K. E., & Sipley, H. J. (2012). Adsorption and desorption of bivalent metals to hematite nanoparticles. *Environmental Toxicology and Chemistry, 31*, 86–92.

Grzegorzek, M., Majewska-Nowak, K., & Ahmed, A. E. (2020). Removal of fluoride from multi-component water solutions with the use of monovalent selective ion-exchange membranes. *Science of the Total Environment, 722*, 137681.

Habuda-Stanić, M., Ergović Ravančić, M., & Flanagan, A. (2014). A review on adsorption of fluoride from aqueous solution. *Materials, 7*(9), 6317–6366.

Hao, Y. M., Man, C., & Hu, Z. B. (2010). Effective removal of Cu (II) ions from aqueous solution by amino-functionalized magnetic nanoparticles. *Journal of Hazardous Materials, 184*(1–3), 392–399.

Hashemifar, M., ShamsKhorramabady, G., Nilufari, N., Mehrabpour, M., & Davoudi, M. (2014). Preparation of nano iron oxide coated activated sludge granules and its adsorption properties for Cd (II) ions in aqueous solutions. *Research Journal of Environmental and Earth Sciences, 6*(5), 259–265.

Jeyaseelan, A., Naushad, M., Ahamad, T., & Viswanathan, N. (2021). Design and development of amine functionalized iron based metal organic frameworks for selective fluoride removal from water environment. *Journal of Environmental Chemical Engineering, 9*(1), 104563.

Khan, A. M., & Ganai, S. A. (2020). Removal and recovery of heavy metal ions using natural adsorbents. *Modern Age Waste Water problems: Solutions Using Applied Nanotechnology*, 251–260.

Khodabakhshi, A., Amin, M., & Mozaffari, M. (2011). Synthesis of magnetite nanoparticles and evaluation of its efficiency for arsenic removal from simulated industrial wastewater. *Journal of Environmental Health Science & Engineering, 8*(3), 189–200.

Lawal, A. A., Hassan, M. A., Zakaria, M. R., Yusoff, M. Z. M., Norrrahim, M. N. F., Mokhtar, M. N., & Shirai, Y. (2021). Effect of oil palm biomass cellulosic content on nanopore structure and adsorption capacity of biochar. *Bioresource Technology, 332*, 125070.

Li, J., Chen, C., Zhu, K., & Wang, X. (2016). Nanoscale zero-valent iron particles modified on reduced graphene oxides using a plasma technique for Cd(II) removal. *Journal of the Taiwan Institute of Chemical Engineers, 59*, 389–394.

Li, Y., Fu, F., Cai, W., & Tang, B. (2019). Synergistic effect of mesoporous feroxyhyte nanoparticles and Fe(II) on phosphate immobilization: Adsorption and chemical precipitation. *Powder Technology, 345*, 786–795.

Lin, S., Lu, D., & Liu, Z. (2012). Removal of arsenic contaminants with magnetic γ-Fe_2O_3 nanoparticles. *Chemical Engineering Journal, 211*, 46–52.

Lingamdinne, L. P., Vemula, K. R., Chang, Y. Y., Yang, J. K., Karri, R. R., & Koduru, J. R. (2020). Process optimization and modeling of lead removal using iron oxide nanocomposites generated from bio-waste mass. *Chemosphere, 243*, 125257.

Lou, T., Yan, X., & Wang, X. (2019). Chitosan coated polyacrylonitrile nanofibrous mat for dye adsorption. *International Journal of Biological Macromolecules, 135*, 919–925.

Maia, L. F., Hott, R. C., Ladeira, P. C., Batista, B. L., Andrade, T. G., Santos, M. S., Faria, M. C., Oliveira, L. C., Monteiro, D. S., Pereira, M. C., & Rodrigues, J. L. (2019). Simple synthesis and characterization of l-Cystine functionalized δ-FeOOH for highly efficient Hg(II) removal from contamined water and mining waste. *Chemosphere, 215*, 422–431.

Mandal, M., & Ghosh, U. (2018). Value addition to horticultural solid waste by applying it in biosynthesis of industrially important enzyme: Cellulase. In *Utilization and Management of Bioresources: Proceedings of 6th IconSWM 2016* (pp. 279–289). Springer Singapore.

Obotey Ezugbe, E., & Rathilal, S. (2020). Membrane technologies in wastewater treatment: A review. *Membranes, 10*(5), 89.

Pang, T., Chan, T. S. A., Jande, Y. A. C., & Shen, J. (2020). Removal of fluoride from water using activated carbon fibres modified with zirconium by a drop-coating method. *Chemosphere, 255*, 126950.

Parham, H., Zargar, B., & Shiralipour, R. (2012). Fast and efficient removal of mercury from water samples using magnetic iron oxide nanoparticles modified with 2-mercaptobenzothiazole. *Journal of Hazardous Materials, 205*, 94–100.

Park, G., Hong, S. P., Lee, C., Lee, J., & Yoon, J. (2021). Selective fluoride removal in capacitive deionization by reduced graphene oxide/hydroxyapatite composite electrode. *Journal of Colloid and Interface Science, 581*, 396–402.

Pillai, P., Dharaskar, S., Shah, M., & Sultania, R. (2020a). Determination of fluoride removal using silica nano adsorbent modified by rice husk from water. *Groundwater for Sustainable Development, 11*, 100423.

Pillai, P., Dharaskar, S., Sinha, M. K., Sillanpää, M., & Khalid, M. (2020b). Iron oxide nanoparticles modified with ionic liquid as an efficient adsorbent for fluoride removal from groundwater. *Environmental Technology & Innovation, 19*, 100842.

Raghav, S., & Kumar, D. (2019). Comparative kinetics and thermodynamic studies of fluoride adsorption by two novel synthesized biopolymer composites. *Carbohydrate Polymers, 203*, 430–440.

Rashmi, S. H., Madhu, G. M., Kittur, A., & Suresh, R. (2013). Synthesis, characterization and application of zero valent iron nanoparticles for the removal of toxic metal hexavalent chromium [Cr(VI)] from aqueous solution. *International Journal of Current Engineering and Technology, 1*, 37–42.

Raul, P. K., Devi, R. R., Umlong, I. M., Banerjee, S., Singh, L., & Purkait, M. (2012). Removal of fluoride from water using iron oxide-hydroxide nanoparticles. *Journal of Nanoscience and Nanotechnology, 12*(5), 3922–3930.

Reddy, M. S., Sivaramakrishna, L., & Reddy, A. V. (2012). The use of an agricultural waste material, Jujuba seeds for the removal of anionic dye (Congo red) from aqueous medium. *Journal of Hazardous Materials, 203*, 118–127.

Sahu, N., Bhan, C., & Singh, J. (2021). Removal of fluoride from an aqueous solution by batch and column process using activated carbon derived from iron infused Pisum sativum peel: Characterization, Isotherm, kinetics study. *Environmental Engineering Research, 26*(4).

Sawangjang, B., Induvesa, P., Wongrueng, A., Pumas, C., Wattanachira, S., Rakruam, P., Punyapalakul, P., Takizawa, S., & Khan, E. (2021). Evaluation of fluoride adsorption mechanism and capacity of different types of bone char. *International Journal of Environmental Research and Public Health, 18*(13), 6878.

Sharma, P. P., Yadav, V., Maru, P. D., Makwana, B. S., Sharma, S., & Kulshrestha, V. (2018). Mitigation of fluoride from brackish water via electrodialysis: An environmentally friendly process. *Chemistry Select, 3*(2), 779–784.

Shirin, S., Jamal, A., Emmanouil, C., Singh, V. P., & Yadav, A. K. (2022). Assessment and characterization of waste material used as backfilling in an abandoned mine. *International Journal of Coal Preparation and Utilization*, 1–9. https://doi.org/10.1080/19392699.2022.2118259

Shirin, S., & Yadav, A. K. (2014). Physico chemical analysis of municipal wastewater discharge in ganga river, Haridwar district of Uttarakhand, India. *Current World Environment, 9*, 536–543. https://doi.org/10.12944/CWE.9.2.39

Sun, R., Zhang, H. B., Qu, J., Yao, H., Yao, J., & Yu, Z. Z. (2016). Supercritical carbon dioxide fluid assisted synthesis of hierarchical AlOOH@ reduced graphene oxide hybrids for efficient removal of fluoride ions. *Chemical Engineering Journal, 292*, 174–182.

Tolkou, A. K., Manousi, N., Zachariadis, G. A., Katsoyiannis, I. A., & Deliyanni, E. A. (2021). Recently developed adsorbing materials for fluoride removal from water and fluoride analytical determination techniques: A review. *Sustainability, 13*(13), 7061.

Wadidi, N. A., Idris, A. A. A., Woldemicheal, S. K., & El Haj, S. I. A. (2022). Properties of natural adsorbent prepared from two local Sudanese agricultural wastes mango seeds and date's stones and their uses in removal of contamination from fluid nutrient. *Bulletin of the National Research Centre, 46*(1), 1–7.

Waghmare, S. S., & Arfin, T. (2015). Fluoride removal from water by various techniques. *International Journal of Innovative Science Engineering and Technology, 2*(3), 560–571.

Wan, K., Huang, L., Yan, J., Ma, B., Huang, X., Luo, Z., Zhang, H., & Xiao, T. (2021). Removal of fluoride from industrial wastewater by using different adsorbents: A review. *Science of the Total Environment, 773*, 145535.

Wan, S., Lin, J., Tao, W., Yang, Y., Li, Y., & He, F. (2019). Enhanced fluoride removal from water by nanoporous biochar-supported magnesium oxide. *Industrial & Engineering Chemistry Research, 58*(23), 9988–9996.

Wang, Y., Zhang, Y., Li, S. Y., Zhong, W. H., & Wei, W. (2018). Enhanced methylene blue adsorption onto activated reed-derived biochar by tannic acid. *Journal of Molecular Liquid, 268*, 658–666.

Xu, L., & Wang, J. (2011). A heterogeneous Fenton-like system with nanoparticulate zero-valent iron for removal of 4-chloro-3-methyl phenol. *Journal of Hazardous Materials, 186*(1), 256–264.

Yadav, A. K., Shirin, S., Hopke, P. K., Pal, D., & Jamal, A. (2022). Process to reduce particulate matter in ambient air using bubbles of sodium palmitate. *Chemical Engineering & Technology, 45*, 1497–1500. https://doi.org/10.1002/ceat.202100454

Zhang, C., Li, Y., Wang, T. J., Jiang, Y., & Fok, J. (2017). Synthesis and properties of a high-capacity iron oxide adsorbent for fluoride removal from drinking water. *Applied Surface Science, 425*, 272–281.

Zhang, H., Wang, J., Zhang, X., Li, B., & Cheng, X. (2019). Enhanced removal of lomefloxacin based on peroxymonosulfate activation by Co_3O_4/δ-FeOOH composite. *Chemical Engineering Journal, 369*, 834–844.

Zhao, J., Lu, Z., He, X., Zhang, X., Li, Q., Xia, T., Zhang, W., & Lu, C. (2017). Fabrication and characterization of highly porous Fe(OH)3@ cellulose hybrid fibers for effective removal of Congo red from contaminated wat201er. *ACS Sustainable Chemistry & Engineering, 5*(9), 7723–7732.

Chapter 9
Electrocoagulation of Fluoride from Water with Fe-Based Ion Electrode

Ram Raj Meena, Sushil Kumar, and Pramod Soni

Abstract In recent years, the environmental sector has become interested in electrochemical-based techniques due to their environmental compatibility, sustainability, versatility, low cost, efficiency, and a small amount of sludge production. Fluoride in drinking water at high levels has been associated with a number of health problems, including skeletal, dental, and various forms of fluorosis. Among the various available de-fluoridation methods, the electrocoagulation procedure was tested experimentally and optimized to increase removal effectiveness while using the least amount of energy possible. The electrocoagulation process with Fe as sacrificial electrodes was used at domestic, industrial, and commercial levels. The Fe ions have a great affinity for fluoride ions, leading to a higher removal efficiency of fluoride from the water. The removal efficiency also depends on factors such as the applied current, the initial fluoride content, the electrode spacing, the electrolysis time, pH, the contact surface area, etc. This chapter deals with the removal of fluoride contents through Fe-based ion electrodes by electrocoagulation (EC). We further discuss the mechanism of CE, the different shapes of Fe electrodes, and the geometry of reactors in contrast to the efficiency of F removal from the water.

Keywords Fluoride removal · Iron electrode · Water purification · Electrocoagulation

R. R. Meena (✉)
Department of Civil Engineering, Motilal Nehru National Institute of Technology Allahabad, Prayagraj 211004, India
e-mail: ram.2020rce10@mnnit.ac.in

S. Kumar
Department of Chemical Engineering, Motilal Nehru National Institute of Technology Allahabad, Prayagraj 211004, India
e-mail: sushilk2006@gmail.com

P. Soni
Department of Civil Engineering, Indian Institute of Technology (Banaras Hindu University), Varanasi 221005, India
e-mail: pramod.civ@iitbhu.ac.in

A. K. Yadav et al. (eds.), *Advanced Treatment Technologies for Fluoride Removal in Water*, Water Science and Technology Library 125,
https://doi.org/10.1007/978-3-031-38845-3_9

9.1 Introduction

The contamination of fluoride ions in the water is possibly due to the high fluoride soil, grasses and grains, forage, anthropogenic reasons, volcanic activity, a deposit of metal processing, glass manufacturing, fertilizers, semiconductors, etc. The discharge of these substances into surface water also leads to fluoride contamination in groundwater. Long-term drinking of high fluoride-contaminated water causes the result of the softening of bones, mottling of the teeth, and osteogenesis of ligaments and tendons. Dental cavities continue to be a serious public health problem in most industrialized countries, influencing 60–90% of children and the vast proportion of adults (Liteplo et al., 2002; Division of Toxicology & Environmental Medicine, 1997). Fluoridated water decreases cavities in children, but its effectiveness in adults is less certain. Currently, 372 million people, representing 5.7% of the global population in around 24 countries, including India, Brazil, Australia, Malaysia, Chile, Canada, the United States, the Republic of Ireland, and Vietnam, receive deliberately fluoridated water. In nations including Sweden, Finland, Zimbabwe, China, Sri Lanka, and Gabon, 57.4 million people receive naturally fluoridated water near or above recommended levels (Liteplo et al., 2002). Whereas in India, 12 m tons of fluoride are dumped out of 85 million in the earth's crust (Tables 9.1 and 9.2).

The use of electrochemistry for sewage treatment dates back to 1889 when the first sewage treatment facility using this technology was established in London. Irrespective of the promising results, the effectiveness of electrochemical (EC) processes has been inconsistent. In recent times, the demand for water treatment solutions has increased, leading to renewed interest in this technology. The advantages of EC processes include adaptability, energy efficiency, automation feasibility, and economic viability. In addition, electrochemically based systems offer controlled and rapid reactions, compactness, and the potential for treating water using electrons instead of chemicals. One of the most efficient methods of treating contaminated water is through the in situ production of active coagulant with the aid of electrically oxidizing the anode material, a process commonly employed in EC.

Among the defluoridation procedures that are now accessible are electrochemical, electrodialysis, chemical precipitation, and adsorption (Lounici et al., 2001; Mameri et al., 1998; Parthasarathy & Buffle, 1985; Toyoda & Taira, 2000). The Nalgonda process and the Activated Alumina (AA) (Susheela et al., 1992) procedure are the two most used defluoridation techniques in India. Due to the massive amounts of sludge produced, the fact that metal hydroxides are now considered hazardous waste, and the excessive prices of chemical treatment and outdated chemical coagulation techniques employed in defluoridation are no longer considered acceptable. The current circumstance demands a solution that addresses the problems described previously. It is necessary to develop a new approach that addresses the drawbacks of current coagulation procedures. Electrocoagulation is a potential replacement for conventional coagulation methods (EC). It has been used to treat sewage and drinkable water with great success (Emamjomeh et al., 2011).

Table 9.1 Fluoride distribution in Indian districts listed

Districts	Fluoride range	Source of contamination	References
Hooghly, West Bengal	0.02–1.28 mg/L	Superphosphate fertilizer	Kundu and Mandal (2009)
Nalgonda, Andhra Pradesh	0.4–20 mg/L	Fluoride-rich granitic rocks	Ramamohana Rao et al. (1993)
Bellary, Karnataka	0.34–7.9 mg/L	Apatite, hornblende, and biotite	Sreenivasan and Wodeyar (1996)
Bhiwani, Haryana	0.14–8.6 mg/L	Rocks	Garg et al. (2009)
Jharkhand	0.1–4 mg/L	Coal ash	Prasad and Mondal (2005)
Guwahati, Assam	0.19–6.89 mg/L	Granite rocks	Das et al. (2003)
Kanpur, U.P	0.15–5.35 mg/L	Industrial waste	Sankararamakrishnan et al. (2008)
Palghat, Kerala	0.2–5.75 mg/L	Biotite gneiss and Hornblende	Shaji et al. (2007)
Visakhapatnam, Andhra Pradesh	0.61–2.11 mg/L	Granitic rocks	Ramamohana Rao et al. (1993)
Kurmapalli watershed, Andhra Pradesh	Up to 21.5 mg/L	Fluoride rich rocks	Mondal et al. (2009)
Delhi	0.1–16.5 mg/L	Irrigation water and brick industries	Datta et al. (1996)
Mehsana, Gujrat	0.95–2.82 mg/L	Granite, pegmatite and gneiss	Salve et al. (2008)
Hanumangarh, Raj	1.02–4.42 mg/L	Rocks	Suthar et al. (2008)
Erode, Tamilnadu	0.5–8.21 mg/L	Host rocks and weathering of fluoride	Karthikeyan et al. (2010)
Yavatmal, Maharashtra	0.3–13.41 mg/L	Biotite, Amphibole, and fluoroapatite	Madhnure et al. (2007)

Table 9.2 Fluoride standards in drinking water by various authorities

Authority	Permissible limit (mg/L)	References
Indian Council of Medical Research (**ICMR**)	1.0	Modi (2013)
Bureau of Indian Standards (**BIS**)	1.0	BIS (2022)
World Health Organisation (**WHO**)	1.5	WHO (2010)
The Committee on Public Health Engineering Manual and Code of Practice, Government of India (**CPHEEO**)	1.00	BIS (2022)

EC is an electrochemical method that effectively removes contaminants from drinking water and wastewater. During this procedure, an electric current (mainly DC) is applied to the electrodes. Metal hydroxides, created when the anode oxidizes, serve as coagulants. In situ coagulant production occurs throughout this procedure. It is possible to compare chemical coagulation methods and the pollutant removal process. EC is evolving as a potential technology that can be used in place of conventional defluoridation techniques. It has been proposed to substitute electrocoagulation (EC) for conventional coagulation. Some benefits of this process include decreased sludge creation, no need for chemical handling, and convenience of operation (Zhu et al., 2007). It can also be observed that the EC method for eliminating fluoride content does not need a huge investment.

9.2 Coagulation Chemistry for Fe Ions

9.2.1 At the Anode

For electrocoagulation iron, stainless steel (SS) or steel (St) anode is used (Zodi et al., 2009), Fe^{2+} is dissolved in water due to Fe ions oxidation.

$$Fe \rightarrow Fe^{2+} + 2e^{-} \tag{9.1}$$

9.2.2 At the Cathode

The water reduction reaction results in the production of H_2 gas and hydroxide ions (Fig. 9.1).

$$2H_2O + 2e^{-} \rightarrow 2OH^{-} + H_2(g) \tag{9.2}$$

During electrolysis, the formation of OH from reaction (9.2) raises the pH of the solution, resulting in the production of various iron hydroxo-complexes. The production of insoluble Fe $(OH)_2$, which aids in the precipitation of coagulation, can be represented as:

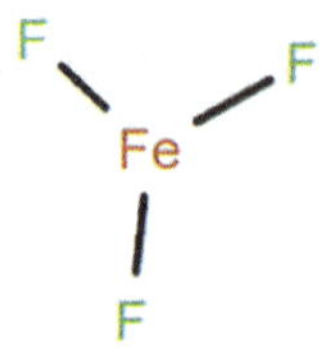

Fig. 9.1 2D structure for ferric fluoride

$$Fe_{2+} + 2OH \rightarrow Fe(OH)_2(s) \quad (9.3)$$

and from all the sequence of reactions for the electrolytic process the following reaction is thru:

$$Fe + 2H_2O \rightarrow Fe(OH)_2(s) + H_2(g) \quad (9.4)$$

Fe(II) species can produce coagulates; however, Fe(III) species have larger charge densities, which further favors the coagulation-flocculation process.

$$Fe \rightarrow Fe^{3+} + 3e^- \quad (9.5)$$

$$4Fe^{2+} + 10H_2O + O_2(g) \rightarrow 4Fe(OH)_3(S) + 8H^+ \quad (9.6)$$

where protons are reduced directly to H_2 gas at the cathode using reactions or neutralized with the OH– generated in reaction.

$$2H^+ + 2e^- \rightarrow 2H_2(g) \quad (9.7)$$

Depending on the characteristics of the pollutant, all of these species with various protective charges and electrostatic attraction promote the development or precipitation to varying degrees. Among all iron (III) species, $Fe(OH)_3$ is expected to be the favored coagulant reagent and the prominent species responsible for pollution removal (Garcia-Segura et al., 2017). The EC setup is shown in Fig. 9.2.

9.3 Parameters Examination for Fluoride Removal

The influence of operational constraints such as initial fluoride concentration, pH, electrolysis time, and inter-electrode distance applied current, which administrates the electrocoagulation method to the removal of fluoride. In addition, more parameters such as charge loading, SEEC, fluoride uptake capacity, and charge loading are also considered for the removal of fluoride.

9.3.1 pH

The pH is a key factor in the chemical and electrochemical coagulation approach. The effectiveness of the reaction process is extremely dependent on the pH of the solution. pH directly affects the specifications of Fe and Al, influencing the defluoridation mechanism.

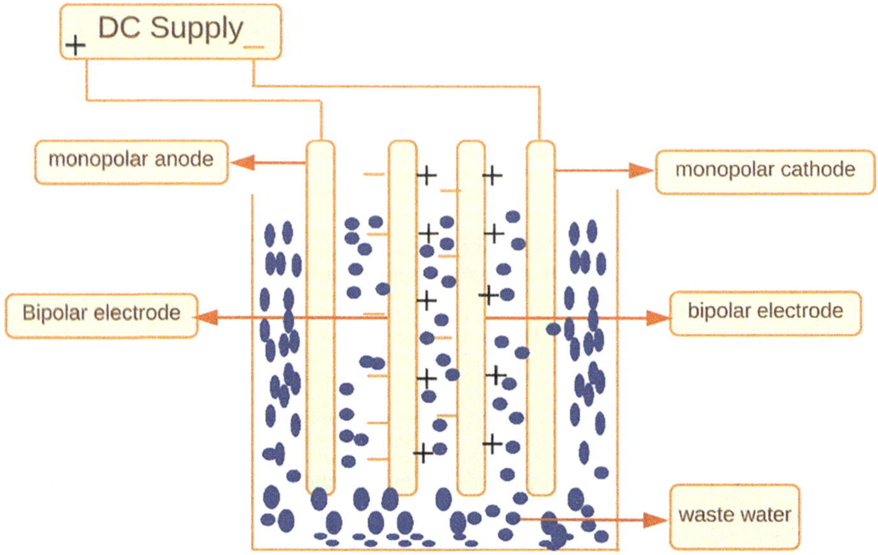

Fig. 9.2 EC cell arrangement

Depending on the pH, several species develop in ionic metals, including ionic metal species, polymeric hydroxide complexes, and monomeric hydroxide complexes. Each exhibits a unique reaction with contaminants, resulting in various coagulation capacities and, as a result, different removal strategies. For aluminium and iron, the species at highly alkaline conditions are $Al(OH)^{4-}$ and $Fe(OH)^{4-}$, respectively. Each of these species has weak coagulation functions. Only slightly acidic conditions are used for coagulation (Fe: 4–5 and Al: 5–6) (Ganesan et al., 2013). The physiochemical characteristics of coagulants include:

1. Electrical conductivity of metal hydroxides.
2. Solubility of metal hydroxides.
3. The size and shape of colloidal particles comprising coagulant complexes are highly sensitive to changes in pH conditions.

Thus, for coagulation, neutral and alkaline media are chosen. It is discovered for aluminium electrodes; defluoridation process is the most proficient between the pH ranging from 6 to 8, so there is no need for re-adjustment after the treatment (Sinha et al., 2015). The influence of the initial and final pH concentration on the defluoridation is shown in Fig. 9.3.

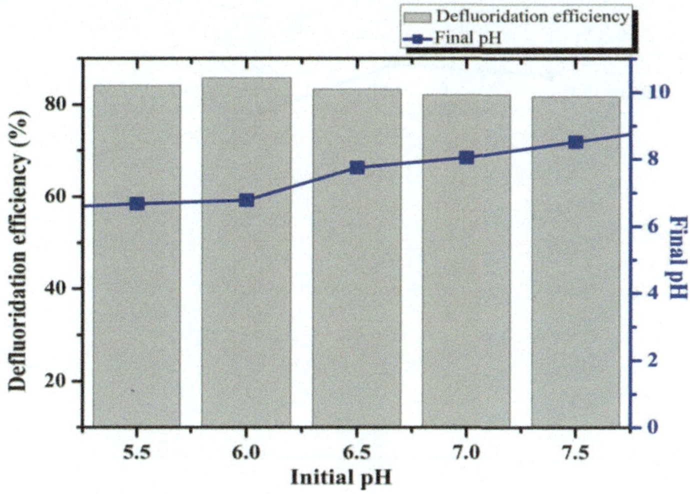

Fig. 9.3 Defluoridation versus pH

9.3.2 Inter-electrode Distance

The IR drop or ohmic potential drop is directly influenced by the interelectrode distance between the anode and cathode and decreases as electrode distance is increased. Short distances between the electrodes can, however, reduce the contaminants' removal efficiency from the water, since several factors can be impacted like coagulation, electroflotation, flocculation, precipitation, etc. Due to the strong electrostatic force that prevents particle collisions, these elements affect the formation and precipitation of flocs and prevent the growth of aggregates (Enciso et al., 2012). On the other hand, a large gap between the electrodes considerably reduces floc development.

An increase in the distance between the electrodes causes increased resistance between the electrodes, which decreases the efficiency of the process. Many studies have been done on the enter-electrode distance. They all found that close to 10 mm is the most efficient distance for better results. Sinha et al. (2016) conducted the study for four electrode distances, that is, 5, 10, 15 and 20 mm. The initial concentration of F, the electrolysis time and applied current were kept constant for all runs. The effectiveness was reported to increase as the distance increases from 5 to 10 mm. But when the distance increases from 10 to 20 mm, a decrease in efficiency was found. The small distance between the electrodes is the reason for the strong air floatation effect in the cathode bubble generation. In addition, the very fewer electrode distances can cause short-circuiting. The efficiency variation with the interelectrode distance is shown in Fig. 9.4.

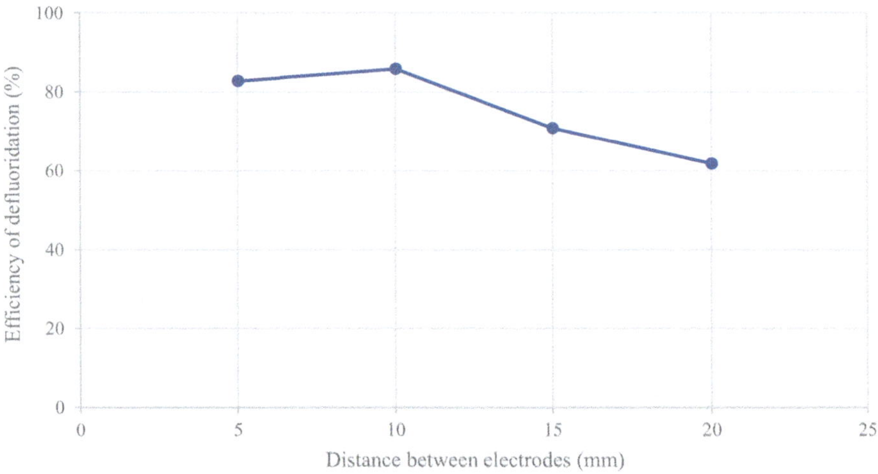

Fig. 9.4 Defluoridation versus electrode distance

9.3.3 Initial Concentration of Fluoride

In India, most F-contaminated regions exhibit fluoride levels ranging from 2 to 8 mg/L. Sinha et al. (2015) revealed that residuals increased from 0.57 to 2.18 mg/L as the initial fluoride grew from 2 to 8 mg/L. under constant electrolysis conditions of a 20-minutes period and an applied current. This may possibly be attributed to the inadequate complex generation of aluminum hydroxide. The authors also noted that when the initial fluoride concentration is high, a high applied current value and a longer electrolysis duration are needed to reduce the fluoride content to acceptable limits. Xu and Zhu (2004) reached the same conclusion.

9.3.4 Current Density and Type

Two crucial factors to consider in the electrocoagulation technique are the applied current and the electrolysis time. The amount of coagulant generated and the rate at which bubbles form are mainly influenced. The electrolysis time and applied current are correlated. More electrolysis time is needed to remove fluoride when the applied current is lower, and vice versa.

The electrochemical reactions that occur in solution (such as the rate of electro-dissolution, water reaction, electro-flotation, gas evolution, etc.), as well as their kinetics and extension, are controlled by the applied current density (j). As a result, the current density specifies (with the applied voltage) the energy required for the electrochemical process to run Colli et al. (2017).

Direct current (DC) is typically used more frequently in EC. Sacrificial anode passivation increases ohmic resistance, which causes the cell potential to increase and increases operating expenses, yet passivation sharply lowers the EC efficiency. Alternately, using an alternating current (AC) might be seen as extending the operating life of the sacrificial anodes by preventing or reducing the production of passivation layers through constant polarity changes (Vasudevan et al., 2009).

9.3.5 *Supporting Electrolyte*

In an electrochemical process, a supporting electrolyte solution is necessary to prevent migration effects, energy consumption, reduce ohmic drop, and boost solution conductivity. As an alternative, the electrolyte significantly impacts the sacrificial anodes' electrodissolution kinetics. It can also affect the ability to produce flocs by creating a double shielding layer. Much research on the EC process with various supporting electrolytes reveals that the various effects are typically linked to anion effects rather than cations (Colli et al., 2017). According to certain authors, sulfate species have a tendency to form multiplexes with aluminum and passivate the anodic surfaces.

9.3.6 *Treatment Time*

Current density refers to a crucial element since it affects the coagulant dose rate, size, bubble formation rate, and floc formation; these affect the efficacy of the electrocoagulation process. The current density has a direct relationship with the anode dissolution rate. The concentration of coagulant created by electrolysis on the electrode is often correlated with the amount of electric charge applied per volume. Treatment time also influences the pollutant removal rate. The efficacy of contaminant elimination increases with the rise in the treatment time. At the beginning of the EC process, fluoride ions are more prevalent. Fluoride was rapidly removed because complexes formed among the produced aluminum hydroxide and the fluoride due to the anode at the time (Zuo et al., 2008). However, as the experiment goes on, the concentration of aluminum hydroxide continues to rise, whereas the amount of fluoride in the aqueous phase continues to decrease.

9.3.7 *Temperature*

When Al and Fe electrodes are used, a high temperature causes the enormous pores in the $Al(OH)_3$ and $Fe(OH)_3$ gel to close, which results in the creation of compressed flocs which are mostly deposited on electrode surface. Without changing the cost or energy usage, increasing the temperature and decreasing the electrolysis duration considerably increase the removal efficiency. In reality, when the temperature rises, particle collision kinetics becomes better as a consequence of enhanced mass transfer. Furthermore, high temperature accelerates flotation and reduces the adhesion of suspended particles while favoring the production of big hydrogen bubbles (Naje et al., 2017).

9.3.8 Conductivity of Solution

The effectiveness of contaminant removal and the rate of treatment of the electrochemical process have direct influence on the conductivity of solution. The conductivity facilitates the movement of electrical current through it. Before treatment, the pH is adjusted by adding salt additions to the solution, such as sodium chloride or sodium sulphate (Naje et al., 2017).

9.3.9 Agitation Speed

Agitation supports the prevention of the establishment of a concentration gradient and maintains uniform circumstances in the electrochemical cell during the electrocoagulation process. Moreover, the electrochemical cell's agitation speeds up the produced ions as they move about. Once the agitation speed reaches the appropriate agitation speed, the effectiveness of pollutant removal increases. This occurs as a result of early floc development linked to a rise in the agility of the produced ions. Furthermore, the effectiveness of pollutant elimination on the specific increase in electrolysis time (Brindha et al., 2018; Yildiz et al., 2007).

9.3.9.1 Electrode Arrangements

The electrode connection mode of an EC cell affects removal efficiency, energy, and financial costs. The three most common configurations are (a) parallel connections for monopolar electrodes, (b) serial connections for monopolar electrodes, and (c) serial connections for bipolar electrodes.

Varying in its electrical polarity, each electrode in a monopolar electrode configuration can function as either a cathode or an anode in an electrochemical cell. Each sacrificial anode in the monopolar parallel electrodes (MP-P) is directly connected to another anode within the electrocoagulation cell using the same setup as for cathodes. In contrast, each pair of anode and cathode of the MP-S (monopolar serial electrodes) design is internally coupled but not with the exterior electrodes (Emamjomeh & Sivakumar, 2006; Goren & Kobya, 2021).

In the case of BP electrodes, each electrode, excluding the exterior ones, which are MP, shows a distinct polarity at every side based on the charge of the anode or cathode in front of it. Bipolar electrode connections are always made serially.

It is important to note that although a serial configuration requires a larger potential difference, all electrodes receive the same amount of current. On the contrary, the parallel mode of the electrochemical reactor divides the electric current among the connected electrodes according to their resistance. The relative efficiencies of these various electrode layouts greatly depend on the operating settings stated previously; therefore, the performance results are only partially conclusive, as well as the nature of the pollutant and the water matrix (Golder et al., 2007; Purkait, 2008; Sahu et al., 2014).

9.4 Other Phenomena's Affecting the Reaction for Fluoride Removal

Electrocoagulation (EC) is a process that has a significant impact on reactor operation, scalability, and performance. The commonly used EC reactor type is an open batch cell with electrodes as plates in which the cathodes and anodes are completely immersed in the solution and the effluent is consistently mixed to ensure homogeneity. Another cylindrical variation of the reactor employs a rotating impeller cathode with two metallic blades, as conveyed by Tezcan Un et al. (2013), which mechanically homogenizes the solution and prevents the coagulants from settling during EC. In addition, for electrocoagulation, the conventional filter press cell is a continuous reactor with rotating screw-type anode and cathode. Recently, a newly created EC technology with continuous reactors and rotating screw-shaped anodes has been developed. These reactors have been used for the purification of wastewater containing groundwater and cheese whey. To achieve a uniform distribution of liquid velocity around the sacrificial rod anode and helical cathode, each of which is located in the center of the electrochemical reactor (involving rotation or without rotation), these cells feature a symmetrical section, as noted by Choudhary and Mathur (2017).

9.4.1 Design of Electrocoagulation Cells

The efficient operation of an electrocoagulation cell is crucial and it depends on various factors. One of these factors is the prevention of the accretion of O_2 and H_2 gas bubbles that nucleate on surfaces. Additionally, for optimum performance, it's also crucial to keep a constant mass transfer across the electrode gap and reduce the IR drop between the anode and cathode. The IR drop can be affected by the geometry of the electrode, the conductivity of the electrolyte solution, and the inter-electrode distances. Therefore, careful consideration of these aspects is necessary to achieve appropriate efficacy in electrocoagulation cells.

The mass transfer overvoltage, equilibrium potential difference, electrode overpotentials, and ohmic potential drop in the solution are all factors that affect the total voltage (U/V). The concentration of the analyte varies close to the electrode surface as a result of the electrode response, which causes a concentration overpotential, sometimes referred to as mass transfer or diffusion overpotential. This results from a difference in the concentration of electroactive species between the electrode surface and the bulk solution. This generally occurs when the mass transfer coefficient is substantially smaller than the reaction rate constant and the electrochemical response to the poor surface concentration over potential is visibly quick. Mass transfer over potential is decreased by an enhancement in the mass of metal ions transferred from the anode surface to the majority of the solution (Khaled et al., 2019; Lekhlif et al., 2014). The turbulence of the electrolyte solution can be increased mechanically to enable this bulk transfer.

9.4.2 Mass Transfer

The mass transfer coefficient is an important limitation on the operation and design of chemical reactors, which is greatly influenced by the flow of fluids, laminar or turbulent. When the mass goes from a high chemical potential to low chemical potential range, mass transfer minimizes differences within the system. These coefficients can be experimentally measured using the current limiting method or the electrochemical methodology. For a rough surface, the evaluation of solid–liquid mass transfer coefficients in chemical reaction systems or complex flow regimes has made great use of this technique. The determination of the maximum current that can pass through an electrode during a specific EC reaction is the basis for mass transfer. The current-limiting method can be utilized to examine how electrode arrangement affects mass transfer rates in different electrochemical reactors.

The mass transfer coefficient can be measured with greater accuracy than with other methods, such as the dissolving method, which calculates the mass transfer coefficient depend on the effective removal of pollutants. Limiting current approaches have two key advantages: the ability to control driving forces and in situ current monitoring (Khaled et al., 2019).

The method of dimensional analysis is helpful in the case of spinning electrodes to consider the mass transfer phenomenon of the EC process. When this analysis is used to describe mass transfer, three non-dimensional groups; Sherwood (Sh), Reynolds (Re), and Schmidt numbers (Sc) are produced,

$$\mathrm{Sh} = \mathrm{K_m d}/D \tag{9.8}$$

$$\mathrm{Re} = \rho N d^2 / \mu \tag{9.9}$$

$$\mathrm{Sc} = \mu / \rho D \tag{9.10}$$

where, ρ, D, d, N and μ symbolizes fluid density (g/cm^3), diffusion coefficient (cm^2/s), effective diameter (cm), rotation speed (rpm), and fluid viscosity (g/cm s), respectively. Rotating electrodes haven't been employed in EC studies before, though. The dimensionless relationship presented below, which is assumed to be applicable for a variation in designs of electrolytic reactor, is used to experimentally correlate mass transfer data as a consequence.

$$\mathrm{Sh} = b\mathrm{Re}^a \mathrm{Sc}^{0.33} \tag{9.11}$$

where, a = exponent of Reynolds number and b = mass transfer correlation constant. Both experimental methods and differential equations can be used to determine these parameters' values. These mass-transfer correlations are beneficial for understanding chemical kinetics and for enhancing reactor design. To understand mass transfer processes, it is required to develop these mass transfer correlations (Khaled et al., 2019; Lekhlif et al., 2014; Naje & Abbas, 2013).

9.4.3 Residence Time Distribution

The features of the mixing that takes place in an electro-chemical reactor are represented by the residence time distribution (RTD) of a chemical reactor. Based on the stimulus–response technique, it is planned to investigate hydrodynamic flows in chemical apparatuses. While the RTD provides information on how long each element has been in the reactor, it does not provide details on the exchange of materials among the fluid particles. Its significance stems from the statistic that it offers a measurable evaluation of the degree of back-mixing into a system (Fogler, 1999), enables precise kinetic modeling of the system, and helps in the design of reactor to achieve required flow patterns. RTD appears to be a key tool in successfully scaling up processes because it allows for a detailed comparison between systems considering various zones or configurations of the reactor.

It offers strong tools and an easy-to-use interface for initially processing experimental data, modeling reactor flow using predefined flow patterns as estimating flow model parameters, building blocks, simulating system response to various input signals, and creating user-defined flow patterns.

The RTD investigation is a standard approach for calculating yields in homogeneous isothermal reactors. However, in some circumstances, nonisothermal or heterogeneous reactors can also benefit from the generalized principle of a reaction history distribution. The yield of a first-order reaction is solely determined by the residence time distribution. At t = 0, a tracer- an inert chemical, atom, or molecule is injected into the reactor. The concentration of the tracer, C, is then measured as a function of time in the effluent stream. This procedure is used to determine the RTD experimentally. The chosen tracer should be completely soluble in the mixture, completely soluble in the reactor liquid, easily detectable, and non-reactive (Fogler, 1999). Pulse input and step input are the two commonly used injection techniques. During pulse input, a specific amount of tracer is fed into the reactor for a minimum period of time.

9.4.4 *Response Surface Methodology (RSM)*

In conventional multi-factor experimental studies, optimization is frequently achieved by varying e factor, whereas the remaining other factors hold constant under predetermined conditions. Because it avoids the interactions between variables, it takes a long time and frequently fails to find the true optimum. In contrast, the RSM has been offered to determine the effects of distinct components and their combined effects. Designing, conceiving, developing, and evaluating new scientific studies and products is dependent on the RSM. It is also effective in enhancing current research and output. The industry has increased its use of RSM as a result of one crucial fact: whether the system comprises a maximum, minimum, or saddle point.

9.5 Results

Different pollutants have been removed from wastewater using electrocoagulation using Fe electrodes. They have been listed as follows: organic pollutants, heavy metals, nonmetallic inorganic species, real industrial effluents, etc.

9.5.1 *Non-metallic Inorganic Species*

The quantity of non-metallic inorganic species in water has increased due to the development of human progress and the excessive use of detergents and fertilizers. Therefore, efforts are necessary to minimize the adverse effects on the ecosystem. When Al, Fe, and SS electrodes were used, the removal of concentration for ammonia, fluoride, nitrite, nitrate, boron, cyanide, phosphate, silica particles, sulfides, powdered activated carbon, and sulphite was greater than 80%. Regarding ammonia, the Al-SS electrode arrangement favors the effectively removing this inorganic molecule from wastewater, while Al–Al only eliminated 80% of it. Although only 15% ammonia was eliminated when using Fe–Fe electrodes, there is still a need for improvement.

In some circumstances, the MP-P reactor was the preferred EC configuration. However, when the supporting electrolyte was altered, no appreciable changes were seen. The effectiveness of the EC technique was increased when a pH between 7.0 and 8.0 was used during the removal of boron in various instances when the EC arrangement was used preferentially according to the pH conditions. On the basis of the amounts of boron in the effluent, the impactful coalition of Fe, Al, or SS electrodes offers a significant decrement in treatment time, even when the nature of the anode and cathode was not an important parameter for boron removal. Using different electrode combinations, such as Al–Al, Al–Fe, and Mg-SS, removal efficiency ranged from 86.3 to 97.3%. The removal efficiency of nonmetallic compounds utilizing various electrode combinations is discussed in Table 9.3.

9.5.2 *Heavy Metals*

Heavy metals are bioaccumulative and not biodegradable, affect the entire food chain process. The combination of electrodes used and the EC reactor are critical factors that affect the removal efficiency attained in several cases, including those involving arsenate, arsenite, manganese, silver, and cadmium. As a result of the simultaneous chemical reactions that occur during the formation of complexes or flocs with the coagulant material, the efficacy of elimination differs depending on the type of effluent (Table 9.4).

9.5.3 *Organic Pollutants*

Organic pollutants have become a major concern and hazardous pollutants during the last decade, according to the UNESCO World Water Report (2012). Harmful, carcinogenic, and mutagenic impacts are dangerous consequences of organic pollution. As a result, pretreatment of effluents before disposal is necessary for the ecosystem. One alternative to reducing the presence and fate of these contaminants in water bodies is considered EC treatment (Yu et al., 2014).

Various EC comparison studies using iron and aluminium anodes have been conducted for synthetic dyes used in the textile industries, such as Crystal Violet Dyes and Reactive Red 43. Fe is preferable to Al to remove the color from various types of dyes. Unlike Al, which only removes days through pure adsorption and coagulation, Fe reduces dyes by adding Fe^{+2} ions to the system, explaining why it is superior in most situations. However, the anions in the solution had a substantial impact on the color decay. For example, when 2.5 mA cm^{-2} of current was applied to 1.8 L of reactive Red 43 at 50 mg/L in the presence of numerous anions at a neutral pH, when the Al–Al electrodes were used (Patel et al., 2011).

Table 9.3 Non-metallic compounds removal efficiency

Compound	Arrangement	Time (min)	pH	[CO]Mg/L	J (mA/cm^2)	Anode–cathode	Removal %	References
Cyanide (CN^-)	MP-P	20	11.5	300	15	Al–Al Al–Fe Fe–Fe Fe–Al	35.0 32.0 87.0 93.0	Moussavi et al. (2011)
Fluoride (F^-)	MP-P	90	3.0	42	5.0	Fe–Fe Al–Al	56.7 87.0	Aoudj et al. (2015)
	MP-P	60	7.0	20	10	Al–Al	91.5	Vasudevan et al. (2011)
Phosphate (PO_4^{3-})	MP-P	60 15	5.0	30	10	Al–Fe	93.0 96.0	Kuokkanen et al. (2015)
	MP-P	30 120	9.0	83	3.0	Fe–Fe Al–Al	100 100	Lacasa et al. (2011)
Sulphides (S^{2-})	MP-P	15	7.0	100 500	32	Fe–Fe	99.0 65.0	Murugananthan et al. (2004)
Sulphite (SO_3^{2-})	MP-P	15	7.0	100 500	62	Fe–Fe	85.0 46.2	Murugananthan et al. (2004)
Sulphate (SO_4^{2-})	MP-P	15	7.0	100 500	62	Fe–Fe	71.3 30.0	Murugananthan et al. (2004)

Table 9.4 Heavy metals removal efficiency

Compounds	Arrangements	Time (min)	pH	[CO]mg/L	J (mA/cm^2)	Anode–cathode	Removal %	References
Arsenate (V) [AsO_4^{3-}]	MP-P	15	7.0	0.10	0.14	Fe–Fe	99.9	Wan et al. (2011)
	MP-P	60	7.0	10	3.0	Al–Al Fe–Fe	99.9 99.9	Lacasa et al. (2011)
Chromium (III) [Cr^{+3}]	BP-S MP-P	50	3.4	1700	10.8 32.5	Fe–Fe	81.5 99.9	Golder et al. (2007)
Arsenite (III) [AsO_3^{3-}]	MP-P	60	2.4	13.4	4.0 4.0 30(AC)	Al–Al Fe–Al Fe–Fe	97.6 99.7 99.6	Gomes et al. (2007)
Chromium (VI) [$Cr2O_7^{2-}$]	MP-P	90	3.0	50	5.0	Al–Al Fe–Fe	25.2 99.8	Aoudj et al. (2015)
Copper [Cu^{2+}]	MP-P	40	5.5	250	25	Fe–Fe	98.0	Al Aji et al. (2012)
Mercury (II) [Hg^{2+}]	MP-P	15 25	7.0	0.4	25	Fe–Fe Al–Al	99.9	Nanseu-Njiki et al. (2009)

9.6 Conclusions and Recommendation

On the basis of the above studies, it is stated that iron (Fe) electrodes can be used for the elimination of different substances which are found in wastewater and drinking water like Fluoride (F^-), Arsenate (V), chlorium (III), Mercury (II), Sulphides, Phosphate, Cyanide, etc. In the case of fluoride removal, Fe–Fe gives less results in comparison of Al–Al electrodes for because of its lower affinity to the fluoride ions. The efficiency of fluoride removal also depends on different factors such as the size of the reactor, the reactor, the geometry of the reactor, arrangements of the electrodes, the current density to electrodes, retention time in the reactor, etc.

To successfully remove fluoride from drinking water, Al–Al electrodes have been preferred over Fe–Fe electrodes because they produce better results and produce less sludge. Furthermore, because Al–Al electrodes produce fewer hazardous residuals, they are a safer and more efficient solution for fluoride removal.

Acknowledgements The authors thank our prestigious institute and other people who are involved in this study for their direct and indirect contributions.

References

Al Aji, B., Yavuz, Y., & Koparal, A. S. (2012). Electrocoagulation of heavy metals containing model wastewater using monopolar iron electrodes. *Separation and Purification Technology*. https://doi.org/10.1016/j.seppur.2011.11.011

Aoudj, S., Khelifa, A., Drouiche, N., Belkada, R., & Miroud, D. (2015). Simultaneous removal of chromium(VI) and fluoride by electrocoagulation-electroflotation: Application of a hybrid Fe–Al anode. *Chemical Engineering Journal*. https://doi.org/10.1016/j.cej.2014.12.081

BIS. (2022). *Bureau Of Indian Standards. 14*(January), 1–23.

Brindha, K., Elango, L., Sen, S., Prajapati, A. K., Bannatwala, A., Pal, D., Kok, M. D. R., Jervis, R., Tranter, T. G., Sadeghi, M. A., Brett, D. J. L., Shearing, P. R., Gostick, J. T., Bhattacharya, D., Mukhopadhyay, J., Biswas, N., Basu, R. N., Das, P. K., De Kubber, D., … Nava, J. L. (2018). Optimization of the process parameters for the removal of boron from drinking water by electrocoagulation—A clean technology. *Electrochimica Acta, 3*(2), 93–127. https://doi.org/10.1039/c9ee00219g

Choudhary, A., & Mathur, S. (2017). Performance evaluation of 3D rotating anode in electro coagulation reactor: Part II: Effect of rotation. *Journal of Water Process Engineering, 19*(August), 352–362. https://doi.org/10.1016/j.jwpe.2017.08.019

Colli, A. N., Toelzer, R., Bergmann, M. E. H., Bisang, J. M., Zhang, T., Shao, T., Zhang, Y., Pan, D., Huang, W., Li, B., Hakizimana, J. N., Gourich, B., Chafi, M., Stiriba, Y., Vial, C., Drogui, P., Naja, J., Yoon, S. K., Fichtl, W., & Kenis, P. J. A. (2017). Electrocoagulation process in water treatment: A review of electrocoagulation modeling approaches. *Desalination, 404*(August 2021), 1–21. https://doi.org/10.1016/j.desal.2016.10.011

Das, B., Talukdar, J., Sarma, S., Gohain, B., Dutta, R. K., Das, H. B., & Das, S. C. (2003). Fluoride and other inorganic constituents in groundwater of Guwahati, Assam, India. *Current Science, 85*(5), 657–661.

Datta, P. S., Deb, D. L., & Tyagi, S. K. (1996). Stable isotope (18O) investigations on the processes controlling fluoride contamination of groundwater. *Journal of Contaminant Hydrology, 24*(1), 85–96. https://doi.org/10.1016/0169-7722(96)00004-6

Division of Toxicology and ENvironmental Medicine. (1997). Public health statement propylene glycol. In *Agency for toxic substances and disease registry*.

Emamjomeh, M. M., & Sivakumar, M. (2006). An empirical model for defluoridation by batch monopolar electrocoagulation/flotation (ECF) process. *Journal of Hazardous Materials, 131*(1–3), 118–125. https://doi.org/10.1016/j.jhazmat.2005.09.030

Emamjomeh, M. M., Sivakumar, M., & Varyani, A. S. (2011). Analysis and the understanding of fluoride removal mechanisms by an electrocoagulation/flotation (ECF) process. *Desalination, 275*(1–3), 102–106. https://doi.org/10.1016/j.desal.2011.02.032

Enciso, R., Padilla, L. A., Ojeda, C., Delgadillo, J. A., & Rodríguez, I. (2012). Computational fluid dynamics characterization of a rotating cylinder electrochemical reactor using an RANS-RNG turbulence model. *International Journal of Electrochemical Science, 7*(12), 12181–12192.

Fogler, H. S. (1999). *Elements of chemical reaction engineering elements of chemical reaction engineering*.

Ganesan, P., Lakshmi, J., Sozhan, G., & Vasudevan, S. (2013). Removal of manganese from water by electrocoagulation: Adsorption, kinetics and thermodynamic studies. *Canadian Journal of Chemical Engineering, 91*(3), 448–458. https://doi.org/10.1002/cjce.21709

Garcia-Segura, S., Eiband, M. M. S. G., de Melo, J. V., & Martínez-Huitle, C. A. (2017). Electrocoagulation and advanced electrocoagulation processes: A general review about the fundamentals, emerging applications and its association with other technologies. *Journal of Electroanalytical Chemistry*. https://doi.org/10.1016/j.jelechem.2017.07.047

Garg, V. K., Suthar, S., Singh, S., Sheoran, A., Garima, M., & Jain, S. (2009). Drinking water quality in villages of southwestern Haryana, India: Assessing human health risks associated with hydrochemistry. *Environmental Geology, 58*(6), 1329–1340. https://doi.org/10.1007/s00254-008-1636-y

Gilbert, N. (2012). Water under pressure: A UN analysis sets out global water-management concerns ahead of Earth Summit. *Nature, 483*(7389), 256–258.

Golder, A. K., Samanta, A. N., & Ray, S. (2007). Removal of Cr^{3+} by electrocoagulation with multiple electrodes: Bipolar and monopolar configurations. *Journal of Hazardous Materials*. https://doi.org/10.1016/j.jhazmat.2006.07.025

Gomes, J. A. G., Daida, P., Kesmez, M., Weir, M., Moreno, H., Parga, J. R., Irwin, G., McWhinney, H., Grady, T., Peterson, E., & Cocke, D. L. (2007). Arsenic removal by electrocoagulation using combined Al–Fe electrode system and characterization of products. *Journal of Hazardous Materials, 139*(2), 220–231. https://doi.org/10.1016/j.jhazmat.2005.11.108

Goren, A. Y., & Kobya, M. (2021). Arsenic removal from groundwater using an aerated electrocoagulation reactor with 3D Al electrodes in the presence of anions. *Chemosphere, 263*, 128253. https://doi.org/10.1016/j.chemosphere.2020.128253

Karthikeyan, K., Nanthakumar, K., Velmurugan, P., Tamilarasi, S., & Lakshmanaperumalsamy, P. (2010). Prevalence of certain inorganic constituents in groundwater samples of Erode district, Tamilnadu, India, with special emphasis on fluoride, fluorosis and its remedial measures. *Environmental Monitoring and Assessment, 160*(1–4), 141–155. https://doi.org/10.1007/s10661-008-0664-0

Khaled, B., Wided, B., Béchir, H., Elimame, E., Mouna, L., & Zied, T. (2019). Investigation of electrocoagulation reactor design parameters effect on the removal of cadmium from synthetic and phosphate industrial wastewater. *Arabian Journal of Chemistry*. https://doi.org/10.1016/j.arabjc.2014.12.012

Kundu, M. C., & Mandal, B. (2009). Assessment of potential hazards of fluoride contamination in drinking groundwater of an intensively cultivated district in West Bengal, India. *Environmental Monitoring and Assessment, 152*(1–4), 97–103. https://doi.org/10.1007/s10661-008-0299-1

Kuokkanen, V., Kuokkanen, T., Rämö, J., Lassi, U., & Roininen, J. (2015). Removal of phosphate from wastewaters for further utilization using electrocoagulation with hybrid electrodes—Techno-economic studies. *Journal of Water Process Engineering*. https://doi.org/10.1016/j.jwpe.2014.11.008

Lacasa, E., Cañizares, P., Sáez, C., Fernández, F. J., & Rodrigo, M. A. (2011). Removal of nitrates from groundwater by electrocoagulation. *Chemical Engineering Journal, 171*(3), 1012–1017. https://doi.org/10.1016/j.cej.2011.04.053

Lekhlif, B., Oudrhiri, L., Zidane, F., Drogui, P., & Blais, J. F. (2014). Study of the electrocoagulation of electroplating industry wastewaters charged by nickel (II) and chromium (VI). *Journal of Materials and Environmental Science.*

Liteplo, R., Howe, P., Malcolm, H., Aggett, P., Belmar, R., Bucher, J., Camargo, J., Cauley, J., Ekstrand, J., Howe, P., Karthikeyan, G., Kierdorf, U., Kurttio, P., Liteplo, R., Nishikawa, A., Stevens, D., Vineis, P., & Yinlong, J. (2002). Environmental health criteria 227: Fluorides. *Environmental Health Criteria.*

Lounici, H., Adour, L., Belhocine, D., Elmidaoui, A., Bariou, B., & Mameri, N. (2001). Novel technique to regenerate activated alumina bed saturated by fluoride ions. *Chemical Engineering Journal, 81*(1–3), 153–160. https://doi.org/10.1016/S1385-8947(00)00213-8

Madhnure, P., Sirsikar, D. Y., Tiwari, A. N., Ranjan, B., & Malpe, D. B. (2007). Occurrence of fluoride in the groundwaters of Pandharkawada area, Yavatmal district, Maharashtra, India. *Current Science.*

Mameri, N., Yeddou, A. R., Lounici, H., Belhocine, D., Grib, H., & Bariou, B. (1998). Defluoridation of septentrional Sahara water of north Africa by electrocoagulation process using bipolar aluminium electrodes. *Water Research, 32*(5), 1604–1612. https://doi.org/10.1016/S0043-1354(97)00357-6

Modi, D. S. (2013). Merits and Demerits of different technologies of defluoridation for drinking water. *IOSR Journal of Environmental Science, Toxicology and Food Technology, 3*(2), 24–27. https://doi.org/10.9790/2402-0322427

Mondal, N. C., Prasad, R. K., Saxena, V. K., Singh, Y., & Singh, V. S. (2009). Appraisal of highly fluoride zones in groundwater of Kurmapalli watershed, Nalgonda district, Andhra Pradesh (India). *Environmental Earth Sciences, 59*(1), 63–73. https://doi.org/10.1007/s12665-009-0004-x

Moussavi, G., Majidi, F., & Farzadkia, M. (2011). The influence of operational parameters on elimination of cyanide from wastewater using the electrocoagulation process. *Desalination.* https://doi.org/10.1016/j.desal.2011.06.052

Murugananthan, M., Bhaskar Raju, G., & Prabhakar, S. (2004). Removal of sulfide, sulfate and sulfite ions by electro coagulation. *Journal of Hazardous Materials.* https://doi.org/10.1016/j.jhazmat.2003.12.009

Naje, A. S., & Abbas, S. A. (2013). Electrocoagulation technology in wastewater treatment: A review of methods and applications. *Civil and Environmental Research, 3*(11), 29–42. http://www.iiste.org/Journals/index.php/CER/article/view/8115

Naje, A. S., Chelliapan, S., Zakaria, Z., Ajeel, M. A., & Alaba, P. A. (2017). A review of electrocoagulation technology for the treatment of textile wastewater. *Reviews in Chemical Engineering, 33*(3), 263–292. https://doi.org/10.1515/revce-2016-0019

Nanseu-Njiki, C. P., Tchamango, S. R., Ngom, P. C., Darchen, A., & Ngameni, E. (2009). Mercury(II) removal from water by electrocoagulation using aluminium and iron electrodes. *Journal of Hazardous Materials.* https://doi.org/10.1016/j.jhazmat.2009.03.042

Parthasarathy, N., & Buffle, J. (1985). Study of polymeric aluminium(III) hydroxide solutions for application in waste water treatment. Properties of the polymer and optimal conditions of preparation. *Water Research.* https://doi.org/10.1016/0043-1354(85)90319-7

Patel, U. D., Ruparelia, J. P., & Patel, M. U. (2011). Electrocoagulation treatment of simulated floor-wash containing Reactive Black 5 using iron sacrificial anode. *Journal of Hazardous Materials.* https://doi.org/10.1016/j.jhazmat.2011.09.064

Prasad, B., & Mondal, K. K. (2005). Leaching characteristics of fluoride from coal ash. *Asian Journal of Water, Environment and Pollution.*

Purkait, D. G. S. K. (2008). *Removal of Fe(II) from tap water by electrocoagulation technique.*

Ramamohana Rao, N. V., Rao, N., Surya Prakash Rao, K., & Schuiling, R. D. (1993). Fluorine distribution in waters of Nalgonda District, Andhra Pradesh, India. *Environmental Geology, 21*(1–2), 84–89. https://doi.org/10.1007/BF00775055

Sahu, O., Mazumdar, B., & Chaudhari, P. K. (2014). Treatment of wastewater by electrocoagulation: A review. *Environmental Science and Pollution Research*. https://doi.org/10.1007/s11356-013-2208-6

Salve, P. R., Maurya, A., Kumbhare, P. S., Ramteke, D. S., & Wate, S. R. (2008). Assessment of groundwater quality with respect to fluoride. *Bulletin of Environmental Contamination and Toxicology, 81*(3), 289–293. https://doi.org/10.1007/s00128-008-9466-x

Sankararamakrishnan, N., Sharma, A. K., & Iyengar, L. (2008). Contamination of nitrate and fluoride in ground water along the Ganges Alluvial Plain of Kanpur district, Uttar Pradesh, India. *Environmental Monitoring and Assessment, 146*(1–3), 375–382. https://doi.org/10.1007/s10661-007-0085-5

Shaji, E., Bindu, Viju, J., & Thambi, D. S. (2007). High fluoride in groundwater of Palghat District, Kerala. *Current Science, 92*(2), 240–245.

Sinha, R., Mathur, S., & Brighu, U. (2015). Aluminium removal from water after defluoridation with the electrocoagulation process. *Environmental Technology (United Kingdom), 36*(21), 2724–2731. https://doi.org/10.1080/09593330.2015.1043958

Sinha, R., Singh, A., & Mathur, S. (2016). Multiobjective optimization for minimum residual fluoride and specific energy in electrocoagulation process. *Desalination and Water Treatment, 57*(9), 4194–4204. https://doi.org/10.1080/19443994.2014.990929

Sreenivasan, G., & Wodeyar, B. K. (1996). Occurrence of fluoride in the groundwaters and its impact in Peddavankahalla basin, Bellary District, Karnataka—A preliminary study. *Current Science Associatio, 70*(1), 71–74.

Susheela, A. K., Das, T. K., Gupta, I. P., Tandon, R. K., Kacker, S. K., Ghosh, P., & Deka, R. C. (1992). Fluoride ingestion and its correlation with gastrointestinal discomfort. *Fluoride, 25*(1), 5–22.

Suthar, S., Garg, V. K., Jangir, S., Kaur, S., Goswami, N., & Singh, S. (2008). Fluoride contamination in drinking water in rural habitations of Northern Rajasthan, India. *Environmental Monitoring and Assessment, 145*(1–3), 1–6. https://doi.org/10.1007/s10661-007-0011-x

Tezcan Un, U., Koparal, A. S., & Bakir Ogutveren, U. (2013). Fluoride removal from water and wastewater with a bach cylindrical electrode using electrocoagulation. *Chemical Engineering Journal*. https://doi.org/10.1016/j.cej.2013.02.126

Toyoda, A., & Taira, T. (2000). A new method for treating fluorine wastewater to reduce sludge and running costs. *IEEE Transactions on Semiconductor Manufacturing, 13*(3), 305–309. https://doi.org/10.1109/66.857940

Vasudevan, S., Kannan, B. S., Lakshmi, J., Mohanraj, S., & Sozhan, G. (2011). Effects of alternating and direct current in electrocoagulation process on the removal of fluoride from water. *Journal of Chemical Technology and Biotechnology*. https://doi.org/10.1002/jctb.2534

Vasudevan, S., Laksmi, J., Sozhan, G. (2009). Studies on a Mg–Al–Zn alloy as an anode for the removal of fluoride from drinking water in an electrocoagulation process.

Wan, W., Pepping, T. J., Banerji, T., Chaudhari, S., & Giammar, D. E. (2011). Effects of water chemistry on arsenic removal from drinking water by electrocoagulation. *Water Research*. https://doi.org/10.1016/j.watres.2010.08.016

WHO. (2010). *Water for Health—WHO guilines for drinking-water quality*. WHO.

Xu, X., & Zhu, X. (2004). Treatment of refectory oily wastewater by electro-coagulation process. *Chemosphere, 56*(10), 889–894. https://doi.org/10.1016/j.chemosphere.2004.05.003

Yildiz, Y. Ş, Koparal, A. S., Irdemez, Ş, & Keskinler, B. (2007). Electrocoagulation of synthetically prepared waters containing high concentration of NOM using iron cast electrodes. *Journal of Hazardous Materials, 139*(2), 373–380. https://doi.org/10.1016/j.jhazmat.2006.06.044

Yu, Y., Wang, X., Yang, D., Lei, B., Zhang, X., & Zhang, X. (2014). Evaluation of human health risks posed by carcinogenic and non-carcinogenic multiple contaminants associated with consumption of fish from Taihu Lake, China. *Food and Chemical Toxicology*. https://doi.org/10.1016/j.fct.2014.04.001

Zhu, J., Zhao, H., & Ni, J. (2007). Fluoride distribution in electrocoagulation defluoridation process. *Separation and Purification Technology, 56*(2), 184–191. https://doi.org/10.1016/j.seppur.2007.01.030

Zodi, S., Potier, O., Lapicque, F., & Leclerc, J. P. (2009). Treatment of the textile wastewaters by electrocoagulation: Effect of operating parameters on the sludge settling characteristics. *Separation and Purification Technology*. https://doi.org/10.1016/j.seppur.2009.06.028

Zuo, Q., Chen, X., Li, W., & Chen, G. (2008). Combined electrocoagulation and electroflotation for removal of fluoride from drinking water. *Journal of Hazardous Materials, 159*(2–3), 452–457. https://doi.org/10.1016/j.jhazmat.2008.02.039

Chapter 10
Fluoride Removal from Water Using Filtration and Chemical Precipitation

Oorv Sumant Devasthali, Arya Johnny Shah, and Sachin Vijay Jadhav

Abstract In precipitation, fluoride-rich water is treated with aluminum salts and lime, which is then taken through filtration or sedimentation. Lime reacts with the free fluorides, originating from salts such as NaF and HF (acids released by industries), forming calcium fluoride (CaF_2), an insoluble impurity that can be eliminated. Since aluminum acts as a coagulant, its salts Aluminium Chloride ($AlCl_3$) or Aluminium sulfate ($Al_2(SO_4)_3$) are used for the viable removal of fluorides. This leads to the formation of Aluminium Hydroxide ($Al(OH)_3$) with the introduction of alum. This leads to the accumulation of all fluoride impurities, which makes it easier to eliminate them. This coagulation process can also be carried out using Ferric salts (Fe salts) as an alternative to aluminum salts. An improvement in this is electrocoagulation (EC), which involves electric current passed through an aqueous medium wherein sacrificial electrodes are used. An aluminum anode is used, which breaks the metal into ions that react with the free fluoride ions in the aqueous solution to form an Aluminium Hydroxide-Fluoride complex. It coagulates, thus increasing the speed of impurity elimination. These complexes need to be filtered to further reduce the fluoride content of the water; carbon, or gravel filters can be used. Here, a layer of a sand filter is used along with layers of gravel. Thus, chemical precipitation (coagulation) is utilized to reduce the fluoride concentration from a high bulk level to a few milligrams per L. This is further reduced below the critical limit of minerals/elements with filtration and sedimentation. Thus, this chapter discusses methods of removal of fluoride impurities through chemical precipitation using coagulation, electrocoagulation, and filtration, including membrane filtration. Explores recent technology and methods that have been developed to improve the efficiency and cost-effectiveness of precipitation and filtration methods.

O. S. Devasthali · A. J. Shah · S. V. Jadhav (✉)
Department of Chemical Engineering, Institute of Chemical Technology, Matunga, Mumbai, Maharashtra 400 019, India
e-mail: sv.jadhav@ictmumbai.edu.in

O. S. Devasthali
e-mail: 20cheos.devasthali@ug.ictmumbai.edu.in

A. J. Shah
e-mail: 20cheaj.shah@ug.ictmumbai.edu.in

A. K. Yadav et al. (eds.), *Advanced Treatment Technologies for Fluoride Removal in Water*, Water Science and Technology Library 125,
https://doi.org/10.1007/978-3-031-38845-3_10

Keywords Chemical precipitation · Coagulation · Filtration · Fluoride · Water

10.1 Introduction

The water scarcity that affects millions of people is discussed in several chapters of this book. Most of the population in developing countries uses groundwater for potable purposes. Groundwater can be contaminated by natural anthropogenic and industrial sources, such as heavy metals, manganese, radioactive materials, nitrates, sulfates, iron, fluorides, and arsenic. However, fluoride is one of the major and most harmful components of groundwater in India. Groundwater has its fluoride sources in fluorspar, sedimentary rocks, granite, fluorapatite, sellaite, dolomite, etc. The simultaneous presence of fluoride in groundwater is connected to volcanic, geothermal, mining, and other activities. Although small amounts of fluorides are beneficial for dental and bone health in humans, fluoride in higher concentrations is detrimental to human health. Fluorides and other contaminants can affect health independently, synergistically, or antagonistically. Thus, it can be interpreted that fluoride contamination is a function of the solubilization of crystalline minerals (Jadhav et al., 2015). Millions of people in India are at risk for fluoride. The following methods that are currently used for fluoride removal are chemical coagulation/precipitation, which also includes electrocoagulation, floatation, and adsorption on special activated solids (Bhatnagar et al., 2011; Gong et al., 2012), ion exchange, and membrane techniques (Richards et al., 2009). Electrocoagulation and floatation techniques are relatively expensive due to their high operational costs. Adsorption methods are the most widespread, but they produce a significant amount of wastewater to recharge the adsorption beds. On the basis of the fluoride concentration in water, these removal techniques are applied sequentially to achieve the desired fluoride concentration that is safe to use. This chapter focuses on various coagulation and filtration techniques in detail.

10.2 Removal by Chemical Precipitation and Coagulation

The coagulation or chemical precipitation technique for fluoride removal first started in Nagpur, India, and is called the Nalgonda process. Since the process is simple and inexpensive, it has been implemented in many Indian towns and villages. Along with this, the Nalgonda process has many other advantages, ranging from the simplicity of operation and construction. Since it is being implemented in towns and villages, the presence of skilled laborers is scarce. This is taken care of as it does not require highly skilled labourers and is simple in design. It also meets the modern sustainability criteria, as it has minimal waste generation, avoids the problem of disposal, and also accounts for minimum water loss. There is little investment and wear and tear due to less mechanical and electrical equipment. The alum used and the sludge produced can be recycled. For the implementation of this process, the only major requirement was hand pumps. It involves the addition of aluminum salts. In addition to this, lime is added to this impurity-concentrated water (fluoride). This process is sequentially succeeded by sedimentation or filtration. To summarize, the basis of this operation is the adsorption of fluoride salts on the aggregations of ammonium hydroxide, called flocs. Using flash mixing, chemical interaction, flocculation, sedimentation, filtration, sludge concentration, and disinfection, the entire process is carried out in solution form. In the above-mentioned process, alum and lime are supplemented with fluoride-rich water, which is constantly churned in a two-bucket deflourinator (Fig. 10.1). The raw water in the bucket is immediately mixed uniformly with the alum and lime mixture by generating turbulence by mixing rapidly for a brief period of time, followed by a slow stirring for five minutes. The minimal rate of addition of the bleaching powder is usually maintained at 3 mg/L, with the main purpose of the bleaching powder being the disinfection of the water. The drawback of this method is that it cannot treat heavy fluoride concentrations without the use of excess alum. This technique utilizes alkalis, chlorine, and aluminum sulfate or aluminum chloride (García-Sánchez et al., 2013). An alternative to the given technique is the usage of both $Al(SO_4)_3$ and AlCl. Therefore, when a large amount of water needs to be treated to eliminate the presence of fluoride, the efficacy of this process is high. However, the Nalgonda technique finds it difficult to control the pH of the treated water and is highly time-consuming. For pH control, lime may be utilized, but it may lead to excess addition of lime and alum, leading to improper usage of chemicals. While the water is being treated, special care must be taken for maintaining the pH of the solution (for effective removal), the pH of the mixture increases to 12, whereas the optimum fluoride is removed at a pH of about 6–7.

When looking objectively at wastewater, especially groundwater in water, the major fluoride impurities that one has to deal with are hydrogen fluoride (HF) and sodium fluoride (NaF). As mentioned earlier, when lime is added to the solution, the fluoride impurities react with the lime to generate an insoluble impurity of calcium fluoride (CaF). This process is represented by a series of reactions given below

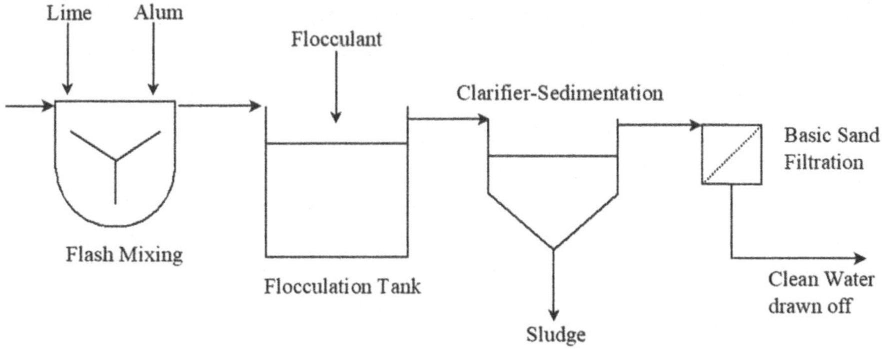

Fig. 10.1 The Nalgonda process

(Jadhav et al., 2015).

$$Ca(OH)_2 + 2F^- \rightarrow CaF_2 + 2OH^-$$

$$Ca(OH)_2(aq) + 2NaF(aq) \rightarrow CaF_2(s) + 2NaOH(aq)$$

To improve calcium hydroxide ($Ca(OH)_2$), several experiments were conducted using calcium chloride ($CaCl_2$) and magnesium hydroxide ($Mg(OH)_2$) to generate data on the maximum reduction of fluoride concentration by reaction (Jadhav et al., 2014). Several parameters were assessed to determine which of these three salts would optimize fluoride removal. These include the solubility of the salt in the solution, the temperature of the solution, which affects the kinetics, the pH generated in the solution, and the particle size of the insoluble salt (generated as a product of the reaction) at different fluoride concentrations. The optimal elimination of fluoride was achieved between the pH range of 5.5–6.5. Calcium fluoride particles were generated from the reaction with $Ca(OH)_2$ and $CaCl_2$. These particles had a size of 1 and 0.5 mm, respectively. As the size of the particle increases, the terminal settling velocity increases. This leads to the particle sinking, and greater energy being required to keep it afloat and active for the reaction. The reaction kinetics was greatly affected by temperature variations, but the particle size remained relatively unaffected. Based on the results that indicated that the fluoride concentration was reduced below 2 mg/L, $CaCl_2$ was found to be more effective than $Ca(OH)_2$. The required molar ratios were optimized by performing experiments at different concentrations of salt and fluoride.

10.2.1 Use of Inorganic Polymeric Coagulants

Recently, a study on the scope of using inorganic polymeric coagulants was used to remove fluoride. This is because the Nalgonda process can be used to efficiently reduce heavy fluoride concentrations to a low level. However, this output might still not be below the permissible limit. Thus, chemical coagulation through inorganic polymeric coagulants is performed to bring the fluoride levels below the permissible levels. At the same time, it has the ability to not remove useful minerals, thus maintaining the quality of water. In a separate study, several polymeric coagulants, along with alum, were tested to optimize the dosage and the type of coagulant to be used. The amount of coagulant to be added is determined on the basis of the Al_2O_3 present in the alum used. This ratio is optimized in the Nalgonda technique. The effect of these coagulants was tested at different pH levels, as well as different initial fluoride concentrations. Therefore, all these grades are titrated to determine the amount of Al_2O_3 content and their basicity. An experiment was carried out to analyze the coagulants, namely IPC-13, IPC-17, IPC-23, IPC-M and IPC-UH (Solanki et al., 2021). The amount of coagulant to be added is determined on the basis of the Al_2O_3

present in the alum used, with IPC-23 yielding maximum basicity and maximum Al_2O_3 concentration. This was fundamental in determining the amount of IPC to be used. As studied in the Nalgonda process, the pH was varied between 5 and 8 (Dubey et al., 2018) to optimize the removal of F^- as the F^- residual was measured using a fluoride-selective electrode (F^-), and the alumina concentration was monitored using spectrometry. Initially, to determine the optimal pH, the initial concentrations of fluoride and alumina were kept constant. As the pH is varied, the fluoride removal efficiency increases by up to 65% at a pH range of 6–7. However, after a pH of 7, the efficiency starts to drop.

One of the main parameters to be accounted for during membrane filtration is the membrane zeta potential. It can be effectively used to clean or alter the properties of the membrane to improve the removal efficiency. The creation of a charge allows for the solute with the opposite charge to be attracted. This phenomenon can be utilized effectively in the cleaning of the membrane. The zeta potentials of a membrane (a number of membranes) were observed and plotted against the pH. It was observed that this potential goes down as the pH increases, and finally, the potential becomes highly negative at very low pH values. If the zeta potential of the membrane is low, it allows the particles to flocculate as the electrostatic repulsion is lower. Here, instead of membranes, coagulants are used, so that the zeta potential being zero initiates faster coagulation and flocculation (López-Maldonado et al., 2014). It is necessary to look for an isoelectric point that would allow effective coagulation, as the coagulants carry no charge.

It was observed that in the case of IPC-17 and IPC-23, the removal was much higher than that of the alum at the same loading. Furthermore, the removal efficiency of IPC-17 improved with an increase in fluoride concentration in the feed. The sample of water also needs to be treated for total dissolved salts (TDS). This happens because as the TDS increases or goes above the permissible limit, the water can become unusable. The result was that IPC-23 showed the lowest increase in TDS from 230 to 385 mg/L; IPC17, IPC-13, and alum showed residual TDS from 241 to 400 mg/L, 270 to 496 mg/L, and 340 to 611 mg/L (Dubey et al., 2018). Here, IPC showed a lower increase in TDS compared to alum. IPC-17 offers water with low fluoride and low TDS. Therefore, it was concluded that IPC-17 is the most effective membrane for fluoride removal, while IPC-23 can also be used with a high fluoride removal efficiency and the lowest TDS.

10.3 Removal by Electrocoagulation

An improvement in the chemical precipitation/coagulation method is the use of the electrocoagulation method. One of the very obvious drawbacks of this method is the utilization of energy, so if the defluorination has to be carried out in rural areas, it is not possible because it is expensive (Hu et al., 2003, 2005). However, fluoride removal is enhanced in this process. Along with retaining the necessary elements in the water, the other benefit of this technique is that it does not lead to the addition

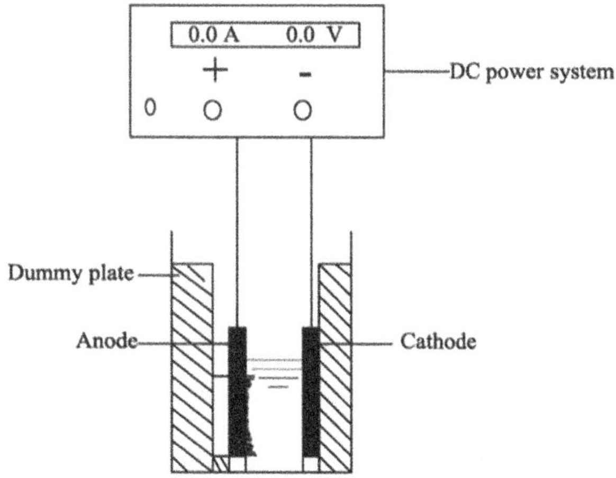

Fig. 10.2 The electrocoagulation cell

of any new impurities to the water. The primary equipment required for performing electrocoagulation is the electrochemical cell. Along with two electrodes, namely the anode and the cathode, an electrolytic cell constitutes an electrocoagulation reactor, as shown in Fig. 10.2. Deterioration of the anode yields the electrons that are carried by the electrolyte to the cathode. They can be made of the same material or constructed from different materials. In most cases, these electrodes can be used in a sacrifice way. This means that the electrodes themselves may become consumed during the reaction.

The process utilizes a destabilizing agent which neutralizes the electric charge to remove pollutants. This leads to coagulation of the charged particles. This accumulation increases the size of the particles and this is where the flocculation comes into play. The flocculating agents interact with these coagulants to make them float on top. The alternative to that is to allow the coagulants with their higher molecular weights to sediment and then separate them. The F^- ion mass transfer is aided by the evolution of hydrogen bubbles at the cathode. Since aluminum is used as a flocculating agent, it forms a complex of ($Al(OH)_3xFx$) with fluorides and it flocs (Emamjomeh & Sivakumar, 2009) to the top of the electrocoagulation reactor. The aluminum complex is isolated periodically to ensure effective removal of F^-. Thus, defluorination is a consequence of the formation of this aluminum complex [$Al(OH)_3xFx$]. The reactions that take place are mentioned below (Zuo et al., 2008).

$$Al \rightarrow Al^{3+} + 3e^- \text{ at the anode}$$

$$Al^{3+} + 3H_2O \rightarrow Al(OH)_3 + 3H^+$$

$$Al(OH)_3 + xF^- \rightarrow Al(OH)_{3-x}F_x + xOH^-$$

$$2H_2O + 2e^- \rightarrow H_2 + 2OH^- \text{ at the cathode}$$

10.4 Removal by Membrane Filtration

The previous sections discussed coagulation and electrocoagulation, which are the traditional ways of fluoride removal. In recent times, newer technologies, such as membrane filtration, have been implemented with continuous improvement (Fig. 10.3). They can ensure much higher degrees of separation, offering potentially clean drinking water without any sort of contaminants.

The basis for separation using membranes is size and electric charge. The fluoride-rich water is guided across a semipermeable membrane. For fluoride removal, the most commonly used processes are Nanofiltration (NF) and reverse osmosis (RO). The RO requires a pressure greater than the normal osmotic pressure to be applied on the heavy pollutant side. Nanofiltration (NF) is a relatively new and modern technology. It is gaining popularity because of its attributes between the RO and Ultrafiltration (UF). It requires or has a lower operating pressure (lower pressure drop). It has the capacity to produce a large permeate flux even at lower pressures, consuming less energy as compared to the RO. However, RO does not discriminate between ions and nonselectively eliminates all monovalent ions. This ensures a 90% removal of fluoride ions from wastewater, whereas NF can only account for 60% of fluoride removal (Malaisamy et al., 2011). Therefore, multiple rounds or iterations of NF would need to be performed to ensure the complete removal of fluorides.

To verify the above claims, two NF membranes were used for an experiment. The two membranes used were NF90 and NF400 (Tahaikt et al., 2007). These were used to treat fluoride-rich groundwater. When assessing and comparing membranes, different parameters need to be taken into consideration, such since the loading capacity, as the elimination/rejection capacity depends on the loading capacity. NF400 was found to be effective at low fluoride amounts. However, when a higher fluoride content, the filtrate (after one pass) needed to be refluxed (double pass) to ensure effective removal. A study further conducted with altered TFC membranes found that with increased positive pressure on fluoride membranes, the removal of fluorides was enhanced.

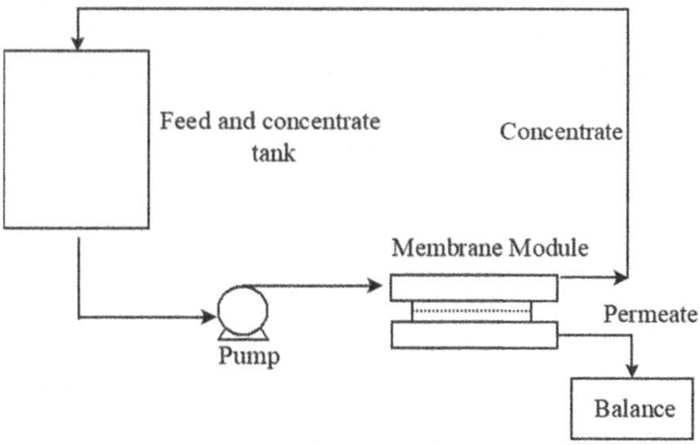

Fig. 10.3 Process flow diagram for membrane filtration

To increase fluoride removal, another experiment was carried out in which a commercially available NF membrane was modified (Malaisamy et al., 2011). To enhance removal and selectivity, this by layer-by-layer dismantling the existing membrane and layer-by-layer reconstructing it using alternating polyelectrolyte thin films. The anionic polyelectrolyte was polystyrene sulfonate (PSS), and the cationic polyelectrolyte was polydiallyldimethylammonium chloride (PDADMAC). After this, these thin (0.5–8.5 μm) PDADMAC/PSS bilayers were deposited on the substrate membrane. Any number of such layers could be generated, with the most common being the 8-bilayer membrane which increased the fluoride removal from 40 to 80% (where 40% was the fluoride removal without the use of modified membranes).

10.4.1 Use of RO and NF in Parallel

This process was implemented in Puntari, Finland (Kettunen & Keskitalo, 2000). The first choice of use was Reverse Osmosis (RO) because of the superior removal efficiency. However, considering that RO requires a large amount of energy in the form of electricity, NF was also considered a viable option. The basic idea behind the NF was that the water here was rich in aluminum (Al^{3+}), so there was a good chance that the fluoride (F^-) to complex with the Al of Al hydroxides. The pH was to be controlled as the water was highly alkaline, so limestone monitoring was a necessity. The pH control also helped remove fluoride. The way this was implemented was using 2-staged RO and NF membranes that function in parallel. There was a cartridge filter for pretreatment to prevent large particulates from mechanically damaging the membranes, and these membranes were chemically scrubbed regularly. The first stage consisted of two vessels, and the concentrate from the second stage was recycled through the first.

Through this experiment, an observation was made to determine that the RO unit led to 95% fluoride removal, while the NF unit led to 76% fluoride removal. However, the F^- removal through NF did reach 90% (Huxstep & Sorg, 1987) when the applied pressure was 1.5 times the usual pressure used. Thus, it can be concluded that as the pressure applied increases, the removal efficiency of the NF membranes increases. If the operating pressures of the RO and NF are reduced, it affects the permeate flow rate (Fox & Sorg, 1987). However, if the temperature of the incoming flow has increased, this effect on the permeate flow is mitigated. However, temperature fluctuations in the incoming flow had adverse effects on the conductivities of the membranes. The conductivity of the RO membrane remained unaffected, but the NF membrane experienced a change in conductivity. The conductivity of the NF increased by 1.8 times for every 10 °C rise in temperature. As a result, at higher temperatures, more ions could pass through the membrane. The RO unit was not affected by the pressure applied only in terms of permeate quality. The salt rejection was lowered if the pressure applied dropped. Similarly, as seen earlier as well, in the

Table 10.1 Fluoride removal efficiency in inorganic polymeric coagulants

Coagulant name	Initial fluoride conc. (mg/L)	Residual fluoride conc. (mg/L)	Removal efficacy (%)
IPC-17	3	1.1	63.3
IPC-23	3	1.1	63.3
IPC-13	3	1.4	53.3
Alum	3	1.3	56.7

Solanki et al. (2021)

case of NF membranes, if the pressure applied increases, the F^- removal efficiency goes up.

The fluoride concentration (4 mg/L) of the raw incoming water had to be reduced to 1.5 mg/L using these membranes. NF plants required twice the capacity of a RO plant (Finnish Standards Association, 1980). Therefore, a cost–benefit analysis was required to determine which was more efficient. This was because the cost saved on RO membranes (by saving up on the installation and energy costs) would be compensated for by the double capacity required by the NF membranes. The cost of the RO plant was 1.7 times that of the NF plant, but the flow rate of the RO plant was 1.3 times that of the NF plant. All of these parameters studied are represented in Table 10.1.

NF membranes utilize semipermeable membranes, but with very low-pressure requirements. These membranes have very large pores, which confirms the fact that the removal efficiency of the NF is lower than that of the RO and uses a lower amount of energy (Maheshwari, 2006). RO can achieve 98% of fluorides, while NF has a slightly lower performance (Mohapatra et al., 2009; Damtie et al., 2019). However, RO and NF membranes prove to be very costly, and so in rural areas, other methods may be required (Maheshwari, 2006).

10.4.2 Use of Modified NF Membranes

The observations of Table 10.2 make it clear that the Nanofiltration (NF) membranes would require a double pass to ensure maximum removal efficiency compared to a RO unit, which only requires a single pass to obtain a similar efficiency. However, NF membranes are cheaper than RO membranes, thus prompting greater exploration in the field of nanofiltration. Another problem with RO membranes is the complete elimination of the minerals, which makes it necessary to remineralize the water which is treated to make it usable for daily use. This further increases the cost of this process. However, NF membranes offer a rejection efficiency lower than that of ROs, so modifications need to be made in these membranes to ensure higher fluoride removal along with a high permeate flux. To investigate this, an experiment was conducted in which modified NF membranes were used. NF 90 or NF270 membranes

Table 10.2 A comparison between the RO and NF units

Stage	RO unit	NF unit
Preliminary pumping		
Operating hours per year	5981	
Energy consumption	53,820	
Energy cost per year, €	2690	
High-pressure pumping		
Operating hours per year	5865	5981
Energy consumption	30,205	17,823
Energy cost per year, €	1510	890

Kettunen and Keskitalo (2000)

were modified for this purpose, as these were the nanofiltration membranes that were easily available (Malaisamy et al., 2011). This modification was carried out by soaking the given membrane in water for 3 h, after which an aqueous solution of 0.02 M polydiallyldimethylammonium chloride (PDADMAC) in 0.5 M sodium chloride (NaCl) was added, after which the membrane was allowed to remain dipped in the solution for a while. After this, the membrane was rinsed and cleaned with water, after which it was exposed to a 0.02 M solution of polystyrene sulfur (PSS) in 0.5 M NaCl (Farhat & Schlenoff, 2001). The membrane was re-rinsed with water, and so a single layer was generated. The NaCl solution was used to increase the charge density on the membrane. Several of such layers can be characterized.

With an increase in the number of polyelectrolyte layers, there was a downfall in the permeate flux. For the 4-bilayer, that reduction was not visible at lower F^- loadings. However, it became deliberately visible at high F^- concentrations. In comparison, the flux loss in 8 bilayers was very high, even for low F^- concentrations. Compared to the unchanged membrane, the 8-bilayer modified membrane experienced a reduction of 50% in the pure water flux (Hollman & Bhattacharyya, 2004). However, this permeate flux was still 30% higher compared to the RO membranes used (BW30). When the F^- loading is low, up to 100 mg/L, the rejection in unmodified and 8 bilayers is approximately the same, around 30%. As the F loading increased up to 1000 mg/L, NF270 gave a rejection of 50%, while the 8-bilayer membrane went up to 70% (Hong et al., 2006).

The objective here is to use polyelectrolyte membranes for the NF. An alternative to this method to increase the rejection of F^- would be to increase the salt concentration of NaCl. As this concentration is boosted, the counterion (Na^+) helps filter out the negative charge on the membrane surface, which helps to neutralize the negative charge and reduce the electrostatic repulsion, thus increasing the rejection capacity of the membrane. Otherwise, the F^- is electrostatically repel the divalent anionic membrane. With 4-bilayers there was a high spike in the F^- flux even in the case of high F^- loading when compared with the unaltered membrane. However, in the case of the 8-bilayer membrane, the F^- flux was reduced in comparison to both the membranes mentioned above. The permeate and F^- flux was lower for it but still higher than those of the RO membranes. Therefore, the 8-bilayer membrane is an improvement over the industry standard.

In general, it was observed that the permeate flux and rejection efficiency were inversely related. The following observation was made: The flux decreased but the rejection efficiency increased with an increase in the number of membrane layers. However, all of the parameters were altered as the fluoride loading was altered.

10.4.3 Use of Thin Film Composite Membranes

Since it has been established in Sect. 10.1, the Nanofiltration (NF) is a cheaper process, and membranes can be modified to achieve a high F rejection rate, another membrane modification was carried out in which thin film composites were used as the NF membranes. Thin-film composite membranes are generally semi-permeable membranes with a negatively charged surface. They are like molecular sieves arranged as two or more layered materials (Hu & Dickson, 2006).

As demonstrated in previous experiments, NaCl solution was added to the membrane for a period of 96 h to ensure the stability of the TFC membranes. This is done by increasing the charge density on the surface. The feed was deionized using carbon adsorption, and the pH of the feed was maintained at 6.4. Every 6 h, the rejection and the solution flux were looked at until both of those quantities became constant. The effective membrane area is to be noted for analysis. Since it has been established that the feed flow rate is much greater than the permeate flux when these samples are connected in series with each other, they can be assumed to be in parallel with the same feed flow rate. A conductivity meter is used to account for salt rejection, and a flowmeter is used to account for the permeate flow rate. The performance of the membrane is characterized in terms of membrane parameters that include pure water permeability (LP), pore radius (rP) and constant surface electrical potential (ψ). The feed consists of a single salt of sodium fluoride (NaF) as the water is deionized to remove any other ions. An important nuance that needs to be taken care of is that the experiment needs to be operated at a low as well as high loading of fluorides in the water being treated.

To offer an example of how the membrane study was conducted, and few TFC membranes were tested, such as DS-5-DL, DS-51-HL, SR-1, TFC-SR3, and SelRO MPF-34 to compare the rejection capacity, pore size, and the zeta potential generated by these membranes. NF membranes usually have a pore size between 4 and 1.5 nm (Bowen & Mohammad, 1998). For the membranes tested here, the pore size was 0.41 nm for HL, 0.7 nm for SR-1, and 0.58 nm for DL. The reason for maintaining such low pore sizes is that for the larger pore sizes leading to higher solution flux, the concentration polarization effect could play a major role in determining the rejection capacity (Wang et al., 1995). The important parameter to be studied in this case was the ratio of the effective thickness to the porosity of the membrane, which is characterized by LP and rP. This is demonstrated by the HL membrane, which has the smallest pore size, which should logically indicate that it will have the lowest pure water flux, but it was actually higher than the DL and SR-1 membranes due to high porosity.

The constant surface electrical potential demonstrates electrical attraction or repulsion between the incoming charged solutes and the membrane. A membrane having the highest surface electric potential (Wang et al., 1995) means that it is able to produce the greatest electric field on the surface. When a potential difference is introduced across a capillary cell that contains the sample, a particle velocity is generated in the sample. Using this velocity, the zeta potential is determined. It is an important factor when it comes to determining flocculation characteristics. This implies that a zeta potential or surface charge of more than 30 mV is needed to ensure membrane stability. It was observed that the flux of the solution that arose depended on the pure water permeability, porosity, and effective thickness of the membrane. Along with this, it was also observed that this flux increased with an increase in the applied pressure. The rejection efficiency for different membranes was 60% for SR-1, 78% for DL, 82% for HL, 90% for TFC-SR3 and 50% for SelRO MPF-34 membranes.

Polymeric membranes such as these are highly susceptible to fouling. Thus, one needs to employ membranes that have greater stability to avoid fouling and which do not undergo rapid coagulation or flocculation. Polymeric membranes are generally recommended to be used to eliminate fluorides at a low loading with very high efficiency.

10.5 Hybrid Fluoride Removal Processes

Part I of this book extensively discusses fluoride contamination or accumulation in groundwater. A cost-effective method developed for fluoride removal is the introduction of sand filters where pumice gravels are used to eliminate fluoride from the incoming raw water. The advantage of pumice usage is that it is easily available and it is non-expensive. Furthermore, pumice has a very large surface area for filtration, allowing greater loading of raw water (Kitis et al., 2007). In general, pumice has very little removal efficiency with fluoride. However, when combined with aluminum oxide coatings, removal efficiency can be increased by 24%. Along with fluorides, it also has a very high removal efficiency for turbidity (Ghebremichael et al., 2012). An alternative to pumice is the use of char and biochar. Its activated form can be used to eliminate fluoride and arsenic for drinking purposes (Gwenzi et al., 2017). Organic matter and fluoride removal have also been demonstrated in homemade biochar filters (Pooi & Ng, 2018).

Another similar study by Devi et al. (2008) on modified homemade filters showed promising results. The system consisted of two metal tanks. One tank was filled with pebbles and gravels of different sizes. The second tank was filled with crushed bricks. These experiments were carried out with a competing ion, arsenic, which is also a

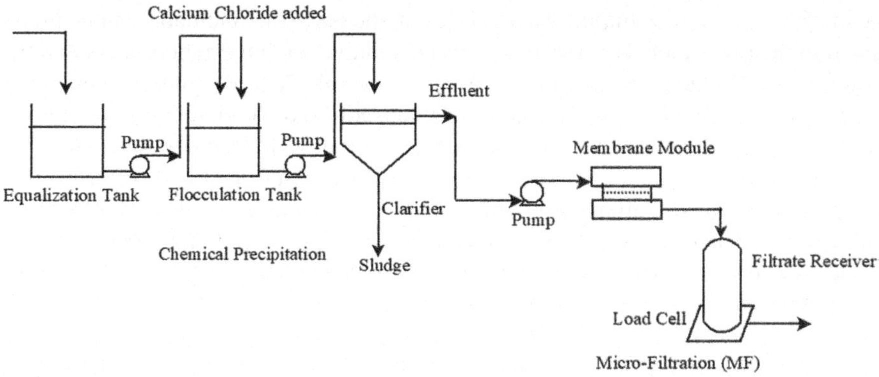

Fig. 10.4 Chemical precipitation and microfiltration combination

lethal contaminant to human health. Starting with 5 mg/L fluoride, this simple set-up was effective in removing fluoride by 86%. Innovations like these allow the larger population to access safe drinking water without the adverse health effects of excess fluoride.

An important nuance to be taken care of is understanding of how fluoride creeps into groundwater. To account for this, groundwater sources were assessed. They include wastewater, rainwater, and agricultural water. Since agricultural water is an important contributor to groundwater, a study was conducted to treat this contaminating source water itself to ensure a reduction in fluorides in groundwater (Lu & Liu, 2010). Another route for phosphate-rich water (which is generally found in agricultural areas) that was studied was the use of the phosphates present in water. As discussed in Sect. 10.2 of this chapter, $CaCl_2$ was added to the wastewater with a molar ratio of ($[Ca^{+2}]$: $[PO_4^3]$: [F]) 1.5:1:0.7 and 2.5:1:0.7 as shown in Table 10.2. During the experiments conducted, the pH was controlled at two different points, 8.5 and 10.5. The removal of fluoride was found to require a higher than stoichiometric amount of calcium. Under alkaline conditions, calcium has a reactive affinity for phosphates rather than fluoride. Thus, in this process, the chemical precipitation was followed by microfiltration (MF), as shown in Fig. 10.4. It was clearly found that the treated water was still active in fluoride, so microfiltration was performed to remove the remaining fluoride to reduce the concentration from 30.87 mg/L to nearly 8.83 mg/L.

10.6 Conclusions

In this chapter, methods like coagulation/chemical precipitation, electrocoagulation, and various types of membrane filtration have been reviewed. Chemical or nonchemical, the precipitation technique remains the conventional process for treating very high concentrations of fluoride. Electrocoagulation is also used, which helps prevent the introduction of new impurities. In addition, the necessary elements can be maintained in the raw water. This makes electrocoagulation an important process, but the requirement of electricity can prove to be very costly. The RO membranes led to great performance, but the permeate flux is very low and the membranes are highly expensive. The NF membranes offer slightly lower removal efficiency, but a very high permeate flux. Modified membranes using polyelectrolytes or TFC show promising results in fluoride mitigation while offering higher flux and removal efficiency. Innovation in these fields is necessary to move forward. This is required to prevent health conditions such as fluorosis (Miretzky & Cirelli, 2011). One such innovation is the use of hybrid technologies, as discussed in the previous sections. This helps in inculcating the advantages of different methods and thus achieving the maximum removal efficiency of fluoride from groundwater. While the Nalgonda process is the oldest process and is used in reducing the heavy fluoride concentration to a low level, this flux can then be passed through the membrane filters to achieve a fluoride concentration less than the permissible drinking water limit. In addition, along with removal efficiency, it is necessary to look at ease of use and cost-effectiveness of the methods that are being implemented.

Acknowledgements The authors declare no conflict of interest. This work did not receive any specific grant from funding agencies in the public, commercial, or not-for-profit sectors. The authors acknowledge the Institute of Chemical Technology, Mumbai, for the resources provided in writing this chapter.

References

Bhatnagar, A., Kumar, E., & Sillanpää, M. (2011). Fluoride removal from water by adsorption—A review. *Journal of Chemical Engineering, 171*(3), 811–840.

Bowen, W. R., & Mohammad, A. W. (1998). Characterization and prediction of nanofiltration membrane performance—A general assessment. *Chemical Engineering Research and Design, 76*(8), 885–893.

Devi, R., Alemayehu, E., Singh, V., Kumar, A., & Mengistie, E. (2008). Removal of fluoride, arsenic and coliform bacteria by modified homemade filter media from drinking water. *Bioresource Technology, 99*(7), 2269–2274.

Damtie, M. M., Woo, Y. C., Kim, B., Hailemariam, R. H., Park, K. D., Shon, H. K., & Choi, J. S. (2019). Removal of fluoride in membrane-based water and wastewater treatment technologies: Performance review. *Journal of Environmental Management, 251*, 109524.

Dubey, S., Agarwal, M., & Gupta, A. B. (2018). Experimental investigation of Al-F species formation and transformation during coagulation for fluoride removal using alum and PACl. *Journal of Molecular Liquids, 266*, 349–360.

Emamjomeh, M. M., & Sivakumar, M. (2009). Review of pollutants removed by electrocoagulation and electrocoagulation/flotation processes. *Journal of Environmental Management, 90*(5), 1663–1679.

Farhat, T. R., & Schlenoff, J. B. (2001). Ion transport and equilibria in polyelectrolyte multilayers. *Langmuir, 17*(4), 1184–1192.

Fox, K. R., & Sorg, T. J. (1987). Controlling arsenic, fluoride, and uranium by point-of-use treatment. *Journal American Water Works Association, 79*(10), 81–84.

García-Sánchez, J. J., Solache-Ríos, M., Martínez-Miranda, V., & Morelos, C. S. (2013). Removal of fluoride ions from drinking water and fluoride solutions by aluminum modified iron oxides in a column system. *Journal of Colloid and Interface Science, 407*, 410–415.

Ghebremichael, K., Wasala, L. D., Kennedy, M., & Graham, N. J. (2012). Comparative treatment performance and hydraulic characteristics of pumice and sand biofilters for point-of-use water treatment. *Journal of Water Supply: Research and Technology-AQUA, 61*(4), 201–209.

Gong, W. X., Qu, J. H., Liu, R. P., Lan, H. C., (2012). Effect of aluminum fluoride complexation on fluoride removal by coagulation. *Colloids and Surfaces A: Physicochemical and Engineering Aspects, 395*, 88–93.

Gwenzi, W., Chaukura, N., Noubactep, C., & Mukome, F. N. (2017). Biochar-based water treatment systems as a potential low-cost and sustainable technology for clean water provision. *Journal of Environmental Management, 197*, 732–749.

Hollman, A. M., & Bhattacharyya, D. (2004). Pore assembled multilayers of charged polypeptides in microporous membranes for ion separation. *Langmuir, 20*(13), 5418–5424.

Hong, S. U., Malaisamy, R., & Bruening, M. L. (2006). Optimization of flux and selectivity in Cl−/$SO_4{}^2$—Separations with multilayer polyelectrolyte membranes. *Journal of Membrane Science, 283*(1–2), 366–372.

Hu, C. Y., Lo, S. L., & Kuan, W. H. (2003). Effects of co-existing anions on fluoride removal in electrocoagulation (EC) process using aluminum electrodes. *Water Research, 37*(18), 4513–4523.

Hu, C. Y., Lo, S. L., & Kuan, W. H. (2005). Effects of the molar ratio of hydroxide and fluoride to Al(III) on fluoride removal by coagulation and electrocoagulation. *Journal of Colloid and Interface Science, 283*(2), 472–476.

Hu, K., & Dickson, J. M. (2006). Nanofiltration membrane performance on fluoride removal from water. *Journal of Membrane Science, 279*(1–2), 529–538.

Huxstep, M. R., & Sorg, T. J. (1987). *Reverse osmosis treatment to remove inorganic contaminants from drinking water*. Charlotte Harbor Water Association Inc Harbour Heights FL (USA).

Jadhav, S. V., Bringas, E., Yadav, G. D., Rathod, V. K., Ortiz, I., & Marathe, K. V. (2015). Arsenic and fluoride contaminated groundwaters: A review of current technologies for contaminants removal. *Journal of Environmental Management, 162*, 306–325.

Jadhav, S. V., Gadipelly, C. R., Marathe, K. V., & Rathod, V. K. (2014). Treatment of fluoride concentrates from membrane unit using salt solutions. *Journal of Water Process Engineering, 2*, 31–36.

Kettunen, R., & Keskitalo, P. (2000). Combination of membrane technology and limestone filtration to control drinking water quality. *Desalination, 131*(1–3), 271–283.

Kitis, M., Kaplan, S. S., Karakaya, E., Yigit, N. O., & Civelekoglu, G. (2007). Adsorption of natural organic matter from waters by iron coated pumice. *Chemosphere, 66*(1), 130–138.

López-Maldonado, E. A., Oropeza-Guzmán, M. T., & Ochoa-Terán, A. (2014). Improving the efficiency of a coagulation-flocculation wastewater treatment of the semiconductor industry through zeta potential measurements. *Journal of Chemistry, 2014*.

Lu, N. C., & Liu, J. C. (2010). Removal of phosphate and fluoride from wastewater by a hybrid precipitation–microfiltration process. *Separation and Purification Technology, 74*(3), 329–335.

Maheshwari, R. C. (2006). Fluoride in drinking water and its removal. *Journal of Hazardous Materials, 137*(1), 456–463.

Malaisamy, R., Talla-Nwafo, A., & Jones, K. L. (2011). Polyelectrolyte modification of nanofiltration membrane for selective removal of monovalent anions. *Separation and Purification Technology, 77*(3), 367–374.

Miretzky, P., & Cirelli, A. F. (2011). Fluoride removal from water by chitosan derivatives and composites: A review. *Journal of Fluorine Chemistry, 132*(4), 231–240.

Mohapatra, M., Anand, S., Mishra, B. K., Giles, D. E., & Singh, P. (2009). Review of fluoride removal from drinking water. *Journal of Environmental Management, 91*(1), 67–77.

Pooi, C. K., & Ng, H. Y. (2018). Review of low-cost point-of-use water treatment systems for developing communities. *NPJ Clean Water, 1*(1), 11.

Richards, L. A., Richards, B. S., Rossiter, H. M., & Schäfer, A. I. (2009). Impact of speciation on fluoride, arsenic and magnesium retention by nanofiltration/reverse osmosis in remote Australian communities. *Desalination, 248*(1–3), 177–183.

Solanki, Y. S., Agarwal, M., Maheshwari, K., Gupta, S., Shukla, P., & Gupta, A. B. (2021). Investigation of plausible mechanism of the synthesized inorganic polymeric coagulant and its application toward fluoride removal from drinking water. *Industrial & Engineering Chemistry Research, 59*(20), 9679–9687.

Solanki, Y. S., Agarwal, M., Maheshwari, K., Gupta, S., Shukla, P., & Gupta, A. B. (2021). Removal of fluoride from water by using a coagulant (inorganic polymeric coagulant). *Environmental Science and Pollution Research, 28*, 3897–3905.

Tahaikt, M., El Habbani, R., Haddou, A. A., Achary, I., Amor, Z., Taky, M., Alami, A., Boughriba, A., Hafsi, M., & Elmidaoui, A. (2007). Fluoride removal from groundwater by nanofiltration. *Desalination, 212*(1–3), 46–53.

Wang, X. L., Tsuru, T., Togoh, M., Nakao, S. I., & Kimura, S. (1995). Evaluation of pore structure and electrical properties of nanofiltration membranes. *Journal of Chemical Engineering Japan, 28*(2), 186–192.

Zuo, Q., Chen, X., Li, W., & Chen, G. (2008). Combined electrocoagulation and electroflotation for removal of fluoride from drinking water. *Journal of Hazardous Materials, 159*(2–3), 452–457.

Chapter 11
Advanced Simulation Technologies for Removal of Water Fluoride

Rashmi Bhardwaj and Inderjeet

Abstract This study shows how fluoride can be removed from drinking water using a lot of techniques. For instance, membrane separation process, ion exchange, adsorption procedures, coagulation-precipitation and the numerically technique magneto-hydrodynamics model. The problem is taken into consideration for two-dimensional planer parallel flow in the magneto-hydrodynamics model. The behaviour of finely dispersed impurities contaminants is illustrated using the drift–diffusion approximation. Because of the high establishment and maintenance cost, membrane and ion exchange techniques are not so common. In India, two more methods are frequently used. The Nalgonda method is one of well-known approaches widely employed in underdeveloped nations, such as India, Kenya, Senegal and Tanzania, for DE fluoridating water. The adsorption technique is widely employed, produces positive results and is undoubtedly a more appealing approach for fluoride removal in the form of cost, simplicity of concept and operation among the several ways for DE fluoridating water. Environmental Protection Agency data show that fluoride is a contaminant that encourages pollution in public drinking water and can be harmful to human health. As per World Health Organization and Water Conservation Agency, maximum pollution of fluoride that has few negative effects has been set at 1.5 mg fluoride per litre. Fluoride can be found both naturally and chemically, however leading sources of fluoride in the human body are pesticides, drinking water, and dental products. Fluoride can influence the human body in a variety of ways, including metabolic and nutritional problems. Fluoride has a number of negative health effects on humans, including dental effects, musculoskeletal effects such as bone fractures and skeletal fluorosis, reproductive and development effects, neurobehavioral and neurotoxicity effects, endocrine effects, as well as some effects on the kidneys and immune system.

R. Bhardwaj (✉) · Inderjeet
University School of Basic and Applied Sciences, Guru Gobind Singh Indraprastha University, Dwarka, New Delhi 110078, India
e-mail: rashmib@ipu.ac.in

Inderjeet
e-mail: yadavinderjeet386@gmail.com

A. K. Yadav et al. (eds.), *Advanced Treatment Technologies for Fluoride Removal in Water*, Water Science and Technology Library 125,
https://doi.org/10.1007/978-3-031-38845-3_11

Keywords Fluoride removal · Water purification · Dental fluorosis · Exchanging ions · Adsorption

11.1 Introduction

In modern time, the problem with drinking water and its substance has emerged by increasing the pollution. One of the main causes behind this is the pollutant called fluoride affecting the lives (Waghmare et al., 2015). A modest amount of fluoride is required for mineralization of bone and defense against dental cavities; excessive consumption results in fluorosis, a form of tooth disease. Fluoride enters the aquatic environment through human activities, such as industrial drains, as the well as weathering of fluoride-rich minerals (Arfin et al., 2015). Fluoride in water sources is a serious problem for tropical countries like Kenya, India, Senegal, and Tanzania. Defluoridation is the most effective method to avoid this problem. Fluoride was removed from the water using techniques, including ion exchange, membrane-based processes, precipitation-coagulation, and adsorption. Precipitation-coagulation methods produce enormous amounts of sludge and can involve the leaching of undesired elements, membrane processes are expensive, and fouling is an inescapable problem. Adsorption methods have owned unique points of such as simplicity and minimal water discharge. A large collection of procedures literature has been compiled for this evaluation. A review of the literature that includes 200 recent articles shows that fluoride removal with little effort approaches has proven to be exceptionally effective.

Throughout human history, the availability of water as a natural resource has been of utmost importance. The total water reserves, including both liquid and frozen water, are 1386 million cubic kilometers. However, saltwater constitutes 97.5% of the world's freshwater supplies. Only 2.5% of it is freshwater. 99.7% of this fresh water comes in the form of ice, persistent snow, or fresh groundwater. On Earth, only 0.3% of fresh water is readily available. No other country has more freshwater than Brazil.

The environment is negatively impacted by the increasing global population, human economic, social and cultural activities, the growth of industry, and new technological advances. The environment is repeatedly destroyed by the current economy's production model "more efficiently" than by historically earlier forms of human life. Human life depends on freshwater, which is also essential for health. In the future, the only source of drinking water available to meet human requirements will be the global ocean. It is important to remove numerous contaminants from water in order to obtain clean water for drinking, medical or industrial uses.

Integration of numerous fields is a specific difficulty for mathematical modelling of water treatment. The movement of water and processes of purification are described using concepts from physics, chemistry, mathematics, and numerical realization on computer systems. It involves understanding the interactions between good water, chemical processes the aquatic ecosystem, and medicine. The development of technological applications as useful tools is just as vital as developing mathematical models that account for various physical aspects.

Life depends on water, and it is also essential for good health and industry. All forms of development, including that of people, animals, and plants, require it. However, other substances that are suspended or dissolved in water may impact how suitable it is for a given function. An example of this is the fluoride impurity of drinkable water. According to the WHO, one of the contaminants in drinkable water is fluoride along with the widely destructive toxins arsenic and nitrate. A major problem that presents a significant risk to human health is groundwater poisoning with fluoride from both anthropogenic and natural sources. Fluoride contamination of water is caused by industrial processes that are likely to discharge effluents, use of pesticides and fertilizers, synthesis of fluorosilicon and fluorocarbon polymers, production of coke, the production of glass and ceramics, production of electronics, the use of electroplating, production of steel and aluminium, the etching of metals (with hydrofluoric acid) and the use of wood preservatives. Several substances, including water, food, air, medications, and cosmetics, contain fluoride that can enter the body. The most popular way that fluoride is made available to people is through drinking water. Depending on the exposure level and time frame, fluoride is known to have both positive and negative results on health. Fluoride levels in drinking water should never exceed 1.5 mg/L. A small amount of fluoride is essential to prevent tooth decay and guarantee that both people and animals have strong bones. It is recognized as a micronutrient for humans because it slows the rate of tooth enamel demineralization, preventing dental caries, or quickens rate of remineralization, stopping the progression of existing disease. Depending on the amount of fluoride in drinkable water, a high amount of fluoride can lead to development of skeletal fluorosis and dental. If the amount of fluoride is less than permitted level, fluoridation units must be constructed at drinking water treatment facilities. More fluoride must be legally removed from the water if the concentration exceeds the allowable amount.

Dental fluorosis, marked by blackened or chalky white teeth, common indication of prolonged exposure to high fluoride water. Adolescents who consume water with a fluoride greater than 4 mg/L over time develop skeletal fluorosis, which can affect bone mineralization and result in severe, long-lasting bone and joint deformations. Fluorosis affects people socially and culturally in addition to affecting their physical health. For example, younger people with extensive tooth pitting and discoloration symptoms of excessive dental fluorosis can find it awkward to smile in public. Many countries in Asia, and Europe, as the well as USA and Australia, struggle with health dangers along with fluoride poisoning of their groundwater. The most affected nations are China and India, however, other nations with major fluoride poisoning issues include Ethiopia, Kenya, Ghana, and Tanzania. Therefore, fluorosis is predicted to affect more than 260 million people worldwide. Most endemic fluorosis cases occur in rural areas of poor nations, where the lack of a developed economy and technological obstacles make the problem of contaminated drinking water worse. Many locations in Ethiopia have reported fluoride concentrations higher than 1.5 mg/L, although the highest amount was found in Rift Valley, the bottomland region with the highest volcanic movement in the nation. Around 14 million residents of the Ethiopian Rift

Valley depend on fluoride-rich water supplies (above 5 mg/L). Therefore, it is crucial to find a technology that removes fluoride from drinkable water that is affordable and efficient.

Fluorosis is a debilitating disease for which there is no known cure, making prevention the only way to deal with this grave problem. Fluorosis can be avoided by using alternate water sources such as rainfall, groundwater with low fluoride levels, and surface water. It can also be avoided by removing an excess amount of fluoride from drinkable water and enhancing nutritional status of the population at risk. Defluoridation of drinkable water appears to be a more direct and useful method of halting fluoride's negative effects. Therefore, it is crucial to develop defluoridation methods that can lower the fluoride content below the WHO-set limit, while also being ecologically beneficial and preferably low-cost.

Fluoride has been removed from drinking water using many technologies, membrane-based, ion-exchange, adsorption, and precipitation-coagulation methods. Fluoride is mixed with lime and alum during the precipitation and flocculation process to create insoluble precipitates. Due to the precipitate's solubility, the fluoride concentration is still around 8.0 mg/L. Large amounts of sludge can also be produced by chemical precipitation. Reverse osmosis and nanofiltration are the two primary components of the membrane separation process. It has a high energy cost and fouling of the membrane, yet it can generate extremely pure. Fluoride ions travel across a semipermeable membrane during electrodialysis because of an electric potential. It is expensive and susceptible to concurrent ion impact. Most of defluoridation methods currently in use are difficult, labor intensive, expensive to instal and maintain, and technically unviable for rural locations. Therefore, it is desirable to find defluoridation media that are locally accessible for safe and simple application at household and small community levels.

Adsorption is the desired method for defluoridation at local level in rural areas due to its affordability and usability, high efficacy, ease of accessibility, environmental friendliness, and lack of need for operational expertise or electricity to operate, making it ideal for use in less developed rural areas. Adsorbents can also theoretically be reused and recycled. It also has the benefit of being adaptable to a decentralized water supply system. They are possible candidates for the defluoridation of remote places, since various adsorbents are readily available and inexpensively priced in large quantities.

11.2 Methods for Defluorodation of Water

There are several defluoridation methods available to combat the harmful effects of fluorosis on human health, including membrane separation technique, coagulation-precipitation, ion exchange, adsorption method, and others. Each technique has advantages and disadvantages, but it was effective in removing fluoride to a more significant range when used under the right circumstances. The benefits and drawbacks of each of the aforementioned methods are briefly discussed.

11.2.1 Coagulation and Precipitation Method

The Nalgonda process for defluorizing water most frequently uses the coagulants alum and lime. Fluoride precipitates as insoluble calcium fluoride as a result of lime expansion, bringing pH between 11 and 12. To ensure optimum fluoride removal, lime is always combined with alum treatment because it leaves 8.0 mg F^{-}/l remaining. The dose of lime causes precipitation as a first step, and then alum is added to cause coagulation as a second phase. Basically, two processes take place when alum adds to water. In the Ist reaction, alum and a certain alkalinity combine to form an insoluble aluminum hydroxide. The second reaction involves alum reacting with fluoride ions in water. The best pH value to remove fluoride is between 5.5 and 7.5 (Razbe et al., 2013). The Nalgonda technique, also known as the coagulation-precipitation method, was created by NEERI and involves expanding bleaching powder, lime, and aluminum salt, followed by quick mixing of sedimentation, flocculation and filtration. When the aluminium salt, fluoride in the water removed. In proportion to its concentration, the amount of fluoride should be determined. Alum is typically administered along with lime at a dose of one twentieth. For quick settling, lime is used to create larger and denser flocs. An addition of 3 mg/l of bleaching powder is made for cleaning (Padmashri, 2001). In communities, it is the defluoridation technique that is most frequently used (Bulusu et al., 1979; Nawlakhe & Rao, 1990; Nawlakhe et al., 1975). Additionally, developed for residential usage is Nalgonda technology based on a bucket defluoridation device (Mjengera & Mkongo, 2002). One day of the procedure can consume 20 L of water. The procedure results in water that has between 1 and 1.5 mg/l of residual fluoride (Dahi, 1996). Nalgonda technique-based fill and draw type defluoridation systems have also been taken into account (Mjengera & Mkongo, 2002). However, there are a few points of interest and limitations with coprecipitation procedures in light of aluminium salts.

Interest

1. A common approach.
2. The process is more effective in contrast to other defluoridation techniques.
3. The approach is straightforward.

Limitations

1. The required chemical amount ($Al(OH)_3$ up to 700–1200 mg/l) is considerable.
2. Sludge transport problem.
3. Can not reach fluoride's usable outermost reaches of fluoride.
4. Requirement of skilled labor.
5. Aluminum release in treated water, which leads to Alzheimer's disease.
6. The final fluoride content in treated water is significantly impacted by the ability of precipitated fluoride, calcium, and aluminum salt to dissolve.
7. Using aluminium sulfate as coagulant significantly increases the sulphate ion concentration, which causes cathartic effects in humans.

Despite initial success, the Nalgonda approach did not catch on because of some fundamental problems, as previously described. Bone char, sodium dihydrogen phosphate, and calcium chloride have been reported to be used in a contact precipitation defluoridation method (Dahi, 1996). Narthasarthy et al. explored the combination of the usage of polymeric aluminum hydroxide and calcium salts for the defluoridation process. In the presence of polyaluminum chloride and polyacrylic acid at low concentrations, Chang and Liu studied the flocculation of coagulation of precipitates of calcium fluoride. Aldaco et al. use granular calcite to capture fluoride contained in water as calcium fluoride in fluidized bed reactor, overcoming the drawback of chemical precipitation (Adhikary et al., 1989).

11.2.2 Membrane Process

Industrial uses of the membrane separation process include the desalination of seawater, wastewater treatment, and defluoridation of groundwater (Maheswari et al., 2002). Particles are isolated during a membrane separation technique on size and form of their molecules, using an incredibly well-built semipermeable membrane. Semi-permeable membranes are made up of porous or nonporous polymeric films, ceramic, liquids or gases, or even liquid or gases. The membrane does not separate, fall off, or disappear (Seadar & Heneley, 2005). RO, nano-filtration, Donnan-dialysis, and electrodialysis are the four most popular membrane separation techniques for removing fluoride.

11.2.3 Reverse Osmosis

Anions are removed physically using reverse osmosis by forcing feed water through a semipermeable membrane under pressure. The increased pressure required for RO results in a more pronounced rejection of dissolved solids. Ions are removed by the membrane while taking their size and electrical charge into consideration. The RO membrane technology forces the solvent filter against a pressure gradient to a low concentration solution by applying hydraulic pressure to the side of the solution with high concentration, which is the opposite of natural osmosis. To balance constant osmotic pressure, reverse osmosis involves applying pressure to solution on one side of a semi permeable membrane using a mechanical pump. Adding salt from saltwater during the desalination process also eliminates soluble and particulate particles (Wimalawansa, 2013). In the 1980s, the RO membrane separation method for treatment of industrial wastewater was primarily used to remove and recover fluoride from its effluents. Regardless of the starting fluoride level, the RO membrane separation process can remove more than 90% of fluoride. Using RO separation technology, Ndiaye et al. defluoridated industrial wastewater, and they found that fluoride ion rejection consistently higher than 98% since the RO membrane fully recovered after all tests (Ndiaye et al., 2005). According to Berhanu Assefa et al., the fluoride retention of the Ethiopian Rift Region RO membrane ranged from 94 to 99% (Alkan et al., 2008). In a Senegalese village where 97 to 98.9% of fluoride was rejected, to extract fluoride and salinity from brackish ground water, Diawara et al. employed low pressure reverse osmosis (Diawara et al., 2011). In the Moradgaon village in Chandrapur district, groundwater was purified with a polyamide RO membrane, according to a study by Gedam et al., to remove 95–98% of the fluoride (Gedam et al., 2012). Schoeman, who had used RO to defluoridate various South African provincial areas, claimed that fluoride could be removed using high pressure RO and low pressure RO, with the latter having a feed water concentration value of 17–0.2 mg/l (Schoeman, 2010). Reverse osmosis used by Briao et al. in southern Brazil to desalt water from the Guarani Aquifer System for drinking. RO rejection efficiencies of RO were 100% for fluoride, 97% for total dissolved solids, and 94% for sulfate ions at 2 MPa pressure and 1.61 m/s cross-sectional flow velocity. Groundwater and permeate were combined to produce mixed water, which had a 93% recovery rate for drinking water (Briao et al., 2014).

The following list includes the RO membrane separation's point of interest and restrictions:

Interest

1. The fluoride elimination process is really persuasive.
2. After each arrangement of examination, the RO membrane was fully recovered.
3. Regardless of the beginning concentration, this method may eliminate fluoride by more than 90%.
4. By using this method, additional dissolved materials can be removed simultaneously.
5. It operates over a broad pH range.
6. Other ions do not interfere.
7. There are no chemical requirements minimal labour requirements, and little operational costs.
8. The procedure allows water to be treated and purified in one step.
9. It provides stable water quality.

Restrictions

1. Rural places are not accessible.
2. This methodology is very costly.
3. Remineralization is required after treatment; remove precious minerals that are vital for healthy development.
4. Salt water waste is a problem as it wastes a lot of water while making saline solutions.
5. The pH needs to be raised since it starts to become acidic.

11.2.4 Nanofiltration Membrane Process

All of the membrane defluoridation methods have seen improvements, but nanofiltration (NF) is the most recent. Reverse osmosis and nanofiltration vary primarily in that reverse osmosis has somewhat larger pores and is less resistant to solute and solvent entry than nano filtration. This results in substantially more thorough solute removal, much low pressure needs, low energy requirements, and fast flow (Mohapatra et al., 2009). Compared to RO, nanofiltration membrane essentially remove the bigger dissolved solids, making the process more responsible. However, nanofiltration membranes perform better in desalination for some brackish water because their permeability is larger than that of RO membranes (Lhassani et al., 2001). Unlike RO membrane separation, which rejects 99% of the salt in the water and eliminates all fluoride ions, NF membrane separation allows for adjustment of operation settings to get the proper fluoride content in water. According to Diawara et al., the fluorine retention rate of the NF membrane varies between 63.3% and 71% at several Senegalese sites. In a single pass for fluoride concentrations below 6 mg/l, Tahaikt et al. found engaging performance of two modules (NF90 and NF400), however double pass was necessary to transfer greater fluoride concentrations as far as feasible (Tahaikt et al., 2007). The South of Morocco brackish ground water was successfully fluoridated, according to Pontie et al., using NF90 (Pontie et al., 2008). Bejaoui et al. use reverse osmosis and nanofiltration to decrease salinity and fluoride ions in an industrial effluent from metal packaging. More than 90% of the fluoride was kept in both membranes. Calculating the reflection coefficient (r) and ion solute permeability coefficient of membrane using the Spiegler-Kedem model (Bejaoui et al., 2014).

Following are the points of interest and restrictions:

Interest

1. A high output rate.
2. No chemicals are required.
3. The existence of other ions does not appear to be interfering.
4. It works as expected over a broad pH range.
5. This method effectively blocks all inorganic toxins, all organic micropollutants, insecticides, and microorganisms.

Restrictions

1. Compared to other defluoridation methods, this method is quite expensive.
2. Simple to scale, pollute, or degrade membrane.
3. Removal of all ions from treated water, which are essential to normal growth, requires remineralization.

11.2.5 Dialysis

Reverse osmosis and nanofiltration hold solutes while water passes through the membrane, whereas dialysis separates solutes by transporting solutes across the membrane. The solute can be forced through membrane pores, which are substantially less constrictive than nanofiltration, using a linked electric field or the Donnan effect (Donnan, 1911). Diffusion dialysis, often referred to as Donnan dialysis, performs functions comparable to an ion exchange membrane but differs from it in that it is driven by chemical potential difference rather than an electric current. The main factor influencing ion transport in Donnan dialysis is the concentration difference. Using the second alkaline stream and Donnan dialysis, which has an anion exchange membrane, a negative ion can be removed from the feed solution. The hydroxide ion must diffuse into the feed solution due to the disparity in hydroxide ion concentration between two solutions. Due to the development of negative-polarity electrical field, feed solution is used to extract negative ions (Hichour, 2000). Hichour et al. study the Donnan dialysis process in the current system with sodium salts and 0.001 M NaF as feed, sodium chloride was loaded onto the anion-exchange membrane. As more ions entered the stream, fluoride moved into the receiver. This technique was used to reduce fluoride in feed to less than 1.5 mg/l after fluoridating solution simulated to high fluoride African groundwater (> 30 mg/l fluoride). Morocco's phosphate mining produced fluoride water with a fluoride level (Hichour, 2000). Grames et al. used the hybrid method of Donnan dialysis with a Neosepta-ACS anion exchanger together with adsorption on aluminum oxide and zirconium oxide to treat water and reduce fluoride content to a suitable level for drinking (Grames, 2002). A Neosepta AHA anion exchanger used by Durmaz et al. to separate fluoride from the diluted solution. Fluoride was removed using a Neosepta AHA anion exchanger membrane, and flux of fluoride changed on its concentration, pH, and accompanying anions. Neosepta AFN, Neosepta AHA, and Polysulfone SB-6407 all had the highest membrane transfer efficiency. Alkan et al. use Donnan dialysis with a plasma Prinstine and AFX anion exchange membrane to remove fluoride (Alkan et al., 2008). It was established that due to advances in wettability and shape, flux values and recovery factors for the plasma modified AFX membrane were higher than those for the Prinstine membrane (Boubakri, 2014).

11.2.6 Elecro-dialysis

In electrodialysis, ions are extracted from aqueous solutions using an ion exchange membrane that is powered by electricity. Using electricity rather than pressure, reverse osmosis and electro-dialysis both remove ionic pollutants from water. Regardless of situation, electro-dialysis is not recommended for remote locations because it requires power. Adhikary et al. defluoridated brackish water with fluoride up to 10 ppm and TDS up to 5000 ppm to bearable peak values of 600 ppm TDS and

1.5 ppm with minimum energy usage 1 kWh/kg of salt remove (Adhikary et al., 1989). Amor et al. defluoridated brackish water with 3 ppm fluoride and 3000 ppm of TDS and produced potable water using an electrodialysis technique (Amor et al., 1988). Annouar et al. tested electrodialysis to remove fluoride from artificial water and groundwater in the city of Youssoufla using CMX-ACS membranes and chitosan, and compared the outcomes to water that had been brought within the WHO permitted range (Annouar et al., 2004). To defluoridate brackish subsurface water in Morocco City, Sahli et al. used chitosan and electrodialysis. Combining these two techniques allowed them to treat defluoridated water with 3 mg/l of fluoride and 3000 mg/l of TDS (Sahli et al., 2007). Kabay et al. investigated the removal of fluoride from aqueous solutions using electrodialysis by various operating parameters, such as applied voltage, flow rate, fluoride concentration, and impacts of sulfate and chloride ions. The efficiency of fluoride removal increases with applied voltage and starting fluoride concentration in feed solution. However, changes in feed flow rate had no impact on performance. The capacity of a chloride ion to distinguish between fluoride and sulfate ions is unaltered (Kabay et al., 2008). Ergun et al. used electrodialysis to reduce the amount of water containing 20.6 mg/l of fluoride—0.8 mg/l that is safe to drink. They used the SB-6407 anion exchanger membrane (Ergun et al., 2008). To remove fluoride, Lahnid et al. evaluated the economic viability of electrodialysis. According to Moroccan standards for rural areas, an industrial facility with a water consumption amount of 2200 m^3/d for 50,000 people per capita and operational cost of € 0.154/m^3 was evaluated with a capital cost of € 833,207 (Lahnid et al., 2008).

The following are electro-dialysis point of interest and restrictions:

Interest

1. Pre and post treatment is costly.
2. Flexibility.
3. Modest chemical request.
4. Increase water recovery.

Restrictions

1. Separation of ions.
2. The electrode rinse may have produced H_2.
3. Power requirements particular to pumps.
4. The necessity of treating concentrates.

11.2.7 Ion-Exchange Technique

Quaternary ammonium functional groups are present in strongly anion-exchange resin. According to the following response, reaction occurs:

$$\text{Matrix NR}_3 + \text{cl}^- + \text{f}^- \rightarrow \text{Matrix NR}_3 + \text{f}^- + \text{cl}^-$$

Fluoride ions are used in place of chloride ions in resin. Until all resin's places are filled, this process is repeated. After that, sodium chloride salt-supersaturated water is used to backwash the resin. The process is restarted when fresh chloride ions replace fluoride ions. Chloride ions from the resin can be replaced by fluoride ions since they have a higher electronegativity. Chikuma et al. investigated the removal of fluoride in batch and column modes using anion exchange resin and newly anion exchange resin with an Alizarin Fluorine Blue lanthanum complex (Chikuma & Okabayashi., 1987). Using Amberlite IRA-400 chloride anion exchanger, Chikuma and Nishimura investigated removal of fluoride. Chloride ions, which were kept on the resin surface in an aqueous solution, were replaced by fluoride ions (Chikuma & Nishimura, 1990). Using a cyclic two-way ion exchange process, Castel et al. look into removal of fluoride (Castel et al., 2000). Dodecylamine was used by Ho et al. to construct titanium oxohydroxide, increasing the exchange amount. Zirconia and silica add to mesoporous titanium oxohydroxide to increase its ion exchange efficiency because it has the lowest particle size and highest degree of homogeneity among all mesoporous materials produced. The process is expensive, and membrane fouling issues have been reported (Ishihaa et al., 2004). Chubar et al. studied the removal of anions from instantaneous solutions, including fluoride, chloride, bromide, and bromate using new ion exchange that used two times hydrous oxide ($Fe_2O_3 \cdot Al_2O_3 \cdot x\ H_2O$) (Chubar et al., 2005). Indion FR 10 and Ceralite IRA 400, two chelating resins, were specifically examined by Meenakshi et al. for their defluoridation potential (CER). Using an anion exchange resin and various equilibrating conditions, we investigated the distinctive evidence of selected sorbent. According to research, chelating resin outperforms anion exchange resin in its ability to remove fluoride (Meenakshi & Viswanathan, 2007). For removal of fluoride, Sundaram et al. used ion exchangers of the organic–inorganic type. $Ce(SO_4)_2 \cdot 4H_2O$, $Al(NO_3)_3 \cdot 9H_2O$ and $ZrOCl_2 \cdot 8H_2O$ were used to modify the polyacrylamide ion exchanger. Ce-Ex (2290 mg F^-/kg) has a somewhat higher defluoridation capacity than the competition. In field trials, the fluoride concentration was reduced from 1.96 mg/l to certain level using 0.25 g of exchangers for 50 ml of samples over 30-min contact period (Sundaram & Meenakshi, 2009).

Following are the Ion-Exchange technique point of interest and restrictions:

Interest

1. High productivity (fluoride elimination of 90–95%).
2. Preserves the excellence of the water.

Restrictions

1. The process is quite expensive.
2. The water that has been treated has low pH and high chloride content.
3. Additional anions, including sulfate, carbonate, phosphate, and alkalinity, can cause interference.
4. Due to the high fluoride waste it generates and the need to be handled prior t final disposal, resin regeneration is difficult.
5. Longer reaction times are required.

11.2.8 Adsorption

Adsorption is a link created by chemical or physical forces between different types of molecules from a bulk solution and solid's surface. Fluoride is removed from the solid bed matrix during the adsorption process in the contact bed through ion exchange or surface chemical reaction. Adsorption is preferred over other defluoridation methods because it is simple and provides broad range of adsorbents. Adsorption onto solid surfaces is an easy, flexible, and useful approach to the filtration of a drinking water system, particularly in small groups. Adsorption is a technique that, over broad pH range, removes ions from a solution while producing less waste than precipitation (Das et al., 2005; Ghorai & Pant, 2005). To assess their potential and techno-economic viability as defluoride experts, some adsorbent materials have previously undergone testing. The literature lists a wide range of adsorbent, including activated alumina, carbon, coated silica gel, calcite, activated sawdust, activated coconut shell powder, activated, groundnut shell, serpentine, tricalcium phosphate, and others (Barbier & Mazounie, 1984; Kariyanna, 1987; Min et al., 1999; Muthukumaran et al., 1995; Rongshu et al., 1995; Wang & Reardon, 2001). The most popular adsorbents are activated alumina and activated carbon. Effectively activated alumina eliminates fluoride and is influenced by pH, hardness, and surface loading. Up to 90% fluoride can be removed using the very practical treatment approach of adsorption. Since the adsorbent's ability to remove fluoride decreases with each cycle, regeneration is required to continue at intervals of 4–5 months. The study by Mckee and Johnston on elimination of fluoride using powdered activated carbon was successful (Mckee & Johnston, 1999). The technique is pH-dependent, producing excellent results only at pH 3 or lower. It costs money to use this material since the pH needs to be changed. Activated alumina defluoridation is being used in a few communities thanks to nonprofits supported by UNICEF or other organizations. By giving a bucket with microfilter at bottom and 5 kg of activated alumina, Sarita Sansthan, Udaipur, is dispersing technique with the practical assistance of UNICEF.

The following are points of interest and restrictions of the adsorption technique:

Interest

1. Simplity of operation.
2. The adsorption process is valuable.
3. High efficacy in the removal of fluoride, removing up to 90% of fluoride.
4. Create superior grade.
5. It is possible to regenerate.

Restrictions

1. The issue of disposed of depleted adsorbents and concentrated regenerates.
2. Competition for active sites in an adsorbent may result from interference caused by the presence of several anions.
3. A decrease in removal effectiveness after the regeneration phase.
4. Extremely subservient to pH.
5. The alumina bed can be fouled by a high concentration of TDS.

11.3 Conclusions and Future Perspectives

The objective was to outline the numerous methods that have been employed up to this point to remove fluoride out of drinking water and industrial wastewater. Fluoride is often removed from drinking water through liming and subsequent fluoride precipitation. However, the majority of these technologies suffer from issues that include high operational and maintenance costs, auxiliary contamination, such as the creation of toxic sludge, and challenging treatment processes. Research was done on several defluoridation technologies. It was suggested that while coagulation techniques have typically been successful in defluoridation and not been successful in providing fluoride at acceptable concentration levels.

Acknowledgements The authors are grateful to the "University Grants Commission (UGC)" and "Guru Gobind Singh Indraprastha University (GGSIPU)" for financial support and research facilities.

Funding This manuscript received no fundings.

References

Adhikary, S. K., et al. (1989). Defluoridation during desalination of brackish water by electrodialysis. *Desalination, 71*(3), 301–312.

Amor, Z., et al. (1988). Optimization of fluoride removal from brakish water by electrodialysis. *Desalination, 120*(3), 263–271.

Andey, S., et al. (2013). Performance evaluation of solar power based electrolytic defluoridation plants in India. *International Journal of Water Resources and Arid Environments, 2*(3), 139–145.

Annouar, S., et al. (2004). Defluoridation of underground water by adsorption on the chitosan and by electrodialysis. *Desalination, 165*, 437.

Alkan, E., Kir, E., & Oksuz, L. (2008). Plasma modification of the anion-exchange membrane and its influence on fluoride removal from water. *Separation and Purification Technology, 61*, 455–460.

Assefa, B. (2006). Defluoridation of Ethiopian rift valley region water using reverse osmosis membranes. *Journal of EEA, 23*, 1–6.

Babu, J. M., & Goel, S. (2013). Defluoridation of drinking water in batch and continuous-flow electrocoagulation systems. *Pollution Research, 32*(4), 727–736.

Barbier, J. P., & Mazounie, P. (1984). Methods of reducing high fluoride content in drinking water. *Water Supply, 2*, 8/1–4.

Bazrafshan, E., Ownagh, K. A., & Mahvi, A. H. (2012). Application of Electrocoagulation process using iron and aluminium electrodes for fluoride removal from aqueous environment. *E-Journal of Chemistry, 9*(4), 2297–2308.

Bejaoui, I., Mnif, A., & Hamrouni, B. (2014). Performance of reverse osmosis and nanofiltration in the removal of fluoride from model water and metal packaging industrial effluent. *Separation Science and Technology, 49*, 1135–1145.

Bennajah, M., et al. (2010). Defluoridation of drinking water by electrocoagulation/electroflotation–kinetic study. *Journal of Urban and Environmental Engineering, 4*(1), 37–45.

Boubakri, A. (2014). Fluoride removal from diluted solutions by Donnan dialysis using full factorial design. *Korean Journal of Chemical Engineering., 31*(3), 461–466.

Briao, V. B., et al. (2014). Reverse osmosis for desalination of water from the guarani aquifer system to produce drinking water in southern Brazil. *Desalination, 344*, 402–411.

Bulusu, K. R., et al. (1979). Fluoride in water, defluoridation methods and their limitations. *Journal of the Institution of Engineers (India), 60*, 1–25.

Castel, C., et al. (2000). Selective removal of fluoride ions by a two-way ion-exchange cyclic process. *Chemical Engineering Science, 55*, 3341–3352.

Chikuma, M., & Nishimura, M. (1990). Selective sorption of fluoride ion by anion exchange resin modified with alizarin fluorine bluepraseodymium(III) complex. *Reactive Polymers, 13*, 131–138.

Chikuma, M., & Okabayashi, Y. (1987). Separation and determination of fluoride ion by using ion exchange resin loaded with alizarin fluoride blue. *Chemical and Pharmaceutical Bulletin, 35*(9), 3734–3739.

Chubar, N. I., et al. (2005). Adsorption of fluoride, chloride, bromide, and bromate ions on a novel ion exchanger. *Journal of Colloid and Interface Science, 291*(1), 67–74.

Cui, H., et al. (2012). Electrochemical removal of fluoride from water by PAOA-modified carbon felt electrodes in a continuous flow reactor. *Water Research, 46*(12), 3943–3950.

Dahi, E. (1996). Contact precipitation for defluoridation of water. *22nd WEDC Conference New Delhi India*, pp. 266–268.

Das, N., et al. (2005). Deflouridation of drinking water using activated titanium rich bauxite. *Journal of Colloid Interface Science, 292*, 1–10.

Diawara, C. K., et al. (2011). Determination Performance of nanofiltration (NF) and low pressure reverse osmosis (LPRM) membranes in the removal of fluorine and salinity from Brakish drinking water. *Journal of Water Resource and Protection, 3*, 912–917.

Donnan, F. G. (1911). Donnan equilibrium and the physical properties of proteins. *The Journal of General Physiology*, 667–690.

Ergun, E., et al. (2008). Electrodialytic removal of fluoride from water: Effects of process parameters and accompanying anions. *Separation and Purification Technology, 64*(2), 147–153.

Gedam, V. V., et al. (2012). Performance evaluation of polyamide reverse osmosis membrane for removal of contaminants in ground water collected from Chandrapur district. *Journal of Membrane Science & Technology, 2*(3), 1–5.

Ghorai, S., & Pant, K. K. (2005). Equilibrium, kinetics and breakthrough studies for adsorption of fluoride on activated alumina. *Separation and Purification Technology, 42*, 265–271.

Grames, H. (2002). Defluoridation of groundwater by a hybrid process combining adsorption and Donnan dialysis. *Desalination, 145*, 287–291.

Hichour, M. (2000). Fluoride removal from water by Donnan dialysis. *Separation and Purification Technology, 18*(1), 1–11.

Ishihaa, T., et al. (2004). Removal of fluoride from water through ion exchange by mesoporous Ti-oxohydroxide. *Journal of Colloid and Interface Science, 272*, 399–403.

Kabay, N., et al. (2008). Separation of fluoride from aqueous solution by electrodialysis: Effect of process parameters and other ionic species. *Journal of Hazardous Materials, 153*(1–2), 107–113.

Kariyanna, H. (1987). Geological and geochemical environment and causes of fluorosis—Possible treatment—A review. In *Proceedings in Seminar on Role of Earth Sciences in Environment Bombay*, pp. 113–122.

Khatibikamal, V., et al. (2010). Fluoride removal from industrial wastewater using electrocoagulation and its adsorption kinetics. *Journal of Hazardous Materials, 179*(1–3), 276–280.

Lahnid, S., et al. (2008). Economic evaluation of fluoride removal by electrodialysis. *Desalination, 230*(1–3), 213–219.

Lhassani, A., et al. (2001). Selective demineralisation of water by nanofiltration application to the defluoridation of brackish water. *Water Research, 35*, 3260–3264.

Maheswari, R. C., Hoelzel, G. (2002). Potential of membrane separation technology for fluoride removal from underground water. *Proceedings of the Water Environment Federation, 17*, 620–636.

Mckee, R., & Johnston, W. S. (1999). Removal of fluorides from drinking water using low-cost adsorbent. *Indian Journal of Environmental Health, 41*(1), 53–58.

Meenakshi, S., & Viswanathan, N. (2007). Identification of selective ion-exchange resin for fluoride sorption. *Journal of Colloid and Interface Science, 308*, 438–450.

Min, Y., et al. (1999). Fluoride removal in a fixed bed packed with granular calcite. *Water Research, 33*(16), 3395–3402.

Mjengera, H., & Mkongo, G. (2002). Appropriate defluoridation technology for use in fluorotic areas in Tanzania. In *3rd Water Net Symposium Water Demand Management for Sustainable Development.*

Mohapatra, M., et al. (2009). Review of fluoride removal from drinking water. *Journal of Environmental Management, 91*, 67–77.

Mumtaz, N., et al. (2012). Evaluation of operational parameters involved in electrolytic defluoridation process. *Environmental and Infrastructure Engineering Research and Development, 2*(4), 23–32.

Muthukumaran, K., Balasubramanian, N., & Ramakrishna, T. V. (1995). Removal of fluoride by chemically activated carbon. *Indian Journal of Environmental Protection, 15*(7), 514–517.

Naim, M., Moneer, M. A. A., & El-Said, G. F. (2015). Predictive equations for the defluoridation by electrocoagulation technique using bipolar aluminum electrodes in the absence and presence of additives: a multivariate study. *Desalination and Water Treatment* 1–13.

Naim, M. M., et al. (2012). Defluoridation of commercial and analar sodium fluoride solutions without using additives by batch electrocoagulation-flotation technique. *Desalination and Water Treatment, 44*, 110–117.

Nava, C., Rios, D. M. S., & Olguin, M. T. (2003). Sorption of fluoride ions from aqueous solutions and well drinking water by thermally heated hydrocalcite. *Separation and Purification Technology, 38*(1), 31–147.

Nawlakhe, W. G., & Rao, A. V. J. (1990). Evaluation of defluoridation plant at Tartatur. *Journal of Indian Water Works Ass, 13*, 287–290.

Nawlakhe, W. G., et al. (1975). Defluoridation of water by Nalgonda technique. *Indian Journal of Environmental Health, 17*, 26–65.

Ndiaye, P. I., et al. (2005). Removal of fluoride from electronic industrial effluent by RO membrane separation. *Desalination, 173*, 25–32.

Padmashri, J. P. (2001). Effectiveness of low cost domestic defluoridation. *International Workshop on Fluoridation Water Strategies, Management and Investigation* (pp. 27–35). Bhopal.

Pontie, M., Dach, H., & Leparc, J. (2008). Nanofiltration as a sustainable water defluoridation operation dedicated to large scale pilot plants for the future. In *13th International Water Resource Association (IWRA) World Water Congress* (Vol. 1–4, pp. 1–6).

Razbe, N., et al. (2013). Various options for removal of fluoride from drinking water. *IOSR Journal of Applied Physics, 3*(2), 40–47.

Rongshu, W., et al. (1995). Study of a new adsorbent for fluoride removal from waters. *Water Quality Research Journal of Canada, 30*(1), 81–88.

Sahli, M. A. M., et al. (2007). Fluoride removal for underground Brakish water by adsorption on the natural chitosan and by electrodialysis. *Desalination, 212*(1–3), 37–45.

Sandoval, M. A., et al. (2014). Fluoride removal from drinking water by electrocoagulation in a continuous filter press reactor coupled to a flocculator and clarifier. *Separation and Purification Technology, 134*, 163–170.

Schoeman, J. J. (2010). Water defluoridation, water denitrification and water desalination in rural areas in South Africa. In *Proceedings of the third IASTED African Conference, Power and Energy System (AfricaPES 2010), Gaborone, Botswana* (pp. 244–247).

Seadar, J. D., & Heneley, J. E. (2005). The separation process principles. *NJ Wiley, 2*, 521–523.

Sundaram C. S., & Meenakshi, S. (2009). Fluoride sorption using organic-inorganic hybrid type ion exchangers. *Journal of Colloid and Interface Science, 333*, 58–62.

Tahaikt, M., et al. (2007). Fluoride removal from groundwater by nanofiltration. *Desalination, 212*, 46–53.

Thergaonkar, V. P., & Nawalakhe, W. G. (1971). Activated magnesia for fluoride removal. *Indian Journal of Environmental Health, 16*, 241–243.

Un, U. T., Koparal, A. S., & Ogutveren, U. B. (2009). Electrochemical process for the treatment of drinking water. In *Thirteenth International Water Technology Conference IWTC Hurghada, Egypt, 13*, 129–137.

Vasudevan, S., et al. (2011). Effects of alternating and direct current in electrocoagulation process on the removal of fluoride from water. *Journal of Chemical Technology and Biotechnology, 86*, 428–436.

Waghmare, S. S., & Arfin, T. (2015). Fluoride removal from water by calcium materials a state of the art review. *International Journal of Innovative Research in Science, Engineering and Technology, 4*(9), 8090–8102.

Waghmare, S. S., & Arfin, T. (2015). Fluoride removal from water by mixed metal oxide adsorbent materials a state of the art review. *International Journal of Engineering Sciences & Research Technology, 4*(9), 519–536.

Waghmare, S. S., et al. (2015). Preparation and characterization of polyalthia longifolia based alumina as a novel adsorbent for removing fluorie from drinking water. *Asian Journal of Advanced Basic Sciences, 4*(1), 12–24.

Wang, Y., & Reardon, E. J. (2001). Activation and regeneration of a soil sorbent for defluoridation of drinking water. *Applied Geochemistry, 16*, 531–539.

Wimalawansa, S. J. (2013). Purification of contaminated water with reverse osmosis—Effective solution of providing clean water for human needs in developing countries. *International Journal of Emerging Technology and Advanced Engineering, 3*(12), 75–89.

Part III
Fluoride Effect on Human Health

Chapter 12
Effect of Fluoride-Contaminated Water on the Living Being and Their Surroundings

Seema Kumari, Harsh Dhankhar, Vikas Abrol, and Akhilesh Kumar Yadav

Abstract Living beings are affected by fluoride (present in soil, water, and air). In the halogen family, fluorine is a naturally abundant element. It has no biological purpose and is exceedingly harmful to humans, animals, and the environment when it is present in large amounts. Fluorine is an extremely reactive element that does not exist in nature in its pure form. It is found as fluoride ion (F^-), which makes up approximately 0.3 g/kg of the Earth's crust. Its pollution has a significant negative impact on 200 million people in 29 nations, including India. It is released into the soil naturally by the breakdown of minerals such as apatite, fluorspar, topaz, and mica. It is transferred to crops, vegetables, and fruits during irrigation using fluoride-contaminated water. This bioaccumulation increases the risk to the population already affected by fluoride poisoning since it adds more fluoride to the food chain in addition to the route through drinking water. Emphasis must be placed on educating people about fluorosis and limiting their use of fluoride-contaminated irrigation water. High

S. Kumari (✉) · H. Dhankhar
Department of Botany, Baba Mastnath University, Rohtak 124001, India
e-mail: kumareeseema2011@gmail.com

H. Dhankhar
e-mail: dhankhar.harsh@gmail.com

V. Abrol
Division of Soil Science, Sher-e-Kashmir University of Agricultural Science and Technology, Jammu 180009, India
e-mail: abrolvics@gmail.com

A. K. Yadav
Department of Environmental Engineering and Management, Chaoyang University of Technology, Taichung 413310, Taiwan

Department of Mining Engineering, Indian Institute of Technology (Banaras Hindu University), Varanasi 221005, India

Environmental Science and Engineering Department, Indian Institute of Technology Bombay, Mumbai 400076, India

A. K. Yadav
e-mail: yadavbasti@gmail.com

A. K. Yadav et al. (eds.), *Advanced Treatment Technologies for Fluoride Removal in Water*, Water Science and Technology Library 125,
https://doi.org/10.1007/978-3-031-38845-3_12

levels of F^- may have a negative impact on human health over time by reducing thyroid function, allowing kidney stones to form, causing bone fractures, skeletal fluorosis, dental fluorosis, and impairment of IQ, especially in young children. In this chapter, we focus on the effects of F^- in the living beings through the water.

Keywords Fluoride · Fluorosis · Humans · Animals · Birds · Environment

12.1 Introduction

Life quality and environmental health are closely linked. Worldwide attention and attempts to identify innovative treatments to better control and sustain the environmental component have been drawn to the increasing occurrence of human health and animal issues caused by industrial pollution and anthropogenic environmental modifications (Ahmed, 2007; Shirin & Yadav, 2014, 2021; Tsiros, et al., 1998).

Air, soil, and water can all emit toxic chemicals into the environment. Pollutants from stacked emissions into the air can be added to the soil, where they may be absorbed by plants through their roots. These contaminants impact both people and wildlife when they build up in the food chain. Variations in anthropogenic and natural activities are causing an increase in global pollution, contaminating the ecosystem with organic chemicals, inorganic compounds, metals and nonmetals (Shirin et al., 2019, 2022). Herbicides, insecticides, sewage disposal, pesticides, and unregulated discharge of waste are the main causes of this contamination. A sizable portion of the population in developed nations is daily exposed to many hazardous chemicals and metals that are bad for human health.

Fluorine-comprising compounds are widely used in practically all biological industries, and the fluoride ion (F^-) is a major environmental pollutant. It is not a necessary element that is not required for the growth of healthy bones and teeth, although it has an antibacterial effect when used topically on teeth (Chouhan & Flora (2010). The health of people, animals, and plants can suffer as a result of consuming too much F^- (Koblar et al., 2011).

This chapter focuses on the significance of fluorine and highlights that high intake of the element from ingested or inhaled sources can be hazardous to humans, birds, animals, and finally the surrounding.

12.1.1 Sources of Fluoride

The main sources of F^- include atmospheric deposition, industrial emissions, and rock weathering, while there are numerous more sources as well. The most frequent origin of F^- ions is geochemical and mineral reserves, and a significant amount of F^- offload on into subsurface water occurs as a result of fluorine-containing rocks degrading (Jacks et al., 2005). It is one of the most common elements in the crust of the earth and may be found at a variety of rocks in concentrations ranging from 100 to 1000 μg/mg, with 625 μg/mg being the average. Granites, quartz, felsic, monzonites, biotite, syenites, and granodiorites contain significant amounts of F^- using various processes, including soil formation and weathering. Rocks containing F such as muscovite, amphibolites, pegmatite, as well as biotite micas contribute F to soil and groundwater (Bhat et al., 2015). Additionally, there are anthropogenic discharge of fluoride, like the removal of hydrogen fluoride in the air or the addition of fluoride to water as a result of a variety of human activities, such as industrialization, the use of pesticides that contain fluoride, fluoridating drinking water supplies, and motorization (Paul et al., 2011). A significant industrial source of F^- is the production of phosphate fertilizers. Fertilizer manufactured from phosphorite has a significant amount of fluorine, such as 3.5%, however, this percentage is decreased during production (Bhat et al., 2015). The contemporary use of chemicals such as sulfur hexafluoride (SF_6), hydrogen fluoride, sodium fluoride (NaF), calcium fluoride (CaF), phosphate manure, etc. is another anthropogenic cause for the introduction of F^- into the earth. As F^- is released into the atmosphere, the wind carries it to the nearby vegetation and soil where it contaminates them. The use of phosphorus fertilizers is the main cause of soil contamination with fluoride ions. Its concentration in soil typically ranges from 150 to 400 g/mg, and in thick clay soil, the figure can rise to 1 g/kg. Humans are ultimately affected by the polluted soil by direct touch, inhalation of soil vapor, and the use of water that has been contaminated with F^- by passing through the soil. Other notable causes of F^- pollution in the environment include the manufacturing and use of phosphate fertilizers, steel, zinc, ceramic industries, and energy plants (Paul, et al., 2011) (Fig. 12.1).

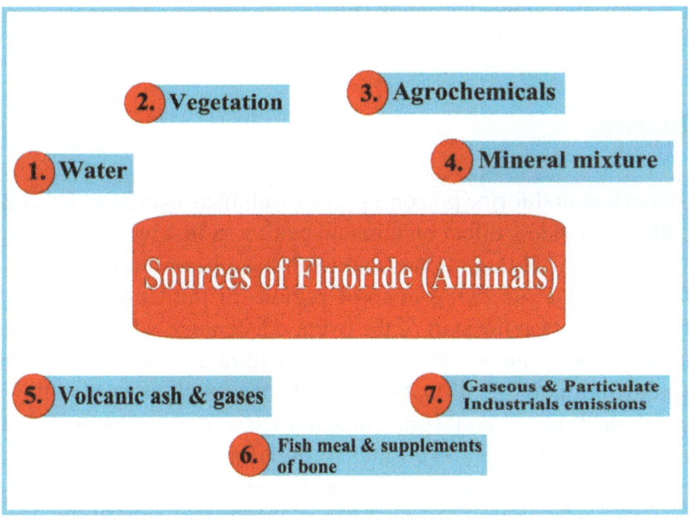

Fig. 12.1 Source of F^- in animals

12.1.2 Spatial Distribution of Fluoride

The spatial distribution is influenced by the soil formation because the parent material typically contributes to the total F^- content within the soil. The average amount of F^- in the soil was calculated to be 329 g/mg worldwide. Rock soil and heavy clay soil often contain the highest amounts of F^-, while sandy soil in very wet conditions typically has the lowest F^- content. The main factors of soil fluoride content are clay, organic carbon content, and soil pH (Kumar et al., 2016). It enters the soil in a variety of ways, including through dry deposition, contaminated litter, and, precipitation where it is easily absorbed. The absorbed F^- can combine with toxic substances such as aluminum and heavy metals, which increase the total soluble content of F^- in the soil. It also affects the pH of the soil. F^- could exist as a free fluoride ion and combined with other elements like calcium, aluminum, boron, iron, and boron. Complexes of Al and F are the most common (Domingos et al., 2003).

12.1.3 Fluoride Pollution Effects

It includes contamination of plants, vegetables, soil, crops, and freshwater bodies. Aerial F^- emissions from industries also contaminate groundwater, which is then contaminated by fluorine-containing mineral deposits. The use of these F^- contaminated goods and resources could have toxic effects on both people's health and domestic animals' health (Choubisa & Choubisa 2016). Although the application of F^- has positive impacts on teeth, the evidence for any positive impacts from systemic absorption is currently viewed as weak (Fig. 12.2; Table 12.1).

12.2 Fluorosis

Fluorosis, or persistent fluorine poisoning, can result from excessive F^- consumption (Ahmed, 2007). Fluoride's effect on human health was first recognized at the end of the nineteenth century. Human bones and teeth have been found to contain a large amount of fluoride. People living in some regions of the United States developed brown dental dark spots at the start of the twentieth century, which were later proven to have a positive association with fluoride contamination in the affected regions' drinking water (Smith et al., 1931). Surprisingly, it was discovered at the same time that adequate fluoride intake could prevent dental caries development without causing orthodontic staining (Dean, 1938). Later in the 1940s, municipal water treatment facilities began fluoridating water to optimally (0.7–1.2 mg/l) in response to the average air temperature of that region (Li et al., 2001; Ozsvath, 2006).

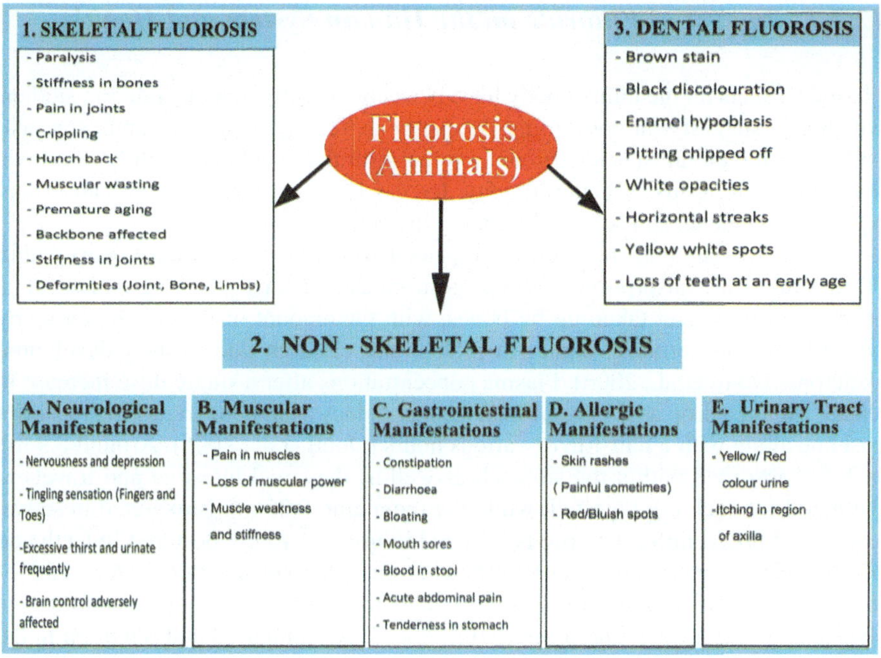

Fig. 12.2 Fluorosis, types, and effects

Table 12.1 Effects of F^- contaminated water on human health

Sl. No	F^- Concentration (mg/l)	Health outcomes\effects
1	Less than 0.5	Dental caries
2	0.5–1.5	Optimum dental health
3	1.5–4.0	Dental fluorosis
4	4.0–10	Skeletal and dental fluorosis
5	More than 10.0	Crippling Fluorosis

12.2.1 Fluoride Absorption in the Body

Fluoride is an electronegative element that can build stronger complexes with Al. B. Be. Fe (III). Although such complexes are extremely rare in natural water bodies, Si, Na, U, and V. The magnesium-fluoride complex is the main component of potable water. According to epidemiology, the biological effects of consuming fluorosilicates and ingesting NaF or fluoride from natural waters are different (Urbansky & Schock, 2000).

12.2.2 *Impact of Fluoride on the Human System and Health*

Fluoride effects on the human body have been consistently demonstrated in studies, which also suggest that consuming F^- can be extremely dangerous and even fatal under specific circumstances. F^- is more harmful than lead and slightly less toxic than arsenic in terms of acute toxicity. Excess F^- concentrations can cause fatal poisoning incidents and serious poisoning incidents.

Chronic toxicity may develop after repeated exposure. Plasma peak occurs within 30 min after fast absorption from the gastrointestinal tract and lungs. Fluoride is excreted in urine and taken up by bones, with the amount of fluoride taken up by bones being substantially higher in infants and children due to their developing skeletons (Doull et al., 2006). Plasma concentrations after a single dose increase to a peak and then drop as F^- is eliminated using kidneys and bone, returning to a brief baseline along with a half-life of various hours (Doull et al., 2006). Around 0.9 mg of F^- is absorbed within the stomach as well as the small intestine and travels to the blood for each mg ingested, while the remaining 0.05 mg is excreted in saliva, sweat, and breast milk, 0.45 mg is excreted in urine, 0.40 mg is stored in teeth and bones, and 0.1 mg is excreted unabsorbed in feces (Doull et al., 2006). About 25% of the F^- was absorbed by the rat's stomach, and 75% came through its small intestine (Nopakun et al., 1989). The primary organs of F accumulation and retention in the human body are bones and teeth, while very modest amounts could also be deposited in the pineal gland, another calcifying organ (Doull et al., 2006; Kumar et al., 2016). The F^- retention period may last for a number of years. The placenta is how F^- gets to the fetus (Ghosh et al., 2013).

Absorption is influenced by a variety of dietary variables, for example the addition of Ca, Mg, and Al salts to the food, which integrates a small amount of fluoride into certain molecules with poor solubility that are eliminated from the body through the excretory process. The absorption of fluoride from the GI tract is also increased by the addition of Mo, sulfates, and phosphates to the diet. More than 90% of the fluoride is in the HPO compound. Gastric acid pH 2 inside the stomach promotes the formation of this complex (Doull et al., 2006). The HPO complex dissociates within the mucosa (which is less acidic) to release fluoride following its absorption from the stomach and small intestine (through the diffusion process) (Whitford et al., 1999).

The growth of teeth and bones quickly absorbs almost half of the fluoride that is ingested, and the other half is excreted in urine (Cerklewski, 1997). Therefore, approximately all fluoride build-up in the body is detected in the teeth and bones. Children are the most effective age group for the fluoride skeleton accumulation process, and lifetime exposure reduces the rate of accumulation (Whitford et al., 1999). Typically, the procedure lasts until age 55 (Nagendra Rao, 2003). Fluoride that has been integrated into hard tissues is recovered, but the process is extremely slow (osteoclastic resorption) (Doull et al., 2006) (Table 12.2).

Table 12.2 Effects observed in human body

Sl. No	Effects on humans		References
1	Effects on reproductive system	Structure and mobility of sperms affected (in males)	Chinoy and Narayana (1994)
		Fertility rate reduced (in females)	Naseem et al. (2010)
		Follicle stimulating hormones, testosterone and inhibin-B quality reduced	Ortiz Perez et al. (2003)
2	Neurobehavioural effects	Interference with enzymatic function, protein structure, brain functions of brain and impaired cognition and memory	Spittle (1994)
		Visuospatial abilities and Bio-chemical activities affected	Calderon et al. (2000)
		Neurotoxicity to the developing brain (in children)	Grandjean and Landrigan (2006)
		Intelligent quotient and thinking ability reduced	Trivedi et al. (2007)
		Dementia, mental and physiological changes, cognitive effects	USEPA
		Interfere with the glycolysis cycle, which ultimately affects the energy requirement of central nervous system	Valdez-Jiménez et al. (2011)
		Mental ability of children affected	Choi et al. (2012)
3	Effects on cardiovascular system	Oxidative stress promotes inflammatory mechanisms, abnormal heart rhythms reduce myocardial function, myocardial cell damage and atherosclerosis vascular stiffness, Bradycardia, diabetes mellitus and obesity hypothyroidism	Xu et al. (1997)
		Hypercalcemia and hypocalcemia	Nureddin
4	Gastro intestinal effects	Hyperaemia, loss of the mucus layer, haemorrhage, oedema, and stomach lining rupture	Pratusha et al. (2011)
		Vomiting, nausea and gastric pain	Nabavi et al. (2013)
5	Effects on endocrine system	Secondary hyperparathyroidism, increased calcitonin and parathyroid activity and impaired glucose tolerance	Doull et al. (2006)
		Dysfunctions and structural changes in thyroid gland	Kheradpisheh et al. (2018)
6	Effects on urinary/renal system	Metabolic, pathological, and histopathological changes in the glomeruli	Bouaziz et al. (2006)
		Increased risk of kidney stones enhanced	Doull et al. (2006)

12.2.3 Beneficial Impacts of Fluoride

It was determined in 1930 that regulated fluoride consumption can prevent the development of tooth cavities and was also responsible for stronger bone growth. The sharpest drop in dental cavities was detected in fluoride contents between 0.7 and 1.2 mg/l, with no additional benefit shown at doses of fluoride above this range (Doull et al., 2006; Nagendra Rao, 2003). Medical studies and epidemiology revealed that consuming fluoride together with the right amounts of Ca and vitamin D could increase bone mineralization but could not significantly decrease the number of fractures. The fluoride level was found to follow a 'U-shaped' curve with the number of bone fractures and the highest potential was found at – 1 mg/l (Li et al., 2001). Fracture incidence was shown to rise with very high fluoride drinking water contamination (4 mg/l) (Sowers et al., 2005). Therefore, further clinical and epidemiological analyzes are needed to examine the efficacy of fluoride in osteoporosis treatment and to optimize the advantages of fluoride while minimizing hazards (Aoba, 1997).

12.2.4 Adverse Effects of Fluoride

These include hemoptysis, vomiting, cramping in the legs and arms, occasionally death, hypocalcemia, hyperkalemia, fixed and dilated pupils, ventricular fibrillation, cardiac arrest, and bronchospasm (Shulman & Wells, 1997; Whitford, 1992). Fluoride toxicity can also result from the ingestion of NaF and dental products. Genetic mutations, allergic diseases, hypersensitivity reactions, birth defects, recurrent bone injury, Alzheimer's disease (AD), and other chronic consequences are caused by long-term exposure to fluoride in humans.

12.2.5 Dental Impacts of Fluoride

The most recognizable sign of dental fluorosis is 'mottled enamel'. When enamel grows, matrix proteins are eliminated and mineralization is increased. The mineralization is disturbed by exposure to fluoride in this enamel growth phase. The result is an increase in porosity (Aoba & Fejerskov, 2002), excessive retention of enamel proteins, and abnormally wide gaps in its crystalline structure. Dental fluorosis at a mild level is characterized by the development of white horizontal striations on the opaque patches or the surface of chalky white discolorations (Nagendra Rao, 2003). When dental fluorosis is severe, opaque patches develop stains that range from 'yellow' to "brown" or even black, and finally, increased porosity of the teeth causes structural damage such as chipping or pitting (Nagendra Rao, 2003).

12.2.6 Skeletal Impacts

These include increased bone mass and density, as well as a variety of skeletal and joint complaints, which is known as skeletal fluorosis. The amount of water consumed, its quality and other dietary parameters all have a significant impact on the threshold concentration of fluoride intake required to cause 'skeletal fluorosis'. Skeletal fluorosis' stages of development are widely known, but the mechanisms underlying it are still little understood. Back pain, hip stiffness, and joint pain are the initial symptoms of osteosclerosis, which is also characterized by increased bone density. The entire spine is made to look like one continuous bone column (a "poker back") due to the gradual rise in rigidity. As time passes, certain ligaments in the spine may calcify.

12.2.7 Fluoride's Impact on Reproduction

Fluoride's effects on human reproduction have not yet been thoroughly investigated. Freni1994 investigated the impact of increased fluoride levels in drinking water (> 3 mg/l) on the decline in birth rates in the US population. Some research findings revealed that increased fluoride intake has detrimental impacts on male reproduction. The morphology and mobility of sperm and the concentrations of inhibin B, testosterone, and follicle stimulating hormones (Ortiz-Perezet al., 2003) are all affected.

12.2.8 Fluoride's Impact on Development

According to certain research, umbilical cord blood plasma and maternal blood plasma fluoride concentrations are positively associated, indicating passive transmission of fluoride from mother to fetus through the placenta (Gupta et al., 1993; Malhotra et al., 1993). When extremely high doses of fluoride were used in clinical trials, negative developmental consequences were discovered. When extremely high dosages of fluoride were used in clinical tests in animals, adverse developmental consequences were discovered; however, the results for humans were inconclusive (Doull et al., 2006). Whiting et al. (2001) demonstrated the possibility of the fluoride-induced predominance of Down syndrome, particularly in infants born to mothers under the age of 30. According to reports, places with excessive fluoride contamination have an unusually high incidence of "*spina bifida occulta*" (Gupta et al., 1995).

12.2.9 Neurological Effects

These include documented effects on children's intellect from dietary fluoride intake (Wang et al., 2007; Xiang et al., 2003). The findings indicated that children who were exposed to higher amounts of fluoride (> 2 mg/l) performed worse on IQ tests compared to children who were exposed to lower levels (1 mg/l). Urinary fluoride levels and IQ were found to be strongly adversely associated in Indian school-age children (Trivedi et al., 2007). The fluoride dose at the neurotoxicity response threshold of children was found to be between 2–4 mg/l. Fluoride can produce a variety of metabolic alterations in enzymes and proteins, interfering with 'normal brain' function and resulting in diminished cognition and memory (Spittle, 1994).

12.2.10 Fluoride's Gastrointestinal Effects

These include a wide range of gastrointestinal symptoms, including abdominal pain, diarrhea, vomiting, and nausea. Fluoride-contaminated locations with undernourished populations tend to have symptoms (Dasarathy et al., 1996; Jenkins, 1991). Fluoride has been shown in clinical trials in animals to be able to accelerate the production of stomach acid, destroy cells of the gastrointestinal tract epithelium, and restrict blood flow to the stomach lining (Doull et al., 2006). When the fluoride level in water is below 4 mg/l, gastrointestinal hypersensitive individuals (less than 1% of the population) will exhibit gastrointestinal symptoms (Doull et al., 2006).

12.2.11 Effects of Fluoride on the Kidneys

Since the kidneys are responsible for excreting excess fluoride from other organs, they are more susceptible to fluoride toxicity than the majority of the body's soft tissues. The use for an extended time has been found to have no carcinogenic impacts on the kidneys (Doull et al., 2006). Juuti and Heinonen (1980) found that people with '*urolithiasis*' were more likely to be hospitalized in areas with groundwater fluoride levels > 1.5 mg/l. By examining a fluoride-contaminated region of India (fluoride content in drinking water: 3.5–4.9 mg/l), Singh found that patients with obvious symptoms of skeletal fluorosis had a 4.6-fold increased risk of kidney stone development.

12.2.12 Carcinogenic Effects

Several researchers (Takahashi et al., 2001; Yang et al., 2000) have shown a connection between the amount of fluoride in 'drinking water' and the occurrence of uterine and colon cancer. Animal studies also revealed increased 'osteoma' (non-cancerous bone tumors) and 'osteosarcoma' (bone cancer) (Doull et al., 2006). Some studies (Grandjean & Olsen, 2004; Takahashi et al., 2001) reported evidence of a link between fluoride exposure and the occurrence of bladder and kidney cancer. The carcinogenic effects of fluoride must be evaluated separately for each type of cancer because epidemiology is required to determine the carcinogenic potential of chronic fluoride toxicity, which is dependent on numerous cancers and potential causal factors (Harrison, 2005).

12.3 Toxicity of Fluoride in Animals and Birds

The fluoride ion (F^-) is toxic to both humans and animals and is known to cause dental, skeletal, and nonskeletal fluorosis when consumed chronically at high levels (more than 6 mg/day). F is not thought to be necessary for human development and growth, 4 including healthy bones and teeth. Fluorosis often manifests itself in 2 ways: (a) endemic fluorosis caused by consuming food or water with a high F^- content, and (b) industrial fluorosis brought on by exposure to air with a high F^- concentration. Fluorosis affects the brain, liver, kidney, thyroid, and spinal cord, among other soft tissues, in people, animals, and birds in addition to the skeletal system. Anjum et al. (2014) studied the impact of the high content of F^- on liver and renal enzymes in 4 groups of domestic chickens, A, B, C, and D, receiving weekly doses of 0, 10, 20 and 30 g/g of NaF by body weight, respectively. Uric acid has been used as a metric for renal function, while AST ('Aspartate Aminotransferase'), ALT ("Alanine Aminotransferase"), ALP ('Alkaline Phosphatase'), as well as bilirubin, have been evaluated as indications of liver function.

Although each animal is vulnerable to large doses of F^-, different species have different levels of tolerance. Drinking water, plants, and soil contaminated with F^- produced by various activities, such as industrial and volcanic eruption activities, are notable sources of F^- for terrestrial species. Animals and humans both have similar F^- metabolisms.

Herbivores are more vulnerable than carnivores and other animals among terrestrial vertebrates. Due to their non-selective eating habits and ability to swallow contaminated food, water, and forage, domestic and wild herbivores are particularly vulnerable to environmental F^- contamination. Researchers from all over the world have focused more on cattle and sheep, perhaps as a result of their vast numbers and greater economic significance. However, F^-toxicity can also occur spontaneously in other animals such as wild cervids, pigs, goats, horses, and water buffalo. Cattle (*Bos taurus*), horses (*Equus caballus*), donkeys (*Equus asinus*), camels (*Camelus dromedarious*), and buffaloes (*Bubalus bubalis*) have all been investigated for skeletal, non-skeletal, and dental fluorosis (Choubisa, 2010).

F^- toxicity in 760 domestic animals from various localities with an average water F range of 1.5 to 4.4 ppm, including 67 sheep (*Ovis aries*), 131 buffaloes (*Bubalus bubalis*), 11 horses (*Equus caballus*), 7 donkeys (*Equus asinus*), 158 goats (*Capra hirus*), and 386 cattle (*Bos taurus*) (Choubisa 2011). Dental mottling was discovered to be present in 371 (487, or 48.7%) of the population. In 325 (42.8%) of the animals, clinical examination found intermittent lameness, periosteal exostoses, deformed hooves, stiffness in the tendons and legs, and wasting of the major mass of the shoulder and hindquarter muscles.

Fluoride contamination of food, water, and air threatens not just humans but also animals. Animals have experienced gastroenteritis, skeletal muscle weakness, lung obstruction, vomiting, nausea, diarrhea, respiratory failure, heart failure, and ultimately death. Dental and bone fluorosis is among the long-term complications. There may be a strong connection between toxic effects on animals and human health since acute and long-term consequences in animals resemble the symptoms that people experience. Fluoride is hazardous to the thyroid hormones, reproductive system, learning and memory functions, blood, growth, and feeding efficiency in scientific tests conducted on animals.

When Camargo and Alonso examined the short and long-term effects of fluoride exposure on the aquatic snail: 'Potamopyrgusantipodarum', they discovered that "the number of embryos with shell was reduced by the highest concentration' and behavioral activity was affected (Alonso & Camargo, 2011). According to this study, fluoride is harmful. Long-term and short-term exposures can both result in mortality, along with negative consequences on behavior and reproduction. High levels of fluoride in drinking water are linked to lower birth rates, according to Gupta et al. (2007). According to their research, male rats exposed to 'sodium fluoride' in drinking water at concentrations of 2, 4, and 6 ppm for a period of six months had testicular problems and experienced negative effects on their reproductive system and fertility.

The fluoride-treated group had significantly lower levels of free triiodothyronine, serum-free thyroxin, and acetylcholine esterase activity. According to the study, iodine absorption may have been inhibited by the interaction of fluoride, which may have contributed to changes in thyroid hormone levels. According to the study findings, exposure to Fluoride has long-lasting transgenerational consequences that lead to decreased thyroid hormone, which is linked to memory and learning problems. According to research on the toxicity of fluoride in animals, "biological reactions of animals to fluoride are connected to dosage and other parameters that influence the animal's physiological and anatomical responses" (Madan et al., 2009).

According to reports, fluoride is poisonous to cattle and can cause chronic fluorosis, thyrotoxic effects, and disruptions in the secretion of thyroxin hormones in mammalian cells (Bouaziz et al., 2004; Cinar & Selcuk, 2005; Rahman & Fetouh, 2013).

More aquatic species living in soft water than those living in hard water are harmed by excessive fluoride. As water hardness increases, fluoride bioavailability decreases, which can have an impact on how quickly organisms grow (Ghosh et al., 2013).

12.4 Conclusion

This chapter discussed about significance of fluorine and makes the case of high intake of element from ingestion and inhalation sources. It can be hazardous to humans, birds, and animals. It finally leads to dental, skeletal, and nonskeletal fluorosis. Depending on the quantity and length of F^- consumption, toxicity can either acute or chronic. According to the available evidence, the high levels of F^- in soil have an impact on terrestrial and aquatic life as well.

Acknowledgements The authors acknowledge the financial support from the CSIR, New Delhi, and other respective institutions also acknowledged for this work.

References

Ahmed, D. (2007). Environmental pollution and its impact on human and animal health: A review. *The Sudan Journal of Veterinary Research, 22*, 37–46.

Alonso, Á., & Camargo, J. A. (2011). Toxic effects of fluoride ion on survival, reproduction and behaviour of the aquatic snail *Potamopyrgus antipodarum* (Hydrobiidae, Mollusca). *Water, Air, & Soil Pollution, 219*(1), 81–90. https://doi.org/10.1007/s11270-010-0685-5

Anjum, K., Mughal, M., Sayyed, U., Yaqub, A., Khalique, A., Rashid, M., Yousaf, M., & Mumtaz, N. (2014). Influence of increasing fluoride dose rates on selected liver and kidney enzymes profile in domestic chicken (*Gallus domesticus*). *The Journal of Animal and Plant Sciences, 24*(1), 77–80.

Aoba, T. (1997). The effect of fluoride on apatite structure and growth. *Critical Reviews in Oral Biology & Medicine, 8*(2), 136–153. https://doi.org/10.1177/10454411970080020301

Aoba, T., & Fejerskov, O. (2002). Dental fluorosis: Chemistry and biology. *Critical Reviews in Oral Biology & Medicine, 13*(2), 155–170. https://doi.org/10.1177/154411130201300206

Bhat, N., Jain, S., Asawa, K., Tak, M., Shinde, K., Singh, A., Gandhi, N., & Gupta, V. V. (2015). Assessment of fluoride concentration of soil and vegetables in vicinity of zinc smelter, Debari, Udaipur, Rajasthan. *Journal of Clinical and Diagnostic Research: JCDR, 9*(10), ZC63. https://doi.org/10.7860/JCDR/2015/13902.6667

Bouaziz, H., Ammar, E., Ghorbel, H., Ketata, S., Jamoussi, K., Ayadi, F., Guermazi, F., & Zeghal, N. (2004). Effect of fluoride ingested by lactating mice on the thyroid function and bone maturation of their suckling pups. *Fluoride, 37*(2), 133–142.

Bouaziz, H., Ketata, S., Jammoussi, K., Boudawara, T., Ayedi, F., Ellouze, F., & Zeghal, N. (2006). Effects of sodium fluoride on hepatic toxicity in adult mice and their suckling pups. *Pesticide Biochemistry and Physiology, 86*(3), 124–130. https://doi.org/10.1016/j.pestbp.2006.02.004

Calderon, J., Machado, B., Navarro, M., Carrizales, L., Ortiz, M., & Diaz-Barriga, F. (2000). Influence of fluoride exposure on reaction time and visuospatial organization in children. *Epidemiology.*

Cerklewski, F. L. (1997). Fluoride bioavailability—Nutritional and clinical aspects. *Nutrition Research, 17*(5), 907–929. https://doi.org/10.1016/S0271-5317(97)00057-2

Chinoy, N. J., & Narayana, M. V. (1994). In vitro fluoride toxicity in human spermatozoa. *Reproductive Toxicology, 8*(2), 155–159. https://doi.org/10.1016/0890-6238(94)90022-1

Choi, A. L., Sun, G., Zhang, Y., & Grandjean, P. (2012). Developmental fluoride neurotoxicity: A systematic review and meta-analysis. *Environmental Health Perspectives, 120*(10), 1362–1368. https://doi.org/10.1289/ehp.1104912

Choubisa, S. (2010). Osteo-dental fluorosis in domestic horses and donkeys in Rajasthan, India. *Fluoride, 43*(1), 5.

Choubisa, S. L., & Choubisa, D. (2016). Status of industrial fluoride pollution and its diverse adverse health effects in man and domestic animals in India. *Environmental Science and Pollution Research, 23*(8), 7244–7254. https://doi.org/10.1007/s11356-016-6319-8

Chouhan, S., & Flora, S. (2010). Arsenic and fluoride: Two major ground water pollutants.

Cinar, A., & Selcuk, M. (2005). Effects of chronic fluorosis on thyroxine, triiodothyronine, and protein-bound iodine in cows. *Fluoride, 38*(1), 65–68.

Dasarathy, S., Das, T. K., Gupta, I. P., Susheela, A. K., & Tandon, R. K. (1996). Gastroduodenal manifestations in patients with skeletal fluorosis. *Journal of Gastroenterology, 31*(3), 333–337. https://doi.org/10.1007/BF02355021

Dean, H. T. (1938). Endemic fluorosis and its relation to dental caries.

Domingos, M., Klumpp, A., Rinaldi, M. C. S., Modesto, I. F., Klumpp, G., & Delitti, W. B. C. (2003). Combined effects of air and soil pollution by fluoride emissions on Tibouchina pulchra Cogn., at Cubatão, SE Brazil, and their relations with aluminium. *Plant and Soil, 249*(2), 297–308. https://doi.org/10.1023/A:1022800225753

Doull, J., Boekelheide, K., Farishian, B., Isaacson, R., Klotz, J., Kumar, J., Limeback, H., Poole, C., Puzas, J., & Reed, N. (2006). Committee on Fluoride in Drinking Water, Board on Environmental Studies and Toxicology, Division on Earth and Life Studies, National Research Council of the National Academies. Fluoride in Drinking Water: A Scientific Review of EPA's Standards. Washington DC.

Ghosh, A., Mukherjee, K., Ghosh, S. K., & Saha, B. (2013). Sources and toxicity of fluoride in the environment. *Research on Chemical Intermediates, 39*(7), 2881–2915. https://doi.org/10.1007/s11164-012-0841-1

Grandjean, P., & Landrigan, P. J. (2006). Developmental neurotoxicity of industrial chemicals. *The Lancet, 368*(9553), 2167–2178. https://doi.org/10.1016/S0140-6736(06)69665-7

Grandjean, P., & Olsen, J. H. (2004). Extended follow-up of cancer incidence in fluoride-exposed workers. *Journal of the National Cancer Institute, 96*(10), 802–803. https://doi.org/10.1093/jnci/djh155

Gupta, R., Khan, T., Agrawal, D., & Kachhawa, J. (2007). The toxic effects of sodium fluoride on the reproductive system of male rats. *Toxicology and Industrial Health, 23*(9), 507–513. https://doi.org/10.1177/0748233708089041

Gupta, S. K., Gupta, R. C., Seth, A. K., & Chaturvedi, C. S. (1995). Increased incidence of spina bifida occulta in fluorosis prone areas. *Pediatrics International, 37*(4), 503–506. https://doi.org/10.1111/j.1442-200X.1995.tb03363.x

Gupta, S., Seth, A., Gupta, A., & Gavane, A. (1993). Transplacental passage of fluorides. *The Journal of Pediatrics, 123*(1), 139–141. https://doi.org/10.1016/s0022-3476(05)81558-6

Harrison, P. T. (2005). Fluoride in water: A UK perspective. *Journal of Fluorine Chemistry, 126*(11–12), 1448–1456. https://doi.org/10.1016/j.jfluchem.2005.09.009

Health, S. C. O., & Risks, E. (2010). Critical review of any new evidence on the hazard profile, health effects, and human exposure to fluoride and the fluoridating agents of drinking water. In European Commission Brussels, Belgium.

Hem, J. D. (1959). Study and interpretation of the chemical characteristics of natural water.

Jacks, G., Bhattacharya, P., Chaudhary, V., & Singh, K. (2005). Controls on the genesis of some high-fluoride groundwaters in India. *Applied Geochemistry, 20*(2), 221–228. https://doi.org/10.1016/j.apgeochem.2004.07.002

Jenkins, G. (1991). Fluoride intake and its safety among heavy tea drinkers in a British fluoridated city. Proceedings of the Finnish Dental Society. *Suomen Hammaslaakariseuran toimituksia, 87*(4), 571–579.

Juuti, M., & Heinonen, O. P. (1980). Incidence of urolithiasis and composition of household water in southern Finland. *Scandinavian Journal of Urology and Nephrology, 14*(2), 181–187. https://doi.org/10.3109/00365598009179558

Kheradpisheh, Z., Mirzaei, M., Mahvi, A. H., Mokhtari, M., Azizi, R., Fallahzadeh, H., & Ehrampoush, M. H. (2018). Impact of drinking water fluoride on human thyroid hormones: A case-control study. *Scientific Reports, 8*(1), 1–7.

Koblar, A., Tavčar, G., & Ponikvar-Svet, M. (2011). Effects of airborne fluoride on soil and vegetation. *Journal of Fluorine Chemistry, 132*(10), 755–759. https://doi.org/10.1016/j.jfluchem.2011.05.022

Kumar, B., Naaz, A., Shukla, K., Narayan, C., Singh, G., Kumar, A., Ramanathan, A. L., & Anshumali. (2016). Spatial variability of fluorine in agricultural soils around Sidhi District, Central India. *Journal of the Geological Society of India, 87*(2), 227–235. https://doi.org/10.1007/s12594-016-0391-z

Li, Y., Liang, C., Slemenda, C. W., Ji, R., Sun, S., Cao, J., Emsley, C. L., Ma, F., Wu, Y., & Ying, P. (2001). Effect of long-term exposure to fluoride in drinking water on risks of bone fractures. *Journal of Bone and Mineral Research, 16*(5), 932–939. https://doi.org/10.1359/jbmr.2001.16.5.932

Liu, X., Wang, B., & Zheng, B. (2014). Geochemical process of fluorine in soil. *Chinese Journal of Geochemistry, 33*(3), 277–279. https://doi.org/10.1007/s11631-014-0688-9

Madan, J., Puri, J. P., & Singh, J. K. (2009). Growth, feed efficiency and blood profile of buffalo calves consuming high levels of fluoride. *Tropical Animal Health and Production, 41*(3), 295–298. https://doi.org/10.1007/s11250-008-9189-3

Malhotra, A., Tewari, A., Chawla, H., Gauba, K., & Dhall, K. (1993). Placental transfer of fluoride in pregnant women consuming optimum fluoride in drinking water. *Journal of the Indian Society of Pedodontics and Preventive Dentistry, 11*(1), 1–3.

Nabavi, S. M., Habtemariam, S., Nabavi, S. F., Sureda, A., Daglia, M., Moghaddam, A. H., & Amani, M. A. (2013). Protective effect of gallic acid isolated from *Peltiphyllum peltatum* against sodium fluoride-induced oxidative stress in rat's kidney. *Molecular and Cellular Biochemistry, 372*(1), 233–239. https://doi.org/10.1007/s11010-012-1464-y

Nagendra Rao, C. (2003). Fluoride and environment—A review. In *Proceedings of the third international conference on environment and health*, Chennai, India.

Naseem, S., Rafique, T., Bashir, E., Bhanger, M. I., Laghari, A., & Usmani, T. H. (2010). Lithological influences on occurrence of high-fluoride groundwater in Nagar Parkar area, Thar Desert, Pakistan. *Chemosphere, 78*(11), 1313–1321. https://doi.org/10.1016/j.chemosphere.2010.01.010

Nopakun, J., Messer, H. H., & Voller, V. (1989). Fluoride absorption from the gastrointestinal tract of rats. *The Journal of Nutrition, 119*(10), 1411–1417. https://doi.org/10.1093/jn/119.10.1411

Organization, W. H. (2002). *Fluorides: Environmental health criteria 227*. World Health Organization (Published under the joint sponsorship of the United Nations Environment Programme, the International Labour Organization and the World Health Organization, and produced within the framework of the Inter-Organization Programme for the Sound Management of Chemicals).

Ortiz-Perez, D., Rodríguez-Martínez, M., Martínez, F., Borja-Aburto, V. c. H., Castelo, J., Grimaldo, J. I., de la Cruz, E., Carrizales, L., & Díaz-Barriga, F. (2003). Fluoride-induced disruption of reproductive hormones in men. *Environmental Research, 93*(1), 20–30.https://doi.org/10.1016/S0013-9351(03)00059-8

Ozsvath, D. L. (2006). Fluoride concentrations in a crystalline bedrock aquifer Marathon County, Wisconsin. *Environmental Geology, 50*(1), 132–138. https://doi.org/10.1007/s00254-006-0192-6

Parsonage, D., Singh, P., & Nikoloski, A. N. (2014). Adverse effects of fluoride on hydrometallurgical operations. *Mineral Processing and Extractive Metallurgy Review, 35*(1), 44–65. https://doi.org/10.1080/08827508.2012.695306

Paul, E., Gimba, C., Kagbu, J., Ndukwe, G., & Okibe, F. (2011). Spectrometric determination of fluoride in water, soil and vegetables from the precinct of river Basawa, Zaria, Nigeria. *Journal of Basic and Applied Chemistry, 1*(6), 33–38.

Pratusha, N. G., Banji, O., Banji, D., Ragini, M., & Pavani, B. (2011). Fluoride toxicity–A harsh reality. *Int Res J Pharm, 2*, 79–85.

Rahman, M. M. A., & Fetouh, F. A. (2013). Effect of sodium fluoride on the thyroid follicular cells and the amelioration by calcium supplementation in Albino rats: A light and an electron microscopic study. *Journal of American Science, 9*(10), 107–114.

Shulman, J. D., & Wells, L. M. (1997). Acute fluoride toxicity from ingesting home-use dental products in children, birth to 6 years of age. *Journal of Public Health Dentistry, 57*(3), 150–158. https://doi.org/10.1111/j.1752-7325.1997.tb02966.x

Shirin, S., Jamal, A., Emmanouil, C., Singh, V. P., & Yadav, A. K. (2022). Assessment and characterization of waste material used as backfilling in an abandoned mine. *International Journal of Coal Preparation and Utilization*:1–9. https://doi.org/10.1080/19392699.2022.2118259

Shirin, S., Jamal, A., Ranjan, P., & Yadav, A. K. (2019). Study on assessment of slope stability and mixed disposal of overburden in voids of singrauli coalfield. *Environmental Quality Management, 28*, 131–139. https://doi.org/10.1002/tqem.21616

Shirin, S., & Yadav, A. K. (2014). Physico chemical analysis of municipal wastewater discharge in ganga river, Haridwar district of Uttarakhand, India. *Current World Environment, 9*, 536–543. https://doi.org/10.12944/CWE.9.2.39

Shirin, S., & Yadav, A. K. (2021). Assessment and suitability for the dumping of flyash in an abandoned mine. *Environmental Challenges, 4*, 100111. https://doi.org/10.1016/j.envc.2021.100111

SK, S. (2014). Determination of Fluoride status in Ground water of Rajasthan. *International Journal of Pharmaceutical, Chemical & Biological Sciences*, 4(3).

Smith, M. C., Lantz, E. M., & Smith, H. (1931). The cause of mottled enamel. *Science, 74*(1914), 244–244.

Sördell, A., & Lundgren, T. (2017). Fluorlack-effekt på kariesprogression.

Sowers, M., Whitford, G. M., Clark, M. K., & Jannausch, M. L. (2005). Elevated serum fluoride concentrations in women are not related to fractures and bone mineral density. *The Journal of Nutrition, 135*(9), 2247–2252. https://doi.org/10.1093/jn/135.9.2247

Spittle, B. (1994). Psychopharmacology of fluoride: A review. *International Clinical Psychopharmacology, 9*(2), 79–82.

Susheela, A., Bhatnagar, M., & Vig, K. (2005). Excess fluoride ingestion and thyroid hormone derangements in children living in Delhi, India. *Fluoride, 38*(2), 98–108.

Takahashi, K., Akiniwa, K., & Narita, K. (2001). Regression analysis of cancer incidence rates and water fluoride in the USA based on IACR/IARC (WHO) data (1978–1992). *Journal of Epidemiology, 11*(4), 170–179. https://doi.org/10.2188/jea.11.170

Trivedi, M., Verma, R., Patel, R., & Sathawara, N. (2007). Effect of high fluoride water on children's intelligence in India. *Fluoride, 40*(2), 178–183.

Tsiros, J., Haidouti, C., & Chronopoulou, A. (1998). Airborne fluoride contamination of soils and olive trees near an aluminum plant. Measurements and simulations. *Journal of Environmental Science & Health Part A, 33*(7), 1309–1324. https://doi.org/10.1080/10934529809376790

Urbansky, E. T., & Schock, M. R. (2000). Can fluoridation affect lead (II) in potable water? Hexafluorosilicate and fluoride equilibria in aqueous solution. *International Journal of Environmental Studies, 57*(5), 597–637. https://doi.org/10.1080/00207230008711299

Valdez-Jiménez, L., Fregozo, C. S., Beltrán, M. M., Coronado, O. G., & Vega, M. P. (2011). Effects of the fluoride on the central nervous system. *Neurología (English Edition), 26*(5), 297–300. https://doi.org/10.1016/S2173-5808(11)70062-1

Wang, S.-X., Wang, Z.-H., Cheng, X.-T., Li, J., Sang, Z.-P., Zhang, X.-D., Han, L.-L., Qiao, X.-Y., Wu, Z.-M., & Wang, Z.-Q. (2007). Arsenic and fluoride exposure in drinking water: Children's IQ and growth in Shanyin county, Shanxi province, China. *Environmental Health Perspectives, 115*(4), 643–647.

Whitford, G. M. (1992). Acute and chronic fluoride toxicity. *Journal of Dental Research, 71*(5), 1249–1254.

Whitford, G. M., Sampaio, F., Arneberg, P., & Von der Fehr, F. (1999). Fingernail fluoride: A method for monitoring fluoride exposure. *Caries Research, 33*(6), 462–467. https://doi.org/10.1159/000016552

Whiting, P., McDonagh, M., & Kleijnen, J. (2001). Association of Down's syndrome and water fluoride level: A systematic review of the evidence. *BMC Public Health, 1*(1), 6. https://doi.org/10.1186/1471-2458-1-6

Xiang, Q., Liang, Y., Chen, L., Wang, C., Chen, B., Chen, X., Zhou, M., & Shanghai, P. (2003). Effect of fluoride in drinking water on children's intelligence. *Fluoride, 36*(2), 84–94.

Xu, R., Wu, D., & Xu, R. (1997). Relations between environment and endemic fluorosis in Hohhot region, Inner Mongolia. *Fluoride, 30*(1), 26–28.

Yadav, A. K., Sahoo, S. K., Mahapatra, S., Kumar, A. V., Pandey, G., Lenka, P., & Tripathi, R. M. (2014). Concentrations of uranium in drinking water and cumulative, age-dependent radiation doses in four districts of Uttar Pradesh, India. *Toxicological and Environmental Chemistry, 96*, 192–200. https://doi.org/10.1080/02772248.2014.934247

Yang, C.-Y., Cheng, M.-F., Tsai, S.-S., & Hung, C.-F. (2000). Fluoride in drinking water and cancer mortality in Taiwan. *Environmental Research, 82*(3), 189–193. https://doi.org/10.1006/enrs.1999.4018

Chapter 13
Health Concerns Associated with the Increased Fluoride Concentration in Drinking Water: Issues and Perspectives

Rashmi Raghav, Rahul Raj, Kamal Kant Tiwari, and Pankaj Kandwal

Abstract The issue of fluoride exceeding permissible limits in drinking water (DW) and its associated problems has been extensively documented, both within and outside India. The excessive presence of fluoride in DW causes detrimental effects on human health. Challenges related to elevated fluoride levels in DW are particularly prevalent in countries such as India. Fluorosis, an illness caused by the assimilation of fluoride, is a serious health problem. This chapter discusses various health issues that arise due to the increased fluoride concentration in water such as impaired joint mobility, neuropathic effects, decreased thyroid function, lower intelligence levels in children, hindered mental growth, elevated risk of kidney stone risk, and potential carcinogenicity. The leaching of fluoride from minerals in the earth's crust is the primary source of fluoride in groundwater. Many rivers in various regions of India have recorded different concentrations of fluoride, ranging from 0.1 to 12.0 ppm. Several countries, including India, China, Ethiopia, Pakistan, Senegal, Sri Lanka, Algeria, Ghana, Kenya, Ivory Coast, Uganda, Tanzania, Mexico, and Argentina, are among the nations severely affected by fluoride contamination. Various solutions have been proposed in the literature to address these issues. Techniques such as ion exchange/adsorption, and coagulation/precipitation are effective in fluoride removal from water or wastewater. The ion exchange/adsorption processes are effective for

---

R. Raghav (✉) · K. K. Tiwari · P. Kandwal
Department of Chemistry, National Institute of Technology Uttarakhand, Srinagar, Garhwal 246174, India
e-mail: raghavrashmi13@gmail.com

K. K. Tiwari
e-mail: kamalkanttiwari@nituk.ac.in

P. Kandwal
e-mail: pankaj.kandwal@nituk.ac.in

R. Raj
Department of Civil Engineering, National Institute of Technology Uttarakhand, Srinagar, Garhwal 246174, India
e-mail: su.rahulraj@gmail.com

A. K. Yadav et al. (eds.), *Advanced Treatment Technologies for Fluoride Removal in Water*, Water Science and Technology Library 125,
https://doi.org/10.1007/978-3-031-38845-3_13

both concentrated and diluted solutions, ensuring complete removal when applied under appropriate conditions. The selection of the most suitable method for a particular situation should be done carefully, considering various factors. This chapter provides a wide-ranging summary of fluoride, mainly its presence in the atmosphere, sources of origin, impact on human health, and methods for defluoridation.

Keyword Fluoride · Groundwater · Health risk assessment · Fluorosis

Environmental Significance

The influence of fluoride on the well-being of individuals is based on the concentration and prolonged ingestion of fluoride. The acceptable range in DW is typically set to a range of 1.0–1.5 ppm. Numerous techniques have been established to eliminate unnecessary fluoride from water through de-fluoridation approaches. Local governments and communities play a crucial role in raising awareness and educating people about the potential health hazards associated with increased fluoride levels in DW. In addition, it is crucial to implement the appropriate remedial technologies in affected areas. This can include methods such as reverse osmosis, activated alumina, bone char, and ion exchange. These technologies must be evaluated and implemented based on factors such as cost effectiveness, sustainability, and suitability for the specific community. Overall, addressing the issue of surplus fluoride in DW requires a comprehensive and multifaceted approach that involves awareness, education, evaluation of various de-fluoridation techniques, and the implementation of suitable remedial technology for affected areas.

13.1 Introduction

Over the past few years, the potential health risks linked with fluoride's elevated levels in DW have gained considerable attention (Ozsvath, 2008; Yadav et al., 2014). Although fluoride occurs naturally in various regions, human actions, such as industrial processes and the seepage of fluoride into groundwater, can also contribute to contamination. Excessive intake of fluoride can result in various health complications, including skeletal and dental fluorosis, neurological and developmental disorders, and increased susceptibility to certain types of cancer (Yiamouyiannis, 1986). History of water fluoridation is well depicted in Fig. 13.1. Contamination can be mitigated through various methods, including water treatment and the regulation of industrial processes. By 2030, every individual will have access to safe and uncontaminated DW, according to a sustainable development target of the United Nations Development Program. Water pollution is a significant global issue that affects aquatic ecosystems, wildlife, and human health. It can be the result of a variety of contaminants, including organic and inorganic chemicals, heavy metals, pathogens, and nutrients.

Fig. 13.1 History of water fluoridation

These pollutants can come from various sources, including industrial activities, domestic sewage, agricultural practices, and urban runoff. Water pollution can have numerous negative impacts, such as the destruction of aquatic habitats, the spread of disease, and the contamination of food supplies (Aoba, 1997). Fluoride is a major contaminant that can be present in groundwater and surface water due to natural procedures or human actions. Natural sources of fluoride include the dissolution of minerals and rocks, while human activities such as trade production and the use of fluoride in water treatment can also be responsible for contamination (Panda et al., 2015; Shirin & Yadav, 2014). The release in the environment can occur through various manufacturing processes, such as the production of aluminum, steel, glass, semiconductors, and fertilizer, as well as electroplating (Kumar et al., 2021). Excessive consumption of fluoride can have harmful health effects, including skeletal and dental fluorosis. It is crucial to implement measures such as water treatment and the regulation of industrial processes to mitigate fluoride contamination. Mixing fluoride ions in consumable water is an effective approach to prevent dental cavities. However, too much ingestion can lead to dental fluorosis. Dental fluorosis is a condition that occurs due to exposure to fluoride during tooth development. Fluoride-based water contamination is a prevalent concern worldwide, including Asia, America, Africa, and Europe (Rafique et al., 2009). This chapter discusses the problem of fluoride contamination in groundwater and the need for effective defluoridation strategies. The focus is on exploring the diverse adsorbent materials developed in recent years for this purpose, which are environmentally friendly and cost-effective for community-level implementation. Provides an overview of fluoride chemistry, its health impacts, and current defluoridation methods, and assesses the benefits, implications, and future perspectives of these methods.

13.2 Various Sources of Fluoride Contamination in the Environment

Fluoride can be found naturally in some water sources, such as wells, springs, and in some foods, for example, seafood and tea. To prevent tooth decay, fluoride is recurrently added to toothpaste, mouthwash, and public water sources. Additionally, fluoride is used in some industrial processes, and as a result fluoride is distributed into the atmosphere. Figure 13.2 shows Indian states with the greatest number of people affected with fluorosis and worldwide distribution of fluoride in groundwater.

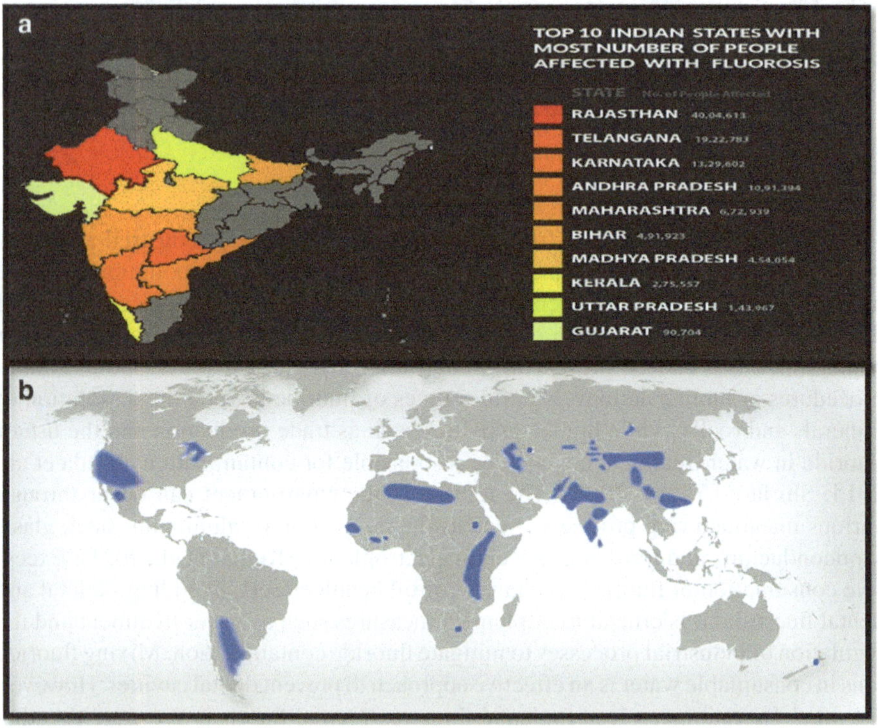

Fig. 13.2 **a** Indian states with the greatest number of people affected with fluorosis, **b** Worldwide distribution of fluoride in groundwater

13.2.1 Fluoride Uptake

Fluoride can be found in numerous sources, and some of the most common sources of fluoride ions include the following:

- **Water**: Fluoride ions are usually mixed with communal water in many countries to prevent tooth deterioration. The amount of fluoride in freshwater can fluctuate depending on the location, but it typically falls within the range of 0.01–0.3 ppm. The amount of fluoride ion in seawater is generally higher, extending from 1.2 to 1.5 ppm. This is because fluoride ions are present in many minerals found in the crust of the earth and these mineral deposits are eroded and weathered and eroded over time, releasing fluoride into the water. Since seawater is constantly replenished by erosion of rocks and minerals, it has a higher concentration of fluoride than freshwater sources (Waghmare et al., 2015).
- **Food**: Fluoride can be found naturally in some foods, such as tea and fish. Fluoride is in fact present in small amounts in many food items, predominantly plant-based foods, namely grains, fruits, and vegetables. Although the amount of fluoride ingested through food and drinks is generally smaller than that ingested through water, it can still contribute to the total fluoride intake. In addition, the body is generally less efficient in absorbing fluoride from food than from water, so the impact of dietary fluoride on overall fluoride exposure is typically lower. It should be noted that some processed foods and beverages, particularly those made with fluoridized water, may contain higher levels of fluoride than unprocessed foods. In addition, certain herbal teas and seafood can be high in fluoride due to their natural fluoride content or their tendency to absorb fluoride from the environment. Overall, while the contribution of fluoride from food to overall intake may be smaller than that from water, it is still worth paying attention to the fluoride content of the foods we eat and making informed choices to minimize our overall fluoride exposure (Susheela et al., 1992).
- **Medicines**: Some medications, such as supplements or certain types of antibiotics, may contain fluoride as an active ingredient.
- **Air**: Fluoride can be found in the air in some industrial areas or near certain sources of pollution.
- **Cosmetics products**: Some cosmetic products may contain fluoride as an ingredient.
- **Humans**: Although the primary route of fluoride absorption is through ingestion, some studies suggest that fluoride can also be absorbed through the skin. Watanabe et al. (1975) reported the absorption of fluoride ions and hydrofluoric acid through the skin membrane in humans and animals. However, the contribution of skin absorption to total fluoride exposure is likely to be minimal compared to the absorption through food ingestion. Fluoride is quickly dispersed throughout the body after ingestion, particularly in the skeletal system and teeth. This is because fluoride has a high affinity for calcium, which is abundant in bones and teeth. More than 99% of the fluoride ingested is deposited in bones and teeth, where it helps protect against tooth decay (Kaminsky et al., 1990). Several sources of fluoride contamination is shown in Fig. 13.3.

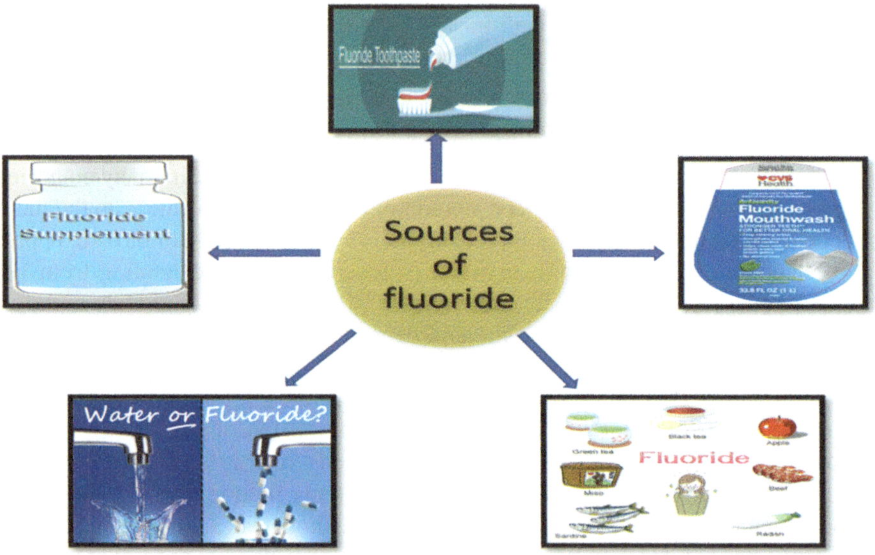

Fig. 13.3 Different sources of fluoride

While fluoride is beneficial for preventing tooth decay, excessive ingestion can cause adverse health effects, mainly skeletal or dental fluorosis. Individuals are recommended to adhere to the safe fluoride consumption guidelines established by their local health authorities.

13.2.2 Contamination in Groundwater

Fluoride contamination of groundwater can be a major global issue. Several factors contribute to groundwater fluoride contamination, including the natural geology of the area, human activities such as mining and agriculture, and improper disposal of industrial waste. Two major factors are responsible for fluoride contamination, that is, geogenic and human-caused or artificial sources (anthropogenic).

13.2.2.1 Geogenic Sources

Naturally, fluoride ions are present in numerous rocks and mineral deposits in the earth's crust, with concentrations varying significantly. The average fluoride concentration in these materials is approximately 625 mg, but it can differ considerably. Water can dissolve some of the fluoride and carry it into groundwater when it meets with fluoride-containing rocks or minerals. Geological processes such as volcanic activity, tectonic activity and weathering can also contribute to fluoride production in groundwater. Geothermal mechanisms can also comprise high fluoride levels due to their contact with rocks and minerals deep within the earth's crust (Vithanage & Bhattacharya, 2015). Fluoride pollution generally occurs due to fluoride leaching from rocks and excessive groundwater use (Sarkar et al., 2016). The amount of fluoride remains high even after the end of volcanic activities in adjacent water and grass for several ages (Araya et al., 1993).

13.2.2.2 Anthropogenic Sources

Fluoride contamination can occur naturally in certain geological formations, but can also be introduced into the environment through anthropogenic (human) activities. Some familiar anthropogenic sources of fluoride contamination include.

- **Industrial Processes**: Several industrial processes can release fluoride into the environment, for example, the production of aluminum, steel, and phosphate fertilizers. These processes often involve the use of fluoride-containing compounds, which can result in fluoride exposure to air and water (Durand & Grattan, 2001).
- **Agricultural practices**: The use of fluoride-containing fertilizers and pesticides can contribute significantly to contamination of soil and groundwater. In addition, the disposal of animal manure from farms can also release fluoride into the environment (Datta et al., 1996).
- **Domestic sources**: Fluoride can be released into the environment from household products such as toothpaste, mouthwash, and cleaning agents. Improper disposal of these products can lead to fluoride contamination of waterways.
- **Mining Activities**: Mining for minerals such as phosphate, coal, and uranium can result in the release of fluoride into the environment. This can occur through the processing of the minerals or the disposal of waste from the mining process.

13.3 Health Effects Due to Fluoride Contamination

The most well-known health effect of fluoride is dental fluorosis, a disorder that causes white or brown stains on the teeth. This condition is generally considered cosmetic and usually does not cause significant health problems. However, in severe cases, dental fluorosis can cause tooth decay and other dental problems. Along with dental fluorosis, extreme fluoride acquaintance remained associated with various other health risks, such as skeletal fluorosis, a condition that causes bones to become brittle and weak. Other potential health effects of excessive exposure to fluoride include neurological problems, such as lower IQ and cognitive impairment, as well as thyroid dysfunction. In addition, young children who ingest large amounts of fluoride toothpaste or mouthwash may also be at increased risk of fluoride toxicity.

Various health issues related to fluoride contamination are as follows.

13.3.1 Issues Related to the Dental and Skeletal System

Tooth decay is a prevalent health problem that affects children around the world. It occurs primarily when the acids produced by oral bacteria lead to the demineralization of the tooth enamel. Scientific research has established that fluoride contributes to the prevention of tooth decay through various mechanisms. It promotes the remineralization process of tooth enamel, inhibits the proliferation of acid-producing oral bacteria, and improves the resistance of tooth enamel to acid erosion. However, excessive exposure to fluoride ions during tooth development can lead to a condition called enamel fluorosis. This condition manifests itself as white or brown stains on the enamel surface, as well as potential pitting. Therefore, it is essential to ensure that children receive the optimal amount of fluoride during their stage of tooth development to prevent dental caries without causing enamel fluorosis. The optimal amount of fluoride ion revelation depends on the individual's age, the risk of dental caries, and other factors. Exposure to fluoride ions (F^-) during the mineralization of teeth can cause a specific disorder characterized by the growth of large gaps in the crystal-like structure, as well as problems with enamel mineralization, extreme retention of enamel proteins, and an upsurge in permeability. This type of disorder is known as a dose-related disorder, meaning that the brutality of the disorder can be associated with the amount of fluoride revelation (Aoba, 1997). Recent findings have suggested a possible link between moderate to severe dental fluorosis and an increased risk of developing esophageal cancer (Menya et al., 2019). Excess fluoride exposure can also result in skeletal fluorosis, identified by symptoms such as joint pain, stiffness, and bone and ligament damage. This condition occurs due to prolonged ingestion of high levels of fluoride, leading to its accumulation in the skeletal system. Although the precise mechanisms underlying the development of skeletal fluorosis are not fully understood, researchers have identified distinct stages associated with its progression (Susheela, 2003; Rao, 2003; Hileman, 1988, etc.). In the initial stage of skeletal fluorosis, fluoride ions are deposited in the bone matrix, increasing bone density and stiffness. As fluoride continues to accumulate, the middle phase is characterized by the formation of bone spurs and osteosclerosis or abnormal hardening of the bones. In the later stages, the bone becomes increasingly brittle and prone to fracture, leading to deformities and chronic pain (Susheela, 2003; Hileman, 1988). The condition known as "poker back" occurs when the entire spine becomes rigid and uncoils into a single column of bone. As this condition worsens, several spine ligaments may start to solidify. In its later stages, fluorosis results in spinal cord compression, crippling irregularities of the vertebral column and major joints, paralysis, neurological problems, muscle atrophy, and paralysis. The exact mechanisms by which fluoride affects bone metabolism are still under investigation, but some research suggests that it may interfere with the activity of osteoblasts, the cells that build and repair bone tissue. Other studies have found that fluoride may also affect the balance of minerals in bone, leading to the formation of abnormal bone tissue. In terms of bone health, fluoride ion can increase bone density and reduce the risk of ruptures (Rao, 2003). This is why fluoride is sometimes used in the treatment of osteoporosis.

13.3.2 Issues Related to the Neural System

Fluoride is considered a neurotoxic substance that can harm developing cognitive abilities. Higher fluoride concentrations exceeding 1 mg/L are associated with an elevated risk of neurotoxicity, leading to cognitive and behavioral impairments (Grandjean & Landrigan, 2006; Choi et al., 2012). The developing brain is particularly vulnerable to the impacts of fluoride, and early exposure can cause long-lasting damage (Grandjean & Landrigan, 2006). Research carried out in China has shown that fluoride ingestion negatively affects children's intelligence quotient (IQ) levels and cognitive abilities (Lu et al., 2000). Research conducted between 1988 and 2008 in various regions of China has linked fluoride concentration to IQ levels, revealing that children who resided in areas with higher fluoride concentrations exhibited lower IQ scores than those with lower fluoride levels (Choi et al., 2012). Excess fluoride concentration can increase lipid peroxidation levels and interfere with essential neuronal enzymes, potentially leading to adverse effects (Shivarajashankara et al., 2002). The direct effects of fluoride on myelin, neurons, and neurotransmitters suggest that it can directly affect brain function (Cheng et al., 2002). Whenever the fluoride concentration exceeds 10 mg/L, it exceeds the safe limit and can contribute to other health problems, such as hypertension and certain types of cancer (Neisi et al., 2018).

13.3.3 Issues Related to the Reproductive System

Some of the consequences of fluoride exposure on the reproductive system are listed below.

- **Reduced fertility**: High levels of fluoride ion revelation have been associated with reduced fertility in both men and women. In men, fluoride exposure has been shown to reduce sperm count, motility, and morphology, which can lead to infertility. In women, high fluoride levels have been associated with changes in the menstrual cycle and reduced fertility (Pati & Bhunya, 2014).
- **Increased risk of miscarriage**: Multiple findings have specified a possible association between exposure to elevated fluoride levels, resulting in a higher risk of miscarriage. In a specific study, women exposed to high levels of fluoride through DW showed a significantly higher risk of miscarriage than those with lower levels of exposure (Solanki et al., 2022).

- **Altered hormone levels**: Fluoride exposure has been shown to alter the levels of certain hormones in the body, including thyroid hormones, which can have an undesirable influence on the reproductive system (Solanki et al., 2022).
- **Birth defects**: Several studies have reported a link between a high intake of fluoride during pregnancy and an increased risk of birth defects in newborns. However, more research is required to establish and confirm this relationship (Guth et al., 2020).
- **Male reproductive system**: The disruption of spermatogenesis, alterations in the structure and functional capabilities of gametes, and instability of several hormone systems are some of the most significant impacts of high fluoride ion consumption on the male reproductive system (Pati & Bhunya, 2014).

13.3.4 Issues Related to Lung Disease

Fluoride exposure is typically associated with dental health and the prevention of cavities. However, excessive exposure to fluoride, particularly from industrial sources, can have negative effects on the lungs. Acute fluoride poisoning, which can occur from accidental ingestion or inhalation of a large amount of fluoride, can be responsible for respiratory failure and lead to death. Inhaling a high amount of fluoride can cause an impairment of the respirational zone, leading to lung damage and inflammation. Workers who are occupationally exposed to fluoride, especially in industries such as aluminum smelting, have been found to have an increased risk of emerging long-lasting bronchitis and chronic obstructive lung disease (COPD) (Soyseth & Kongerud, 1992). This association may be attributed to the inhalation of dust and fumes.

13.3.5 Issues Related to the Immune System

The influence of fluoride on the immune system remains a topic of ongoing research and discussion. Various studies have indicated that high fluoride has adverse effects on the immune system, potentially increasing susceptibility to infections and other health problems. For example, Wang and Bian (1988) explained that high exposure to fluoride ion concentration reduces white blood cell activity, which plays a crucial role in combating infections.

13.3.6 Issues Related to the Developmental System

Fluoride ions have been found to easily pass through the placenta and enter the fetal circulation. Research has indicated a direct connection between the fluoride ion concentration in maternal blood plasma, fetal blood plasma and umbilical cord samples. These findings suggest that fluoride can traverse the placenta and reach the developing fetus (Malhotra et al., 1993). Some studies have suggested that high exposure to fluoride during fetal development could have detrimental effects on neurodevelopment and cognitive function (Collins et al., 1995; Takahashi, 1998; Whiting et al., 2001).

13.3.7 Issues Related to the Renal System

The unnecessary ingestion of fluoride ions can potentially cause antagonistic effects on the kidneys. The kidneys help filter and remove excess fluoride ions from the body. When the body's fluoride ion levels exceed the excretion capacity, there is a potential for fluoride accumulation, leading to potential damage and damage. Acute fluoride toxicity can cause acute kidney injury and renal failure, especially in individuals with pre-existing kidney disease or in those who ingest large amounts of fluoride at once (Malin et al., 2019). However, acute fluoride toxicity is relatively rare and usually occurs only in accidental or intentional ingestion of large amounts of fluoride-containing products, such as toothpaste, mouthwash, or insecticides.

13.3.8 Issues Related to the Endocrine System

Fluoride can have endocrine effects, particularly in the thyroid gland. High levels of fluoride intake can disrupt the synthesis and secretion of thyroid hormones, leading to thyroid dysfunction (Jooste et al., 1999; Shirin et al., 2019). The following are some of the ways fluoride can affect the endocrine system.

- **Thyroid gland**: Fluoride can accumulate within the thyroid gland and alter the uptake and utilization of iodine. Iodine is a crucial element necessary for the synthesis of thyroid hormones. Therefore, fluoride interference can potentially affect the proper functioning of the thyroid gland. This can be responsible for a condition known as hypothyroidism; there is an underproduction of thyroid hormones.
- **Parathyroid gland**: Fluoride can also affect the parathyroid gland, which regulates calcium levels in the body. High levels of fluoride can cause hyperparathyroidism, which can lead to weakened bones, kidney stones, and other health problems (Collins et al., 1995).
- **Pineal gland**: This gland is responsible for the formation of melatonin, a hormone that regulates the sleep–wake cycles. Fluoride ions accumulate in the pineal gland and disrupt its function, which can affect sleep patterns and circadian rhythms.
- **Pancreas**: Some studies suggest that fluoride can affect insulin secretion and glucose metabolism, increasing the risk of emerging type 2 diabetes (Fig. 13.4).

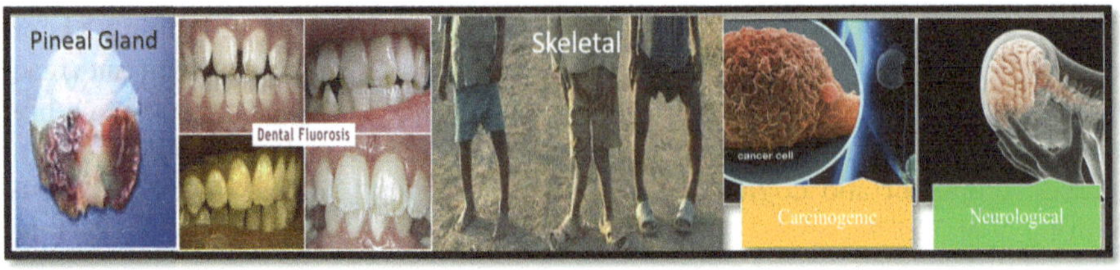

Fig. 13.4 Shows effect of fluoride on different organs

13.3.9 Issues Related to the Gastrointestinal System

Fluoride exposure can have gastrointestinal effects, particularly if ingested in large amounts. Acute fluoride poisoning can cause nausea, vomiting, abdominal pain, and diarrhea (Sidhu & Kimmer, 2002). These symptoms can occur within minutes to hours of exposure and can last for several days. Long-lasting revelation that high levels of fluoride ion can be responsible for gastrointestinal problems, such as gastrointestinal irritation, inflammation, and ulceration (Dasarathy et al., 1996). Gastrointestinal effects based on fluoride exposure can vary with dose amount, duration, and frequency of exposure. People who work with high fluoride exposure or who live in areas where the water or soil has high fluoride concentrations may be more likely to develop gastrointestinal problems.

13.3.10 Issues Related to Carcinogenicity

Although fluoride is generally considered safe when used appropriately, excessive exposure to fluoride can have harmful effects, including potential carcinogenic effects. The International Agency for Research on Cancer (IARC) has classified fluoride as "possibly carcinogenic to humans" on evidence derived from animal and human studies. Specifically, the IARC investigation found a connection among fluoride revelation and an elevated risk of bone cancer in male rats. Furthermore, certain epidemiological studies have indicated an increased risk of osteosarcoma (a type of bone cancer) in young men exposed to high levels of fluoride (Bassin et al., 2006; IARC, 1982).

13.3.11 Issues Related to Hair and Fingernails

Excess fluoride intake may cause changes in the structure of hair and nails, resulting in brittleness and discoloration. A common method to determine the body's total fluoride load is to measure the amount of fluoride in hair and nails (Parimi et al., 2013).

Although fingernails are short-term indicators of exposure that occurred during the last three to six months, damaged fingernails inherently signify high fluoride concentrations (Idowu et al., 2021).

13.3.12 Issues Related to Urine

High levels of fluoride in the body can lead to increased urine volume, as the body attempts to eliminate excess fluoride. Furthermore, fluoride can increase the concentration of certain substances in the urine, such as calcium and magnesium. The renal defecation process of fluoride has remained well documented (Jarnberg et al., 1981).

13.3.13 Issues Related to Saliva

Saliva can be a valuable biomarker for assessing fluoride exposure. The fluoride concentration in saliva is about 75% that of plasma, making it a reliable indicator of the overall fluoride levels. Individuals, unexposed to excessive amounts of fluoride, the typical range of salivary fluoride values is 0.01–0.06 mg L^{-1}. However, after ingesting fluoride, the concentration in saliva can increase rapidly, reaching around 15 times its typical value in just 15 min. This increase is due to the rapid transfer of fluoride from the bloodstream to saliva. In general, measuring fluoride concentration in saliva can provide a fast and non-invasive way to assess fluoride exposure in humans (Toth et al., 2005).

13.3.14 Genetic Erythrocyte Disorder

A study was carried out to assess long-term acquaintance with fluoride and its associated health risks in tribal communities residing in different parts of the scheduled areas of Rajasthan, India. This study revealed that the tribal population in Rajasthan is affected by various blood-related genetic disorders such as β-thalassemia, glucose-6-phosphate dehydrogenase (G-6-PD) enzyme deficiency, and sickle cell anaemia (Choubisa & Choubisa, 2021).

13.4 Fluoride Mitigation

Fluoride mitigation refers to the process of reducing or eliminating excessive amounts of fluoride ions from water or other sources to prevent harmful health effects. An unnecessary amount of fluoride ion can be responsible for dental fluorosis, a disorder that causes white or brown spots on the surface of teeth, as well as skeletal fluorosis, which affects bones and joints and can cause pain and stiffness. There are several techniques that are used to eliminate fluoride ions from groundwater, including.

1. **Reverse Osmosis**: It is a water treatment process that involves passing water through a semi-permeable membrane. This membrane has pores that are small enough to eliminate fluoride ions from the water but allow water molecules to pass through. Reverse osmosis is one of the most effective procedures for the elimination of fluoride (Dubey et al., 2021).
2. **Activated Alumina**: It is a permeable material that can remove fluoride from water by a process called adsorption. Fluoride ions are attracted to and adhere to the surface of activated alumina, where they become trapped (Waghmare et al., 2015).
3. **Ion Exchange**: In this technique, water is passed through a resin bed containing ion-exchange resin beads. The resin beads are functionalized with NR_3X^- groups ($X^- = OH^-$, Cl etc.) where X^- is exchanged with fluoride ion (Ojekunle et al., 2016).
4. **Distillation**: Distillation is a process in which water is boiled, and the steam is collected and condensed into pure water. Fluoride is removed along with other impurities in the water.
5. **Electrocoagulation**: Electrocoagulation is a process that uses an electrical current to remove fluoride ions from water. Electrical current causes metal ions to be released into the water, which react with fluoride ions and precipitate out that can be easily removed (Mollah et al., 2001). Various advantages and disadvantages of several procedures that can be adopted for the elimination of fluoride content from water are shown in Table 13.1.

Table 13.1 Listed the advantages and disadvantages of several procedures that can be adopted for the elimination of fluoride content from water

Methods	Advantages	Disadvantages
Ion exchange	Remove up to 90–95% fluoride	Limited loading capacity
Reverse osmosis	Easy and cheaper	Generation of lethal slush, biofouling
Electrodialysis	No chemical requirement	Power consumption is high
Electrochemical	Efficiency is good	High-cost factor, power consumption is high
Precipitation and coagulation	Widely used	Non-selective
Adsorption	High efficiency	pH dependent, ionic competition

13.5 Conclusion

In conclusion, contamination of aquatic medium by fluoride ions is a serious problem that requires urgent attention to protect human health. It is imperative to implement a comprehensive approach that includes water de-fluoridation, reducing fluoride ingestion through foods and consumable foods, and nutritional supplements. Increasing awareness through the media and social media can help prevent and control fluorosis. Ultimately, prevention is better than treatment, and a holistic approach is necessary to address this problem. Further research should adopt a comprehensive and comprehensive method to investigate fluoride concentrations and their impact on human health. It is time to act and protect our communities from fluoride contamination.

Acknowledgements The authors thank NIT Uttarakhand for supporting this work.

References

Aoba, T. (1997). The effect of fluoride on apatite structure and growth. *Critical Reviews in Oral Biology and Medicine, 8*, 136–153.

Araya, O., Wittwer, F., & Villa, A. (1993). Evolution of fluoride concentrations in cattle and grass following a volcanic eruption. *Veterinary and Human Toxicology, 35*, 437–440.

Bassin, E., Wypij, D., Davis, R., & Mittleman, M. (2006). Age-specific fluoride exposure in drinking water and osteosarcoma (United States). *Cancer Causes and Control, 17*, 421–428.

Bulusu, K. R., Sundaresan, B. B., Pathak, B. N., Nawlakhe, W. G., Kulkarni, D. N., & Thergaonkar, V. P. (1979). A study on removal of fluorides from drinking water by adsorption onto low-cost materials. *IE(I) Jl., 60*, 1–25.

Cheng, X. T., Wang, S. X., & Gao, J. G. (2002). Arsenic and fluoride exposure in drinking water: Children's IQ and growth in Shanyin County, Shanxi Province, China. *Chinese Journal of Endemiology, 2*, 357–358.

Choi, A. L., Sun, G., Zhang, Y., & Grandjean, P. (2012). Association of lifetime exposure to fluoride and cognitive functions in Chinese children: A pilot study. *Environmental Health Perspectives, 120*, 1362–1368.

Choubisia, S. L., & Sompura, K. (1996). Dental fluorosis in tribal villages of Dungarpur district (Rajasthan). *Pollution Research, 12*(1), 45–47.

Choubisa, S. L., & Choubisa, A. (2021). A brief review of ideal bio-indicators, bio-markers and determinants of endemic of fluoride and fluorosis. *Journal ISSN, 2766*, 2276.

Collins, T. F., Sprando, R. L., Shackelford, M. E., Black, T. N., Ames, M. J., Welsh, J. J., Balmer, M. F., Olejnik, N., & Ruggles, D. I. (1995). Developmental toxicity of sodium fluoride in rats. *Food and Chemical Toxicology, 33*, 951–960.

Dasarathy, S., Das, T. K., Gupta, I. P., Susheela, A. K., & Tandon, R. K. (1996). Gastroduodenal manifestations in patients with skeletal fluorosis. *Journal of Gastroenterology, 31*, 333–337.

Datta, P. S., Deb, D. L., & Tyagi, S. K. (1996). Assessment of groundwater contamination from fertilizers in the Delhi area based on 18O, NO_3/− and K+ composition. *Journal of Contaminant Hydrology, 24*, 85–96.

Doull, J., Boekelheide, K., Farishian, B. G., Isaacson, R. L., Klotz, J. B., & Kumar, J. V. (2006). *A standard defluoridation capacity test* (p. 530). National Academies Press.

Dubey, S., Agarwal, M., & Gupta, A. B. (2021). Fluoride removal using Alum & PACl in batch & continuous mode with subsequent microfiltration. *Membrane Water Treatment, 12*, 83–93.

Durand, M., & Grattan, J. (2001). Effects of volcanic air pollution on health. *Lancet, 357*, 164.
Edmunds, W. M., & Smedley, P. L. (2005). *Fluoride in natural waters* (pp. 301–329). Elsevier Academic Press.
Faraji, H., Mohammadi, A. A., Adergani, B. A., Saatloo, N. V., Lashkarboloki, G., & Mahvi, A. H. (2014). Distribution of fluoride contamination in drinking water resources and health risk assessment using geographic information system, northwest Iran. *Iranian Journal of Public Health, 43*, 1664–1668.
Grandjean, P., & Landrigan, P. J. (2006). Developmental neurotoxicity of industrial chemicals. *Lancet, 368*, 2167–2178.
Guth, S., Huser, S., Roth, A., Degen, G., Edlund, P. K., Eisenbrand, G., Engel, K. H., Epe, B., Grune, T., Heinz, V., Henle, T., Humpf, H. U., Jager, H., Joost, H. G., Kulling, S. E., Lampen, A., Mally, A., Marchan, R., Marko, D., … Hengstler, J. G. (2020). Toxicity of fluoride: Critical evaluation of evidence for human developmental neurotoxicity in epidemiological studies, animal experiments and in vitro analyses. *Archives of Toxicology, 94*(5), 1375–1415.
Harrison, P. T. C. (2005). Determination of total fluoride content in teas by using fluoride ion-selective electrode. *Journal of Fluorine Chemistry, 126*(11–12), 1448–1456.
Hileman, B. (1988). Mineralogy and human welfare: Applications in geology, petrology. *Chemical & Engineering News, 1*, 26–42.
Idowu, O. S., Duckworth, R. M., Valentine, R. A., & Zohoori, F. V. (2021). Biomarkers for the assessment of exposure to fluoride in children. *Caries Research, 55*, 292–300.
International Agency for Research on Cancer 1982, IARC monographs on the evaluation of carcinogenic risks of chemicals to humans, Volume 27.
Jarnberg, P. O., Ekstrand, J., & Irestedt, L. (1981). Renal fluoride excretion and plasma fluoride levels during and after enflurane anesthesia are dependent on urinary pH. *Anaesthesiology, 54*, 48–52.
Jolly, S. S., Singh, L. D., Prasad, S., Sharma, R., Singh, B. M., & Mathur, O. C. (1969). Research priorities in maternal, newborn, & child health & nutrition for India: An Indian Council of Medical Research-INCLEN Initiative. *Indian Journal of Medical Research, 57*, 1333–1346.
Jooste, P. L., Weight, M. J., Kriek, J. A., & Louw, A. J. (1999). Endemic goitre in the absence of iodine deficiency in schoolchildren of the Northern Cape Province of South Africa. *European Journal of Clinical Nutrition, 53*, 8–12.
Kaminsky, L. S., Mahony, M. C., Leach, J., Melius, J., & Miller, J. M. (1990). Fluoride: Benefits and risks of exposure. *Critical Reviews in Oral Biology and Medicine, 1*, 261–281.
Keshavarzi, B., Moore, F., Esmaeili, A., & Rastmanesh, F. (2010). The source of fluoride toxicity in Muteh area, Isfahan, Iran. *Environmental Earth Science, 61*, 777–786.
Kumar, R., Sinha, R., Sharma, P. K., Ivy, N., Kumar, P., Kant, N., … & Prasad P. V. (2021). Bioaccumulation of fluoride in plants and its microbially assisted remediation: A review of biological processes and technological performance. *Processes, 9*(12), 2154.
Lu, Y., Sun, Z. R., & Wu, L. N. (2000). Effect of high-fluoride water on intelligence in children. *Fluoride, 33*, 74–78.
Malhotra, A., Tewari, A., Chawla, H. S., Gauba, K., & Dhall, K. (1993). Placental transfer of fluoride in pregnant women consuming optimum fluoride in drinking water. *Journal of the Indian Society of Pedodontics and Preventive Dentistry, 11*, 1–3.
Malin, A. J., Lesseur, C., Busgang, S. A., Curtin, P., Wright, R. O., & Sanders, A. P. (2019). Fluoride exposure and kidney and liver function among adolescents in the United States: NHANES, 2013–2016. *Environment International, 132*, 105012.
Menya, D., Maina, S. K., Kibosia, C., Kigen, N., Oduor, M., & Some, F. (2019). Oral health and gastrointestinal neoplasia: Narrative review. *International Journal of Cancer, 145*(1), 99–109.
Mollah, M. Y. A., Schennach, R., Parga, J. R., & Cocke, D. L. (2001). Electrocoagulation (EC)–science and applications. *Journal of Hazardous Materials, 84*, 29–41.
Mullenix, P. J., Denbesten, P. K., & Schunior, A. (1995). Neurotoxicity of sodium fluoride in rats. *Neurotoxicology and Teratology, 17*, 169–177.

Neisi, A., Mirzabeygi, M., Zeyduni, G., Hamzezadeh, A., Jalili, D., Abbasnia, A., Youse, M., & Khodadadi, R. (2018). Data on fluoride concentration levels in cold and warm season in City area of Sistan and Baluchistan Province, Iran. *Data Brief, 18*, 713–718.

Ojekunle, O. Z., Ojekunle, O. V., Adeyemi, A. A., Taiwo, A. G., Sangowusi, O. R., Taiwo, A. M., & Adekitan, A. A. (2016). Evaluation of surface water quality indices and ecological risk assessment for heavy metals in scrap yard neighbourhood. *Springerplus, 5*, 560.

Ozsvath, D. L. (2008). Fluoride and environmental health: A review. *Environmental Science and Biotechnology, 8*, 59–79.

Panda, L., Kar, B. B., & Patra, B. B. (2015). Exploring the role of excess fluoride in chronic kidney disease: A review. *IOSR Journal of Electronics and Communication Engineering (IOSR-JECE), 10*, 79–91.

Parimi, N., Viswanath, V., Kashyap, B., & Patil, P. U. (2013). Hair as biomarker of fluoride exposure in a fluoride endemic area and a low fluoridated area. *International Journal of Trichology, 5*, 148–150.

Pati, P. C., & Bhunya, S. P. (2014). Genotoxic effect of an environmental pollutant, sodium fluoride, in Mammalian in vivo test system. *Caryologia, 40*, 79–87.

Poureslami, H., Khazaeli, P., Mahvi, A. H., Poureslami, K., Poureslami, P., Haghani, J., & Aghael, M. (2016). Changes in the concentration of ions in Saliva and dental plaque after application of CPP-ACP with and without fluoride among 6–9-year-old children. *Fluoride, 49*, 485–494.

Prystupa, J. (2011). Fluorine—A current literature review. An NRC and ATSDR based review of safety standards for exposure to fluorine and fluorides. *Toxicology Mechanisms and Methods, 21*, 103–170.

Rafique, T., Naseem, S., & Usmani, T. H. (2009). Geochemical controls of high fluoride groundwater in Umarkot Sub-District, Thar Desert, Pakistan. *J. Hazardous Mate., 171*(1), 424–430.

Rao, N. C. R. (2003). Fluoride and environment—A review. In *Proceedings of third International Conference on Environment and Health*, York University, Chennai, India.

Reddy, C. K. V. C., Golla, V., Badapalli, P. K., & Reddy, N. B. Y. (2022). Evaluation of groundwater contamination for fluoride and nitrate in Nellore Urban Province, Southern India: A special emphasis on human health risk assessment (HHRA). *Applied Water Science, 12*, 32.

Sarkar A., Ali, S., Kumar, S., Shekhar, S., Rao, S. V. N., Pandey, V. P., & Thatikonda, S. (2016). Groundwater system of National Capital Region Delhi, India. *Groundwater of South Asia, 5*, 77–108.

Shirin, S., Jamal, A., Ranjan, P., & Yadav, A. K. (2019). Study on assessment of slope stability and mixed disposal of overburden in voids of singrauli coalfield. *Environmental Quality Management, 28*, 131–139. https://doi.org/10.1002/tqem.21616

Shirin, S., & Yadav, A. K. (2014). Physico chemical analysis of municipal wastewater discharge in ganga river, Haridwar district of Uttarakhand, India. *Current World Environment, 9*, 536–543. https://doi.org/10.12944/CWE.9.2.39

Shivarajashankara, Y. M., Shivashankara, A. R., Bhat, P. G., & Rao, S. H. (2002). Lipid peroxidation and antioxidant defense systems in liver of rats in chronic fluoride toxicity. *Fluoride, 35*, 197–203.

Shivarajashankara, Y. M., & Shivashankara, A. R. (2012). Lipid peroxidation and antioxidant defense systems in liver of rats in chronic fluoride toxicity. *Journal of Clinical and Diagnostic Research, 6*, 740–744.

Sidhu, K. S., & Kimmer, R. O. (2002). Fluoride overfeeds at a well site near an elementary school in Michigan. *Journal of Environmental Health, 65*, 16–21.

Solanki, Y. S., Agarwal, M., Gupta, A. B., Gupta, S., & Shukla, P. (2022). Fluoride occurrences, health problems, detection, and remediation methods for drinking water: A comprehensive review. *Science of the Total Environment, 10*, 807.

Soyseth, V., & Kongerud, J. (1992). Prevalence of respiratory disorders among aluminium potroom workers in relation to exposure to fluoride. *British Journal of Industrial Medicine, 49*, 125–130.

Susheela, A. K. (2003). A treatise on fluorosis—Revised 2nd edition 2003. Delhi, India, 137.

Susheela, A. K., Das, K. T., Gupta, I. P., Tandon, R. K., Kacker, S. K., Ghosh, P., & Deka, R. C. (1992). Fluoride ingestion and its correlation with gastrointestinal discomfort. *Indian Academy of Science, 25*, 5–22.

Susheela, A. K., & Jethanandani, P. (1996). Circulating testosterone levels in skeletal fluorosis patients. *Journal of Toxicology Din Toxicol, 34*, 183–189.

Takahashi, K. (1998). Fluoride-linked down syndrome births and their estimated occurrence due to water fluoridation. *Fluoride, 31*, 61–73.

Toth, Z., Gintner, Z., & Banoczy, J. (2005). The effect of ingested fluoride administered in salt, milk, and tablets on salivary and urinary fluoride concentrations. *Fluoride, 38*, 199–204.

Vithanage, M., & Bhattacharya, P. (2015). Fluoride in the environment: Sources, distribution and defluoridation. *Environmental Chemistry Letters, 13*, 131–147.

Waghmare, S. S., & Arfin, T. (2015). Fluoride removal from water by aluminium based adsorption: A review. *Journal of Biological and Chemical Chronicles, 1*, 01–11.

Waghmare, S., Aran, T., Manwar, N., Lataye, D., Labhsetwar, N., & Rayalu, S. (2015). Fluoride removal from water by aluminium based adsorption: A review. *Asian Journal of Advanced Basic Science, 4*, 12–24.

Wang, J. X., & Bian, Y. M. (1988). Microplastic pollution in the surface waters of the Bohai Sea, China. *Environmental Pollution, 52*, 11–18.

Watanabe, M., Yoshida, Y., Shimada, M., & Kurimoto, K. (1975). Effect of hydrofluoric acid on glucose metabolism of the mouse studied by whole-body autoradiography. *British Journal of Industrial Medicine, 32*, 316–320.

Whiting, P., MacDonaghm, M., & Kleijnen, J. (2001). Association of down's syndrome and water fluoride level: A systematic review of the evidence. *BMC Public Health, 1*, 1–8.

Yadav, A. K., Sahoo, S. K., Mahapatra, S., Kumar, A. V., Pandey, G., Lenka, P., & Tripathi, R. M. (2014). Concentrations of uranium in drinking water and cumulative, age-dependent radiation doses in four districts of Uttar Pradesh, India. *Toxicological and Environmental Chemistry, 96*, 192–200. https://doi.org/10.1080/02772248.2014.934247

Yiamouyiannis, J. (1986). *Fluoride, the aging factor: How to recognize and avoid the devastating effects of fluoride* (3rd ed.). Health Action Press.

Chapter 14
Human Nutritional Condition and Dental Fluorosis in Populations with Varying Concentrations of Fluoride in Their Water Sources

Shruti Chaudhari, Himakshi Parmar, and Prakash Samnani

Abstract Fluoride is largely consumed by people all around the world through drinking water. Fluorosis is caused by largely through high level of fluoride in consumable water, lowering the quality of human life. This becomes crucial considering that fluoride intake primarily comes from water. According to estimates, bones and teeth hold 99% of the fluoride that is absorbed by humans. Children retain 80% of the fluoride they have received, compared to adults who retain only 50%. The typical dietary fluoride in adults ranges from 0.020 to 0.048 mg/kg (living in places with water fluoride values of 1.0 mg/L). At a water fluoridation level of 1 mg/L, the prevalence of dental fluorosis has remained calculated to be 48% in fluoridated areas and 15% in non-fluoridated parts. Due to its global proliferation into further than 40 countries, mainly in mid-latitude zones, fluorosis has turned out to be endemic in numerous areas of the world. Typically, fluorosis in tooth tissues is considered a more severe form as a result of the increased intake of fluoride through water. Additionally, consuming vegetables and cereals full-grown in fluoridated zones increase the incidence of fluorosis in kids among the ages of 3 and 14 years. Established on different sources of water consumption between different communities, nutritional deficiency can be seen that leads to dental fluorosis in different concentration can be seen. The community fluoride index is used to measure the load of dental fluorosis in designated population. Nutritional status in children can be improved through preventive programs that are targeted at subpopulations that live in regions with high

S. Chaudhari · H. Parmar
Department of Environmental Studies, Faculty of Science, The Maharaja Sayajirao University of Baroda, Vadodara 390002, India
e-mail: shruti2chaudhari@gmail.com

H. Parmar
e-mail: parmarhimakshi@gmail.com

P. Samnani (✉)
Department of Chemistry, Faculty of Science, The Maharaja Sayajirao University of Baroda, Vadodara 390002, India
e-mail: prakash.samnani-chem@msubaroda.ac.in

A. K. Yadav et al. (eds.), *Advanced Treatment Technologies for Fluoride Removal in Water*, Water Science and Technology Library 125,
https://doi.org/10.1007/978-3-031-38845-3_14

concentrations of fluoride in water. These programs are designed to prevent children from consuming an excessive amount of fluoride. Policymakers in the field of public health could target these subpopulations.

Keywords Fluoride · Dental fluorosis · Nutritional deficiency · Community fluoride index · Fluoride management

14.1 Introduction

Many populations in developing countries are plagued by serious health problems due to fundamental variables such as inadequate diet and poor water quality. During the last few decades, tremendous pressure has been exerted on water quality, particularly in developing countries, as a result of rapid population growth and the advent of industrialisation.

Because of its high reactivity and electronegative characteristics, fluorine (F) is almost never found in nature in its basic form. Fluorides are created when it combines with all other elements-besides oxygen and the noble gases. Commercially important fluorine-containing products are chlorofluorocarbons (freons) used as refrigerants, polytetrafluoroethylene (Teflon) and fluoride-containing toothpastes, among many others. Chlorofluorocarbons have unique properties: chemical and thermal stability. Another important application area of fluorine-containing organic compounds is pharmaceuticals and agrochemicals.

About 30% of the pharmaceutical drugs available in the market at present are organic molecules containing one or more fluorine atoms. Although fluorine does not have a large biochemical role, its applications are increasing, contributing to fluorine pollution, especially affecting water.

Due to the widespread presence in the crust of the Earth, all water supplies comprise a significant amount of fluoride, although at varying levels. Concentrations of 1.2–1.4 mg/L can be found in seawater (Moss & Kumar, 2021), up to 67 mg/L in subsurface water, and less than 0.1 mg/L in most cases in surface waters. Hence, whether we like it or not, all of our water comes from fluoride-rich sources.

In the structures of several natural resources, containing fluorspar, rock phosphate, cryolite, apatite, mica, and hornblende, there are significant levels of fluoride. Fluorite, which has the chemical formula CaF_2, is a fluoride mineral that can be found in both igneous and sedimentary rocks. These sources are linked to geological and geochemical processes in a region. Human activities contribute significantly to atmospheric fluoride levels.

It has traditionally been believed that consumers receive certain health benefits from fluoride (F) in drinking water, such as a reduced incidence of dental cavities; however, taking too much of this anion or consuming it at high concentrations (more than 1.5 mg/L) can result in dental and skeletal fluorosis (Kumar et al., 2018). Furthermore, recent data have indicated that the use of fluoridated water may not actually be as beneficial as consumption of defluorinated water. Very few developed countries fluoridate community waters, except the United States of America (Water Fluoridation Status in OECD Nations, 2012).

Fluoridated water, meals and beverages, toothpastes, and other dental products contain sufficient levels of fluoride for human consumption. It should be understood, however, that fluoride has only topical effect, that is, it works on the surface of the teeth for the effects cited above; and can exhibit toxic effects on its ingestion (Samnani & Bhattacharya, 2020).

14.1.1 Fluoride Containment in Air

Human activity is the largest contributor of fluoride to the atmosphere. Fluoride is present in the atmosphere in both gaseous and particle forms, released by natural and man-made sources (Lewandowska et al., 2013).

Natural sources such as volcanoes, rock dust, and the ocean barely discharge any of this substance into the atmosphere. Aluminium smelters, fertilizer plants, and other industrial operations such as blocks, tiles, art pottery and concrete works, ceramics industries and glass manufacturing are the main human-made sources. Hydrogen fluoride (HF), carbon tetrafluoride (CF_4), hexafluoroethane (C_2F_6) and silicon tetrafluoride (SiF_4) are examples of the gaseous fluorides. Cryolite (Na_3AlF_6), chiolite ($Na_5Al_3F_{14}$), calcium fluoride, aluminum fluoride and sodium fluoride are all examples of fluoride particles.

Hydrogen fluoride is easily formed when fluorine is released by volcano eruptions, forest fires, and marine aerosols. Fluorine can only change its form in the environment; it cannot be destroyed. Thus fluorine reacts with water in the atmosphere. This forms typically nonvolatile, stable fluorides upon reaction with numerous materials (in both the vapor phase and aerosols) (ATSDR, 2003). Hydrogen fluoride and particle fluorides are also human-caused fluoride pollutants. Fluoride in the atmosphere can be deposited in different places and to different depths depending on several characteristics, including emission intensity, particle size distribution, and species' chemical reactivity. When it comes to long-distance transport of fluorides, atmospheric circulation is key, both for larger fluoride aerosol particles (diameter, > 10 μm) and for smaller fluoride particles (diameter, 10 μm) (Slooff et al., 1989).

Because the F compounds in the atmosphere are so stable for extended periods of time, their ability to go further is greatly enhanced. These substances can be removed from the air via wet deposition with atmospheric precipitation (rain, snow, fog, etc.), leading to cloud washout and atmospheric scavenging (Chate et al., 2003).

As long as the source of the emission is not directly adjacent to the atmosphere, the average fluoride concentration in the air is less than 0.1 g/m^3 (World Health Organization, 2000).

Indoor air is frequently contaminated by pollutants found in outside air. Indoor pollution is becoming more of a concern as rules for these contaminants tighten (Lin & Liu, 2020). According to WHO research from 2007, there were estimated to be 420,000 premature losses due to ambient air contamination every year, much more than the 300,000 deaths due to outdoor air pollution (Zhang & Smith, 2007). Indoor air pollution is caused by the heating and cooking of houses using renewable energy sources (coal and biofuels).

14.1.2 Fluoride Concentration in Groundwater Globally

The quantity of aquatic fluoride varies throughout time and between different geographic areas. The concentration in groundwater is exaggerated by a number of influences, as well as the geological, chemical and physical features of aquifers; the acidity and porosity of top soil and rocks; local temperature; the distance of wells; and the impact of other chemicals. The primary reasons for the presence in groundwater are the processes of weathering and leaching, as well as the percolation of water through the soil and sediments. Fluoride pollution of groundwater is a major problem in several nations, including India, China, Ethiopia, Kenya, Mexico, Argentina, etc. The concentration in ground water varies globally, and the possible causes are comprehensively reported by Yadav et al. (2019). The situation is critical because there is continuous decline in the availability of clean surface water and an increasing dependence on groundwater.

14.1.3 Fluoride Concentration in Different Foodstuffs

As a trace element, fluoride is not considered an essential nutrient, although it is considered to contributes a key role in the emergence of teeth and bones. Fluoride can be found in practically all foods, although in extremely trace amounts. Soil, water, and fertilisers that are used for irrigation and agriculture are the primary factors determining the quantity of fluoride that is present in food. Fluoride is absorbed into the human body through the consumption of a variety of food items, including wheat, cabbage, tea, carrots, spinach, and some beverages. Table 14.1 shows fluoride concentration in various foodstuff by the references.

14.1.4 Fluoride Concentration in Dental Care Products

Fluoride can be taken in through toothpaste and other products used to care for teeth. For children under the age of 6, toothpaste can provide up to 87% of their fluoride consumption every day (Zohoori et al., 2013). It is mainly due to children's lack of control over their swallowing reflex. The fluoride content of toothpastes usually ranges between 1000 and 1450 mg/kg, so 1 g of toothpaste contains between 1 and 1.5 mg. Children ingest fluoride each time they brush their teeth in an average amount of 0.13–0.59 mg, depending on how old the child is, how much toothpaste is used

Table 14.1 Fluoride concentration in different foodstuffs

Items		Conc (mg/L)	References
Animal product	Cow milk	0.05–6.87	Bhargava and Bhardwaj (2009)
	Buffalo milk	0.04–6.85	
	Fermented milk	1.76–93.68	Lodi et al. (2011)
	Chicken	0.08–8.63	
	Fish	1.45–2.55	Chowdhury et al. (2018)
	River water fish	0.17–1.56	
	Sea water fish	0.77–4.29	Ganta et al. (2015)
Milk	Soya milk	0.015–0.964	Lal et al. (2014)
	Human breast milk	0.005–0.025	Koparal et al. (2000)
		0.003–0.011	Şener (2007)
	Infant milk	0.49–1.53	Nohno et al. (2011)
		0.15–1.24	
		0.06–1.09	Zohoori et al. (2012)
Cereals	Breakfast cereals	0.08–1.86	Vanessa Eid da Silva Cardoso (2003)
	Wheat	0.51–14.3	
	Corn	0.3–5.9	Chandravanshi and Zewge (2014)
	Rice	0.51–5.52	
	Maize	0.0–5.6	
Vegetables	Lettuce	2.12–5.76	Dagnaw (2017)
	Swiss chard		
	cabbage		
	Abyssinian cabbage		
	Cabbage	0.36–0.71	
	Onion		Bhat et al. (2015)
	Tomato		
Beverages	Alcoholic drinks	0.30–0.199	Goschorska et al. (2016)
	Carbohydrate drinks	0.12–0.42	Bansal and Gupta (2015)
		0.02–2.77	Syazleen et al. (2021
Fruit	Apple	0.11–1.14	Bethencourt-barbuzano et al. (2022)
	Peach	0.06–0.72	
	Pineapple	0.2–1.1	
Tea leaves	–	0.45–8.80	Chan et al. (2013)
	–	0.25–3.55	Koblar (2012)

Table 14.2 Daily adequate intakes (AI) for fluoride and daily tolerable upper intake levels for fluoride (National Institutes of Health 2020)

Age	Daily adequate intakes (AI) as mg F/kg body's weight				Daily tolerable upper intake as mg F/kg body's weight			
	Male	Female	Pregnancy	Lactation	Male	Female	Pregnancy	Lactation
Birth to 6 months	0.01	0.01	–	–	0.7	0.7	–	–
7–12 months	0.5	0.5	–	–	0.9	0.9	–	–
1–3 years	0.7	0.7	–	–	1.3	1.3	–	–
4–8 years	1.0	1.0	–	–	2.2	2.2	–	–
9–13 years	2.0	2.0	–	–	10.0	10.0	–	–
14–18 years	3.0	3.0	3.0	3.0	10.0	10.0	10.0	10.0
19+ years	4.0	3.0	3.0	3.0	10.0	10.0	10.0	10.0

and how often the child rinses. Some young children use fluoridated mouth rinses, and it is not a good idea. For kids less than 6 years old, parents should only put a small amount of toothpaste on their young children's toothbrushes and watch them brush their teeth to make sure they do not swallow too much toothpaste. The adequate daily intake for fluorides is shown in Table 14.2 along with the tolerable upper intake levels.

14.2 Health Consequences of Consuming Groundwater and Food with Fluoride Contamination

The Bureau of Indian Standards (BIS) has developed standards for the concentrations of fluoride in drinking water. The acceptable limit in the absence of a backup supply has been established at 1.5 mg/L, whereas the extreme desired concentration has been established at 1.0 mg/L. Fluorosis is a devastating disease that progresses slowly but ultimately affects most of the body's connective tissue, cell membranes, and body parts, resulting in a number of health complications (Dutta et al., 2020). Fluorosis in the skeletal and dental cavities has received a lot of research and has been extensively discussed because they are the two most common fluorosis-related disorders that are caused by drinking fluoride contaminated groundwater.

14.2.1 Dental Fluorosis

Human dental fluorosis is caused by consuming too much F-contaminated water, during tooth growth, often between birth and 6–8 years of age. The excessive incorporation of F into these structures is associated with dental fluorosis, similarly known as hypoplasia or hypo mineralization of tooth coating. Development of dental fluorosis can be triggered by long-term exposure to higher-than-ideal quantities of fluoride in the body system during crucial phases of the process known as amylogenesis. The following characteristics describe dental fluorosis: In mild cases, opaque white patches appear on the teeth; in severe cases, the teeth become discoloured and pitted as a result of increased subsurface porosity (Denbesten & Li, 2011). During tooth development does dental fluorosis occur. Early childhood, especially the first three years between the ages of 6 and 24 months, permanent teeth and the initial set of permanent molars are developing at a rapid rate, making this a critical time for dental fluorosis prevention, while the lasting canines, premolars, second molars are still forming, high fluoride intake between the ages of 3 and 6 can negatively affect their development.

14.2.2 Skeletal Fluorosis

According to estimates, roughly only half of the fluoride ingested is removed from the human body. The remaining becomes absorbed in calcified tissues such as bones and teeth. Bioaccumulation of fluoride is known to occur in the case of continued intake. The hydroxyl ion in hydroxyapatite is changed with fluoride, fluorohydroxyapatite or simply fluoroapatite is formed on teeth and bones. However, this binding is reversible. Therefore, when fluoride intake is stopped, fluoride can go back to plasma from bones during bone remodelling and can be excreted by kidneys (Samnani & Bhattacharya, 2020).

First manifesting as joint stiffness and pain, skeletal fluorosis is characterized by weaker-than-normal bones. Osteophytic, mineralized and bridging between the margins of vertebral bodies formed within the periosteal sleeves of improperly constructed osseous tissue contribute to muscular weakness and an irregularly thickened skeleton in severe cases.

14.2.3 Acute Toxicity

When fluoride is consumed excessively, either in one dose or repeated doses, fluoride toxicity can occur, just as excessive amounts of many other minerals can cause toxicity. The stomach is the foremost structure to be impacted by systemic severe exposure, hence experimental systemic poisonousness starts with gastric signs and indications. These signs and indications can range from some grade of sickness to stomach aching, haemorrhagic gastroenteritis, unsettled stomach, and diarrhoea. The toxic levels are estimated to be 5 mg/kg body weight. Fluoride intake of approximately 16 mg/kg body weight is dangerous (World Health Organization: Environmental health criteria, 1984).

14.3 Dental Fluorosis Among the Different Regions

Several studies have been conducted around the globe monitoring fluoride concentrations in water, food items, and effects of fluoride ingestions through various routes in the human population. The following sections describe the data reported for dental fluorosis and other fluoride toxicity in several states of India and in other countries.

14.3.1 Agra

A study was carried out to: (a) estimate F exposure; and (b) evaluate the non-carcinogenic threat associated with groundwater drinking among urban residents of various ages in the city of Agra. F levels ranged from 0.90 to 4.12 mg/L, with 1.88 mg/L representing the average. The exposure dose (ED) and hazard quotient (HQ) metrics were used to conduct an investigation on the quantitative nature of the physical risk associated with the consumption of F through groundwater, for three different age groups, namely new-borns (aged 0 to 6 months), children (aged 7 to 18 years), and adults (aged 19 years and older). It was based on three factors: the mean weight of the participants (kg), the amount of water consumed (L/d) and the level of F in groundwater (mg/L). For children, infants, and adults, the highest predicted exposure levels were correspondingly 0.69, 0.31, and 0.12 mg/kg/day. The gathered results disclose that babies and adolescents were more visible to the danger of F than adults were when matched to the oral orientation dose of 0.06 mg/kg/day advised by IRIS 2003 (The US Environmental Protection Agency, 2012). For all the sites in the study area, the HQ is greater than 1, although only eight sites recorded a high HQ for adults. Adults at around 71% of the sampled sites, on the other hand, may be exposed to noncarcinogenic risk (Yadav et al., 2019).

14.3.2 Haryana

For the Jhajjar district of Haryana, India, Yadav et al. published their findings on their investigation on the connection among water fluoride ranks and the frequency of dental fluorosis among schoolchildren. 60 villages were selected after the preliminary study, with 10 villages in each block having fluoride values above 1.5 mg/L and 2 villages from each block having fluoride values up to 1.0 mg/L. The water samples comprised fluoride levels ranging from 1.52 to 4.0 mg/L. Fluoride-rich villages (1.5–4.0 mg/L) and low villages (0.3–1.0 mg/L) Under these two distinct groups, the communities were divided to better understand how fluoride affects dental health, fluorosis data was collected from schoolchildren (aged 7–15) in the communities.

Dental fluorosis phases amongst the individuals viz., chalky white, yellowish brown, and brownish black, with horizontal streaks and spots, were present. The proportion of children with dental fluorosis ranged from 30.00% to 94.85% in high-fluoride villages and from 8.80% to 28.20% in low/normal-fluoride villages (Yadav et al., 2008).

14.3.3 Jammu Kashmir

School children between the ages of 8 and 15 were subjected to case–control research to determine nutritional intake, renal health, thyroid function and bone metabolic markers, furthermore, fluorosis intensity and prevalence in the mouth. 824 children between the ages of 8 and 15 participated in this research. Fluoride levels in this research region wide-ranging from 1.43 to 3.84 mg/L. Children of school age were split into two communities depending on the concentration of fluoride in their water: control and affected. Boys and girls were equally affected out of 379 people from the affected region, 48% affected by various grades of dental fluorosis. Results of the research indicated that children in the impacted region had much higher dental fluorosis than the children in the control group. The alteration in the pH ratio caused by low oxygen levels in high-altitude residents of high elevation was the proposed cause, which increases the retention of fluoride due to decreased urine elimination and increases the risk of fluorosis (Khandare et al., 2017).

14.3.4 Mysore

The occurance and extremity of dental fluorosis among presecondary school students who were raised after birth in three villages in the Mysore district were studied. This study included Nerale (2.0 mg/L water fluoride), Belavadi (1.2 mg/L), and Naganahally (0.4 mg/L) villages; 405 students from schools in the age variety of 10–12-years (204 males, 201 females) were picked for study. Dental fluorosis was found to be 41.73% using Dean's index. Areas with high water fluoride had higher CFI increases. The fluoride content of CFI and drinking water was positively correlated. Optimal fluoride consumption reduced dental cavities. Excessive ingestion, especially during development, can be observed to dental and skeletal fluorosis (Sebastian et al., 2016).

14.3.5 Rajasthan

In the Bhilwara region of central Rajasthan, research on the distribution and health issues associated with fluoride pollution in groundwater was carried out in 1030 communities. These communities had access to water containing 0.2–13.0 mg/L fluoride. In this research, an in-depth assessment of fluorosis was performed in 60 villages out of which 41 towns had fluoride higher than 5.0 mg/L. In addition, information on job possibilities, gender, and age was acquired. Dental fluorosis was studied in 4252 people over 5 years of age, while skeletal fluorosis was examined in 1998 people over 21 years of age. Dental and skeletal fluorosis was 3270/4252 (76.9%) and 949/1998 (47.5%), respectively. Maximum 23.9% (1016) were placed in mild Dean's categorization, 374 (8.8%) had significant dental fluorosis. The examined area's Dean's Community Fluorosis Index was 1.62. Surajpura's CFI was 3.0. 566 individuals (28.3%) had Grade I skeletal fluorosis, while 12 (0.6%) had Grade III. Fluorosis became more common and severe as fluoride concentrations increased and in economically impoverished areas. Fluorosis was higher in male workers (Hussain et al., 2009).

14.3.6 Tamil Nadu

Fluoride in drinking water, common foods, rice that has been cooked, vegetables, and cow's milk was a subject of this investigation. In this study, local drinking water sources provided around 98% of the fluoride that new-borns consume each day comes from their diet. All drinking water tests from 22 fluoride-endemic villages showed fluoride levels greater than 1.5 mg/L. Cow milk had 0.043–0.147 mg/L fluoride. Moringa oleifera had 7.68 mg/kg fluoride and Sesbania grandiflora 4.47 mg/kg. The black-eyed bean had 4.21 mg/kg of fluoride. Water provided 70% of the consumption of fluoride by children. Children and adults consume a lot of fluoride from prepared rice. Adults in low, medium and highly fluorotic areas consumed rice that had been cooked at rates of 0.34, 0.38, and 0.50 mg/d, respectively (Amalraj & Pius, 2013).

14.3.7 Telangana

The objective was to quantify the prevalence of dental caries and fluorosis were among school-age children in the Mahbubnagar district, as well as F levels in drinking water since various locations of the district. The study involved 2000 kids amongst the ages of 6–12 years. Dental caries affected 64.2% of primary teeth and 26.6% of the permanent dentition, respectively indicating that dental caries was more prevalent in children aged 7–8 and less prevalent in children aged 11–12. Dental fluorosis affected 15% of children in the key dentition and 70% of children in the everlasting dentition; it affected children of 9–10 years older than 6 years' children did. The northern and eastern Mahbubnagar district had 2 mg/L Fluoride, while home-grown communities had 1.2 mg/L (Kola et al., 2019).

14.3.8 West Bengal

A cross-sectional investigation was conducted in the Simlapal block of Bankura District, West Bengal, to establish the relationship between fluoride (F) exposure as exposure dose (ED), dental fluorosis (DF), urine fluoride levels (UF), intelligence quotient (IQ), as well as a person's body mass index (BMI). From that target area, 50 groundwater samples were taken. 149 children aged 6 and 18 were taken into account for this study. The daily average consumption of water was between 0.35 and 1.00 L. Due to water restrictions, the children in the research consumed less water daily. Fluoride in drinkable water can sort from 0.02–0.22 mg/kg/day for newborns from places with a range of fluoride endemicities. Normal children are exposed to 0.02–0.05 mg/kg/day of fluoride. The average daily fluoride exposure for children aged 6 to 8, 8 to 10 years10 years and over 10 is 0.05, 0.08, and 0.1 mg/day, correspondingly (Das & Mondal, 2016).

14.3.9 Mexico

Consequences of fluoride in water in some regions of Mexico were studied. Evaluations of the amount of fluoride present in teenage students' urine, their diets, also their dental fluorosis were the main goals of the study. Fluoride levels in water was 4.42 mg/L. A total of 307 participants were included, with women accounting for 59.9% ($n = 184$) and men accounting for 40.1% ($n = 121$). The average age was 15.6 ± 1.6 years. The Thylstrup and Fejerskov index (TFI) criteria were used to measure the existence and harshness of dental fluorosis. 61.6% of those surveyed had TFI > 4, and 91.9% had dental fluorosis. Most of the group under study had mild to severe dental fluorosis. The severity and nutritional health were correlated with the concentration of fluoride in urine. Children who were underweight had more fluoride in their urine and severe dental fluorosis. Thus, overall poor nutritional health has more prominent effects of fluoride toxicity (Del Carmen et al., 2016).

14.3.10 The USA

Researchers examined statistics from the National Health and Nutrition Examination Survey (NHANES), which was conducted between 2015 and 2016. Data from 2098 kids and youngsters with Dean's Index scores as well as water and plasma fluoride measurements were examined. Dental examiners used Dean's fluorosis classification system to calculate Dean's Index score. Fluoride levels in plasma and water for consumption were measured. This investigation displayed that although the incidence of dental fluorosis remained 70%, the amount of fluoride level in water beyond advised amount of 0.7 mg/L was only 25%. This study revealed that even very little exposure to water fluoride or plasma fluoride increased the probability of dental fluorosis (Dong et al., 2021).

14.3.11 Brazil

Among children aged 8–12 years in southern Brazil, a longitudinal investigation evaluated exposure factors associated with dental fluorosis (DF). Twenty schools' worth of students were chosen at random ($n = 1196$). Their parents responded to a questionnaire given home, and they were also questioned. The average incidence of DF was 8.53% by modified Dean's criteria, while incidence of serious DF existed 0.17%. Multiple logistic regression studies demonstrated a link between DF and more frequent brushing of teeth, as well as the initial application of fluoride toothpaste at the eruption of the initial tooth (Azevedo et al., 2014).

14.3.12 Canada

In Canada, researchers looked into the relationship between children's intellectual ability and fluoride exposure during infancy in cities with and without fluoridation. Within the cohort of Maternal-Infant Research on Environmental Chemicals, researchers looked at 398 mothers-child pairs who admitted to consuming tap water. 38% of the mother–child pairs resided in fluoridated areas. Each time the fluoride in water rises by 0.5 mg/L, children who were breastfed or given formula showed a 9.3 and 6.2-point reduction in their performance intelligence score (IQ), respectively. This decrease was almost like the modification among fluoridated and non-fluoridated areas. While allowing prenatal fluoride exposure in fed formula and breast-fed kids, the link between the water fluoride levels and behavioural IQ remained substantial. When baby formula was reconstructed with appropriately fluoridated water according to the latest research, up to 59% of new-borns under four months of age exceeded the maximum allowed level (0.1 mg/kg/day), and 33 and 14.3% of new-borns between six and nine months old did so. If new-borns under the age of six months only consume formula made with fluoridated tap water, their fluoride intake may exceed the upper tolerated limits. Fluoride exposure during infancy was shown to be related with a reduction in non-verbal cleverness in children after adjustment for fetal exposure. The study found that if there is no advantage in ingesting fluoride in the initial six months, by utilising non-fluoridated or low fluoride water as a formula diluent, it is recommended to maintain fluoride exposure to a minimal (Till et al., 2020).

14.3.13 China

The study was carried out to determine whether fluorosis in teeth occurs frequently and severely in kids between 12 and 13 who had permanent teeth, as well as the potential effects of clay pots (which are used to store water) on caries with dental fluorosis in permanent teeth in China. According to the fluoride satisfied of the waters 0.4, 1.0, 1.8, 3.5, and 5.6 mg/L, respectively, 477 kids were separated into 5 sets (A to E). The investigation revealed that dental fluorosis is significantly more severe in 13-year-older than in 12-year-olds across all fluoride levels. In both of the groups, the variations were not significantly different, most likely due to the comparatively small age range. By using TF score and the DMF-T index, dental fluorosis and caries were evaluated. An average TF scores and water fluoride levels were found to be positively correlated. In groups B and D, 13-year-olds had a higher TF score than 12-year-olds. Between 2.6% and 0.03 (group E) and 22.1% and 0.38 (group A), respectively, were the mean of DMF-T and the prevalence of caries. Fluorosis in the teeth can become more severe if water is kept in local clay pots for storage (Ruan et al., 2005).

14.3.14 Japan

To determine whether there was a link between fluoride concentrations in drinking water and the frequency of dental caries and fluorosis in seven Japanese groups with varying fluoride levels in their drinking water, a study was conducted. 1060 residents in the age bracket 10- to 12-years were inspected. The incidence of dental caries and fluorosis was directly and inversely correlated with the fluoride ranks in water used for drinking, respectively. The mean DMFS in fluoride-rich communities with fluoride ranks among 0.8 and 1.4 mg/L was amongst 53.9 and 62.4% lower than it was in the societies with insignificant fluoride levels. According to a multivariate analysis, the amount of fluoride in the water had the greatest impact on the DMFS results (Tsutsui et al., 2000).

14.3.15 South Africa

A study compared the occurrence and sternness of dental fluorosis in kids and teenagers who were vegetarian and non-vegetarian and who lived in a region where dental fluorosis was common in South Africa. Children ($n = 165$) from five schools in Arusha town, ranging in age from 6 to 18, were investigated. The children were exposed to 3.6 mg/L of water throughout their entire life. The Thylstrup and Fejerskov index was generated to determine the degree of dental fluorosis (TFI). The rate of fluorosis in teeth (TFI score ≥ 1) was 67% in the vegan group ($n = 24$), and 21%

had severe fluorosis (TFI score 5). The frequency of fluorosis and severe fluorosis was 95 and 35%, individually, in the non-vegetarian group (n = 141). According to a multiple logistic regression analysis, nonvegetarians had a seven-fold higher risk of acquiring dental fluorosis than vegetarians do (Awadia et al., 1999).

14.3.16 Nigeria

12- to 15-year-old lifetime residents were chosen using the multistage sample technique to study the fluorosis of the teeth is prevalent and severe in high- along with low-altitude zones of the Province of Plateau's Central Parliamentary District. The Thylstrup and Fejerskov index was used as the primary outcome measure to detect whether there is dental fluorosis and how severe it is. A total of 1000 children were tested, with 554 (50.4%) arriving from the high-altitude area and 546 (49.6%) arriving from the low-altitude area. Fluorosis was 12.9 penetrance common across the area, although it was substantially more common in high-altitude regions (22.2%) rather than in low-altitude regions (3.5%). The buccal/labial surfaces of the primary premolar (tooth 1.4) and the rightmost permanent central incisor (tooth 1.1) buccal/labial surfaces were examined for the existence and ruthlessness of dental fluorosis with the TF index. Most serious fluorosis in the district affected teeth 1.4 and 1.1, with scores of TF 6 and TF 5, respectively. The occurrence and ruthlessness of dental fluorosis are significantly more common and severe in the high-altitude portions of the district than in the low-altitude areas. To solve the problem, more investigation and work is needed (Akosu et al., 2009).

14.4 Fluoride Remediation

The traditional and well-proven way of providing safe water to populations impacted by fluorosis is defluoridation. According to the definition, it is "the decrease of fluoride concentrations in drinking water to an ideal level." One way to defluoridate water is to treat it centrally (at the source); another way is to treat it locally (at the home level) (point of use treatment). Most wealthy nations choose treatment at the source because it may be done on a large scale under the direct supervision of trained personnel. In contrast, it could be advised in less established territories to treat water at home or viewpoint of benefit level. Therapy at the point of use is superior to treatment at the community level in a number of ways. The following section categorizes various defluoridation methods (Jamwal & Slathia, 2022).

(a) Adsorption technique

This technique relies on fluoride ions attaching to the surface of an active substance and forming bonds. Running through a bed of defluoridating material is how the raw water used in the adsorption process gets defluoridated. Fluoride is retained in the substance through ion exchange, chemical, or physical adsorption processes. After some use, the adsorbent becomes saturated and needs to be recharged. Activated alumina, bone char, calcined clay, natural absorbents (drumstick seed, roots of Vetiver grass, Tamarind seeds, tea ash, egg shell powder, etc. have been used successfully).

(b) Ion-exchange technique

Ion-exchange materials, which are used to exchange ions from solutions, are insoluble in water and hold the replaceable ions loosely. Ion exchange materials can be divided into two categories: natural and synthetic. Bone, bone char, anion and cation exchange resins such as carbon, defluoron-1, defluoron-2, etc. are some of the several ion exchange materials that have been researched.

(c) Precipitation technique

Chemicals are add into the raw water, the fluoride salt precipitates as inexplicable fluorapatite, which is then removed from the water. Materials recycled frequently in the precipitation procedure include brushite, poly-aluminium chloride, lime, poly aluminium hydroxy sulphate, and aluminium salts (such as alum). The Nalgonda technique, contact precipitation, the Indian Institute of Science (IISc) method, etc. are the techniques used in precipitation of fluoride.

(d) Other techniques

Physical techniques that have been tried and tested for defluorinating water include reverse osmosis, electrolysis, and electro dialysis, and distillation. Although they are efficient in removing fluoride salts from water, they are not widely used because of a number of procedural issues.

14.5 Conclusion

- According to the literature, fluoride may be a necessary ingredient for both animals and humans. However, the importance of fluoride in humans has not yet been proven conclusively. As a result, data on the minimal dietary need for fluoride are unavailable.
- Numerous epidemiological studies have shown that consuming fluoride over an extended period of time can have adverse effects and result in health problems. These studies clearly reveal that fluoride affects primarily skeletal tissues, particularly bones and teeth.
- The Public Health Service (PHS) recommended fluoride concentrations, which ranged from 0.7 to 1.2 mg/L dependent on regional atmospheric high temperature. A maximum fluoride conc. of 1 mg/L is generally recommended by regulatory authorities for drinking water. This is done to provide maximum dental caries protection and to prevent the spread of dental fluorosis.

- Water management and environmental protection must be extremely effective due to the serious threat that fluoride-contaminated water poses to human health. It is a serious health hazard since anthropogenic processes cause the release of fluoride into the environment, and most of the problems arise from the consumption of fluoride from natural resources.
- Dental cavities are affected by an ecological inequity between the physical stability of tooth minerals and oral microorganisms biofilms.
- Dental fluorosis, which is apparent to the naked eye, could be treated as an early symptom of fluoride attack. It causes irreversible toxicity in tooth-forming cells.
- The substantial literature suggests that the concentrated allowable limits of fluoride in drinking water established by various regulatory authorities should be revised in areas where significant fluoride ingestion occurs through the food chain.
- Due to excessive fluoride interaction from the consumption of water, the occurrence of dental fluorosis was high in children aged 6 to 12.
- The vegetarian group has a much lesser incidence and harshness of dental fluorosis than non-vegetarians.
- It is challenging to draw any conclusions about this complex fluorosis problem or to suggest a widely accepted solution that would be "good" for everyone. Fluorosis is still a widespread issue. Fluorosis is a disease that is constantly expanding its geographic reach across the world. Studies conducted by numerous researchers around the world have made it clear that fluoride in groundwater has the potential to cause problems for human society.

Acknowledgements The authors thank our prestigious institute and other people who are involved in this study for their direct and indirect contributions.

References

Akosu, T. J., Zoakah, A. I., & Chirdan, O. A. (2009). The prevalence and severity of dental fluorosis in the high and low altitude parts of Central Plateau, Nigeria. *Community Dental Health, 26*(3), 138.

Amalraj, A., & Pius, A. (2013). Health risk from fluoride exposure of a population in selected areas of Tamil Nadu South India. *Food Science and Human Wellness, 2*(2), 75–86.

ATSDR. (2003). Toxicological profile for Flourides, Hydrogem Flouride, And Flourine. In *US Department of Health and Human Services, Public Health Service Agency for Toxic Substances and Disease Registry; Atlanta, GA.* (Issue September).

Awadia, A. K., Haugejorden, O., Bjorvatn, K., & Birkeland, J. M. (1999). Vegetarianism and dental fluorosis among children in a high fluoride area of northern Tanzania. *International Journal of Paediatric Dentistry, 9*(1), 3–11.

Azevedo, M. S., Goettems, M. L., Torriani, D. D., & Demarco, F. F. (2014). Factors associated with dental fluorosis in school children in southern Brazil: A cross-sectional study. *Brazilian Oral Research, 28*, 1–7.

Bansal, M., & Gupta, N. (2015). Estimation of fluoride levels in various commercially available carbonated soft drinks in Chandigarh city, India. 514–516.https://doi.org/10.4103/2319-5932.171182

Bethencourt-barbuzano, E., Torre, H., Paz-, S., Gutiérrez-, Á. J., Alejandro-vega, S., & Rubio-armendáriz, C. (2022). Evaluation of dietary exposure to fluoride from fruit juices and nectars in children and adolescents. *10*(4), 115–120. https://doi.org/10.15406/ppij.2022.10.00372

Bhargava, D., & Bhardwaj, N. (2009). Study of fluoride contribution through water and food to human population in fluorosis endemic villages of North-Eastern Rajasthan. *African Journal of Basic & Applied Sciences, 1*, 55–58.

Bhat, N., Jain, S., Asawa, K., Tak, M., Shinde, K., Singh, A., Gandhi, N., & Gupta, V. V. (2015). Assessment of fluoride concentration of soil and vegetables in vicinity of zinc smelter. *Journal of Clinical and Diagnostic Research,* 63–66. https://doi.org/10.7860/JCDR/2015/13902.6667

BIS. (2012). Indian Standard Drinking Water Specification (Second Revision). *Bureau of Indian Standards, IS 10500*(May), 1–11. http://cgwb.gov.in/Documents/WQ-standards.pdf

Chan, L., Mehra, A., Saikat, S., & Lynch, P. (2013). Human exposure assessment of fluoride from tea (Camellia sinensis L.): A UK based issue? *Food Research International, 51*(2), 564–570. https://doi.org/10.1016/j.foodres.2013.01.025

Chandravanshi, B. S., & Zewge, F. (2014). Levels of fluoride in staple cereals and legumes produced in selected areas *July*.

Chate, D. M., Rao, P. S. P., Naik, M. S., Momin, G. A., Safai, P. D., & Ali, K. (2003). Scavenging of aerosols and their chemical species by rain. *Atmospheric Environment, 37*(18), 2477–2484. https://doi.org/10.1016/S1352-2310(03)00162-6

Chowdhury, C., Khijmatgar, S., Kumari, D., Chowdhury, A., Grootveld, M., Hegde, C., & Lynch, E. (2018). Fluoride in fish flesh, fish bone and regular diet in south-coastal area of Karnataka state of India. *Indian Journal of Dental Research, 29*(4), 414–417. https://doi.org/10.4103/ijdr.IJDR_653_16

Das, K., & Mondal, N. K. (2016). Dental fluorosis and urinary fluoride concentration as a reflection of fluoride exposure and its impact on IQ level and BMI of children of Laxmisagar, Simlapal Block of Bankura District, W.B., India. *Environmental Monitoring and Assessment, 188*(4). https://doi.org/10.1007/s10661-016-5219-1

Dagnaw, L. A. (2017). Fluoride content of leafy vegetables, irrigation water, and farmland soil in the rift valley and in non-rift valley areas of ethiopia. *Fluoride, 50*(December), 409–429.

Del Carmen, A. F., Javier, F. H., & Aline, C. C. (2016). Dental fluorosis, fluoride in urine, and nutritional status in adolescent students living in the rural areas of Guanajuato, Mexico. *Journal of International Society of Preventive & Community Dentistry, 6*(6), 517–522. https://doi.org/10.4103/2231-0762.195510

Denbesten, P., & Li, W. (2011). Chronic fluoride toxicity: Dental fluorosis. *Monographs in Oral Science, 22*, 81–96. https://doi.org/10.1159/000327028

Dong, H., Yang, X., Zhang, S., Wang, X., Guo, C., Ma, X., Niu, X., Zhang, P., & Chen, T. (2021). Associations of low level of fluoride exposure with dental fluorosis among US children and adolescents, NHANES 2015–2016. *Ecotoxicology and Environmental Safety, 221*, 112439.

Dutta, D., Chowdhury, S., & Hassan Mullick, T. (2020). Prevalence of dental and skeletal fluorosis in relation with the fluoride levels in drinking water among the primary school children at Laxmisagar Grampanchayat of Bankura District, West Bengal India. *IP Journal of Nutrition, Metabolism and Health Science*, *2*(2), 64–67. https://doi.org/10.18231/j.ijnmhs.2019.011

Ganta, S., Yousuf, A., Nagaraj, A., Pareek, S., Sidiq, M., Singh, K., & Vishnani, P. (2015). Evaluation of fluoride retention due to most commonly consumed estuarine fishes among fish consuming population of Andhra Pradesh as a contributing factor to dental fluorosis: A cross-sectional study. *Journal of Clinical and Diagnostic Research*, *9*(6), ZC11–ZC15. https://doi.org/10.7860/JCDR/2015/12271.6035

Goschorska, M., Gutowska, I., Baranowska-bosiacka, I., Ewa, M., & Chlubek, D. (2016). Fluoride content in alcoholic drinks. *Biological Trace Element Research*, 468–471. https://doi.org/10.1007/s12011-015-0519-9

Hussain, J., Hussain, I., & Sharma, K. C. (2009). Fluoride and health hazards: Community perception in a fluorotic area of central Rajasthan (India): An arid environment. *Environmental Monitoring and Assessment, 162*(1–4), 1–14. https://doi.org/10.1007/s10661-009-0771-6

Jamwal, K. D., & Slathia, D. (2022). A review of defluoridation techniques of global and Indian prominence. *Current World Environment: An International Research Journal of Environmental Sciences, 17*, 41–57.

Khandare, A. L., Gourineni, S. R., & Validandi, V. (2017). Dental fluorosis, nutritional status, kidney damage, and thyroid function along with bone metabolic indicators in school-going children living in fluoride-affected hilly areas of Doda district, Jammu and Kashmir, India. *Environmental Monitoring and Assessment, 189*(11). https://doi.org/10.1007/s10661-017-6288-5

Koblar, A. (2012). Fluoride in teas of different types and forms and the exposure of humans to fluoride with tea and diet. *Food Chemistry, 130*, 286–290. https://doi.org/10.1016/j.foodchem.2011.07.037

Kola, S. R., Mallela, M. K., Puppala, R., Kethenaeni, B., Tharasingh, P., & Reddy, V. S. (2019). Prevalence of dental caries and dental fluorosis among 6–12 years old school children in relation to fluoride concentration in an endemic fluoride belt of Mahabubnagar district, Telangana state, India. *Journal of Dr. NTR University of Health Sciences, 8*, 29.

Koparal, E., Ertugrul, F., & Oztekin, K. (2000). Fluoride levels in breast milk and infant foods. *Journal of Clinical Pediatric Dentistry, 24*(4), 299–302.

Kumar, S., Venkatesh, A. S., Singh, R., Udayabhanu, G., & Saha, D. (2018). Geochemical signatures and isotopic systematics constraining dynamics of fluoride contamination in groundwater across Jamui district, Indo-Gangetic alluvial plains, India. *ECSN*. https://doi.org/10.1016/j.chemosphere.2018.04.116

Lal, H., Zohoori, F. V., Omid, N., Valentine, R., & Maguire, A. (2014). The fluoride contents of commercially-available soya milks in the UK. *British Dental Journal, 217*(4), 1–4. https://doi.org/10.1038/sj.bdj.2014.736

Lewandowska, A., Falkowska, L., & Jóźwik, J. (2013). Factors determining the fluctuation of fluoride concentrations in PM10 aerosols in the urbanized coastal area of the Baltic Sea (Gdynia, Poland). *Environmental Science and Pollution Research, 20*(9), 6109–6118. https://doi.org/10.1007/s11356-013-1592-2

Lin, Y., & Liu, F. (2020). Indoor air quality and health: Empirical evidence from fluoride pollution in China. *China EconomicReview, 63*(October 2018), 1–16. https://doi.org/10.1016/j.chieco.2019.03.001

Lodi, C. S., Manareli, M. M., Sassaki, K. T., Delbem, A. C. B., & Martinhon, C. C. R. (2011). Fluoride concentration of some brands of fermented milks available in the market. *European Journal of Dentistry, 5*(2), 139–142. https://doi.org/10.1055/s-0039-1698870

Moss, M. E., & Kumar, J. (2021). Fluoride and human health. Flouride and human health. Burt and Eklunnd's Dntistry, Dental Practice, and the Community, 266–276. https://doi.org/10.1016/B978-0-323-55484-8.00024-1

National Institutes of Health. (2020). Fluoride—Health professional fact sheet. In *Fact Sheet for Health Professionals* (pp. 1–9). https://ods.od.nih.gov/factsheets/Fluoride-HealthProfessional/

Nohno, K., Zohoori, F. V., & Maguire, A. (2011). Fluoride intake of Japanese infants from infant milk formula. *Caries Research, 45*(5), 486–493. https://doi.org/10.1159/000330604

Ruan, J. P., Yang, Z. Q., Wang, Z. L., Astrøm, A. N., Bårdsen, A., Bjorvatn, K. (2005). Dental fluorosis and dental caries in permanent teeth: rural schoolchildren in high-fluoride areas in the Shaanxi province, China. *Acta Odontologica Scandinavica, 63*(5), 258–65. https://doi.org/10.1080/00016350510019991. PMID: 16419430.

Samnani, P. B., & Bhattacharya, P. K. (2020). *Metal ions in biochemistry*. CRC Press.

Sebastian, S. T., Soman, R. R., & Sunitha, S. (2016). Prevalence of dental fluorosis among primary school children in association with different water fluoride levels in Mysore district, Karnataka. *Indian Journal of Dental Research, 27*(2), 151.

Şener, Y. (2007). Fluoride levels of human plasma and breast milk. *European Journal of Dentistry, 1*(January), 21–24.

Slooff, W., Eeerens, H. C., Janus, J. A., & Ros, J. P. (1989). National institute of public health nd environmnetal protection bilthoven the netherlands. *Integr Electr Resour Plan.*

Syazleen, A., Azlida, N., Nor, M., & Afiqah, N. (2021). Assessment of fluoride and pH levels in a range of ready-to-drink children ' s beverages marketed in Malaysia. *Archives of Orofacial Sciences, 16*(2), 177–189. https://doi.org/10.21315/aos2021.16.2.8

Till, C., Green, R., Flora, D., Hornung, R., Martinez-Mier, E. A., Blazer, M., Farmus, L., Ayotte, P., Muckle, G., & Lanphear, B. (2020). Fluoride exposure from infant formula and child IQ in a Canadian birth cohort. *Environment International, 134*, 105315.

Tsutsui, A., Yagi, M., & Horowitz, A. M. (2000). The prevalence of dental caries and fluorosis in Japanese communities with up to 1.4 ppm of naturally occurring fluoride. *Journal of Public Health Dentistry, 60*(3), 147–153.

Vanessa Eid da Silva CARDOSO. (2003). *Breakfast Cereals and Snacks found in Brazil, 11*(4), 306–310.

Water Fluoridation Status in OECD Nations. (2012). https://fluoridealert.org/content/oecd_nations/

World health Organization: Environmental health criteria. (1984). Fluorine and fluorides. *Environmental Health Criteria, NO. 36.*

World health Organization: Environmental health criteria. (2000). Fluorine and fluorides. *Environmental Health Criteria, NO. 36.*

Yadav, J. P., Lata, S., Kataria, S. K., & Kumar, S. (2008). Fluoride distribution in groundwater and survey of dental fluorosis among school children in the villages of the Jhajjar District of Haryana, India. *Environmental Geochemistry and Health, 31*(4), 431–438. https://doi.org/10.1007/s10653-008-9196-3

Yadav, K. K., Kumar, V., Gupta, N., Kumar, S., Rezania, S., & Singh, N. (2019). Human health risk assessment: Study of a population exposed to fluoride through groundwater of Agra city, India. *Regulatory Toxicology and Pharmacology, 106*(April), 68–80. https://doi.org/10.1016/j.yrtph.2019.04.013

Zhang, J., & Smith, K. R. (2007). Household air pollution from coal and biomass fuels in China: Measurements, health impacts, and interventions. *Environmental Health Perspectives, 115*(6), 848–855. https://doi.org/10.1289/ehp.9479

Zohoori, F. V., Moynihan, P. J., Omid, N., Abuhaloob, L., & Maguire, A. (2012). Impact of water fluoride concentration on the fluoride content of infant foods and drinks requiring preparation with liquids before feeding. *Community Dentistry and Oral Epidemiology, 40*(5), 432–440. https://doi.org/10.1111/j.1600-0528.2012.00688.x

Zohoori, F. V., Walls, R., Teasdale, L., Landes, D., Steen, I. N., Moynihan, P., Omid, N., & Maguire, A. (2013). Fractional urinary fluoride excretion of 6–7-year-old children attending schools in low-fluoride and naturally fluoridated areas in the UK. *British Journal of Nutrition, 109*, 1903–1909.

Chapter 15
Bioaccumulation of Fluoride Toxicity in Plants and Its Effects on Plants and Techniques for Its Removal

Seema Kumari, Harsh Dhankhar, Vikas Abrol, and Akhilesh Kumar Yadav

Abstract Fluorine is an element that can be discovered in water, soil, rocks, minerals, and other natural surroundings. It is a naturally occurring substance. Fluoride's effects on plant health may be both beneficial and detrimental depending on how it's used. Gaseous hydrogen fluorides (HF) are the most phytotoxic air pollutants because they accumulate in the leaves of fragile plants against a concentration gradient and harm them at very low concentrations. This is because they harm plants even at extremely low concentrations. The most damaging effect that HF has on plants is caused when it enters their systems in the form of gas and disrupts many physiological processes. As HF accumulates in plant leaves, it has the potential to negatively impact human and animal health as it moves up the food chain. Fluoride is transferred from the fluoridated water used for irrigation to the crops, vegetables, and fruits that are being grown. Because additional fluoride is added to the food chain in addition to entering the body through drinking water, the risk of fluoride

S. Kumari (✉) · H. Dhankhar
Department of Botany, Baba Mastnath University, Rohtak 124001, India
e-mail: kumareeseema2011@gmail.com

H. Dhankhar
e-mail: dhankhar.harsh@gmail.com

V. Abrol
Division of Soil Science, Sher-e-Kashmir University of Agricultural Science and Technology, Jammu 180009, India
e-mail: abrolvics@gmail.com

A. K. Yadav
Department of Environmental Engineering and Management, Chaoyang University of Technology, Taichung 413310, Taiwan

Department of Mining Engineering, Indian Institute of Technology (Banaras Hindu University), Varanasi 221005, India

Environmental Science and Engineering Department, Indian Institute of Technology Bombay, Mumbai 400076, India

A. K. Yadav
e-mail: yadavbasti@gmail.com

A. K. Yadav et al. (eds.), *Advanced Treatment Technologies for Fluoride Removal in Water*, Water Science and Technology Library 125,
https://doi.org/10.1007/978-3-031-38845-3_15

poisoning to the population already at risk from fluoride poisoning increases as a result of bioaccumulation. The condition known as fluorosis should be brought to people's attention, and it should be prevented from using fluoridated irrigation water. The purpose of this review, in its entirety, is to bring attention to the following points:

1. The origins of fluoride ions in the environment, their geochemistry, and the toxicity of these ions.
2. The absorption of fluoride by plants and their subsequent accumulation in their tissues.
3. Techniques or approaches for the elimination of fluoride.

Keywords Fluoride · Bioaccumulation · Water · Irrigation

15.1 Introduction

Fluorine is a naturally occurring element that is found in various places, such as rocks, minerals, water, and soil. Fluorine is the halogen with the highest electronegative property in the periodic table (Lee et al., 2021) and is an essential component (number 13) in the Earth's crust with an average concentration of 625 mg kg^{-1} (Buck et al., 2016; Camargo & Alonso, 2006). The environment can be exposed to fluoride from geogenic and human-made sources.

15.1.1 Fluoride Sources and Geochemistry in the Environment

Both the prevalence of fluorosis and the severity of its symptoms are influenced by the amount of fluoride that occurs naturally in the soil, plants, and water systems (such as feed, grains, and grasses). A researcher suggests that fluoride-rich minerals that can be discovered in the subsurface zone have the potential to cause damage to natural water supplies when present in large numbers (Bera & Ghosh, 2019). This occurs when fluoride ions are released into the soil and groundwater from areas of the structure that are deemed to be weak. The existence of fluorine in the crust of the Earth is what enables the formation of groundwater feasible. This is due to the fact that rocks that are rich in fluoride minerals comprise huge fluoride reservoirs. Considering that groundwater moves via fissures and voids in granitic rocks and other types of cemented materials, this makes sense (Chabukdhara et al., 2019). Fluoride-containing minerals such as biotite, hornblende, and muscovite have been found in close proximity to volcanic rocks and other igneous rocks. These minerals have the ability to leak fluoride into groundwater (Amiri & Berndtsson, 2020). Tables 15.1 and 15.2, which give a list of the most important rocks and minerals that contain fluorine (Hem, 1985; Pickering, 1985; Yadav et al., 2018).

Table 15.1 Concentration of F^- in different rocks

Rocks	F^- range (ppm)	Average (ppm)
Basalt	20–1060	360
Gneisses and granite	20–2700	870
Clay and shale	10–7600	800
Limestone	0–1200	220
Sandstone	10–880	180
Phosphorite	2400–41,500	31,000
Coal (ash)	40–480	80

Table 15.2 Major geogenic sources of fluoride

Mineral	Chemical formulae
Fluorspar/fluorite	CaF_2
Cryolite	Na_3AlF_6
Fluorapatite	$Ca_5(PO_4)_3F$
Villiaumite	NaF
Topaz	$Al_2SiO_4(F,OH_2)$
Apatite	$Ca_5(Cl,F,OH)\ (PO_4)_3$
Mica	$(AB_{2\text{-}3}[X,Si]_4)_{10}(O,F,OH)_2$
Amphiboles	$A_{0\text{-}1}B_2C_5T_8O_{22}(Cl,F,OH)$
Sellaite	MgF_2

Fluoride-bearing mineral rocks are the primary contributors to groundwater fluoride contamination, as well as anthropogenic pollution, desorption of the surface desorption, ion exchange, and dissolution. The hydrochemistry of groundwater resources is essentially determined by four processes: the dissolution of soluble salts, the weathering of silicate minerals, the nitrate oxidation of organic carbon, and the dissolution of sulfate minerals in aquifers. According to a study, the process of silicate minerals breaking down due to weathering in these aquifers has a direct influence on fluoride enrichment (Yidana et al., 2012). The main cause of fluoride pollution in groundwater is the disintegration of minerals in laterite-sheeted granite, gneiss, and basalt such as biotite and fluorite. Natural seeps and springs are two more sources of fluoride pollution in the environment. Fluoride concentrations are regulated not only by the groundwater depth of the level of groundwater but also by the temperature of the groundwater (Hossain & Patra, 2020).

According to the study, geothermal temperature is one of the variables that leads to the high concentrations of fluoride in groundwater that comes from deep aquifers and geothermal springs (Onipe et al., 2020). This is the case since deep aquifers and geothermal springs are both sources of groundwater. Fluoride concentrations in East and Sub-Saharan Africa have been found to be very high as a direct result of volcanic activity that has occurred in the region. Volcanic ash is formed in a natural process, contains a high fluoride concentration, and dissolves easily in water (Onipe et al., 2020).

Human activities and companies are the main sources of fluoride input into the subsurface aquatic system (Luo et al., 2018). These primary sources include household waste, the use of pesticides, and the application of fertilizer in agricultural settings. Coal combustion and the use of fertilizers in agriculture are two of the most significant sources of fluoride that are induced by human activity (Kaur et al., 2020; Mikkonen et al., 2018). According to a research, the use of phosphate fertilizers can have an effect on the fluoride levels of irrigated crops (Yu et al., 2020). Quite a few factories and commercial establishments already use brick kilns to burn coal. The high fluoride level in groundwater is due to inefficient disposal of fly ash on the earth's surface, which is a contributing factor. When particulate fluoride from an airborne emission disperses to the surface, it first percolates through precipitation and then eventually penetrates the groundwater zone when it rains (Brindha & Elango, 2011; Masood et al., 2022).

Fluoride concentrations in the bicarbonate HCO_3^- fluids found in deep aquifers are higher than those found in shallow aquifers. The length of residence and physicochemical processes such as breakdown, dissociation, and subsequent dissolution may be responsible for the leaching of fluoride into groundwater. There is a connection between the mineralogy of metamorphic rocks and granitoid and the geogenic source of fluoride pollution in groundwater (Madhnure et al., 2007).

The majority of fluoride found in rain comes from three main sources: human activities, volcanic emissions, and marine aerosols. Fluoride amounts found in rain generally range from 0.02 to 0.2 mg L^{-1} when there are clear continental showers are present (Edmunds & Smedley, 2005). Rainfall has a very minimal effect on groundwater (Amor et al., 2001). Fluoride concentrations in seawater are quite high, with values ranging between 1.0 and 1.4 mg L^{-1}, and this is because erosion is responsible for absorbing fluoride and then transferring it to saltwater via streams or rivers.

15.1.2 Fluoride Flow from Irrigation Water to Cultivated Plants and Bioaccumulation

There are two primary methods that are utilized in the application of fluoride to plants (Fig. 15.1). According to Davison and Weinstein 1998, plants move fluoride from one organ to another via xylematic flow, with the leaves being particularly important in this process. Stomata, the openings in plant cells that allow air to enter, are the initial point of contact with the environment. According to Domingos et al. (2003), fluoride is carried by the transpiration stream to the leaf's tip and margins, where it accumulates and has the potential to induce morphological, biochemical, and physiological alterations, as well as occasional cell death. When fluoride is present in the soil, plants have the potential to absorb it via their roots and be exposed to it

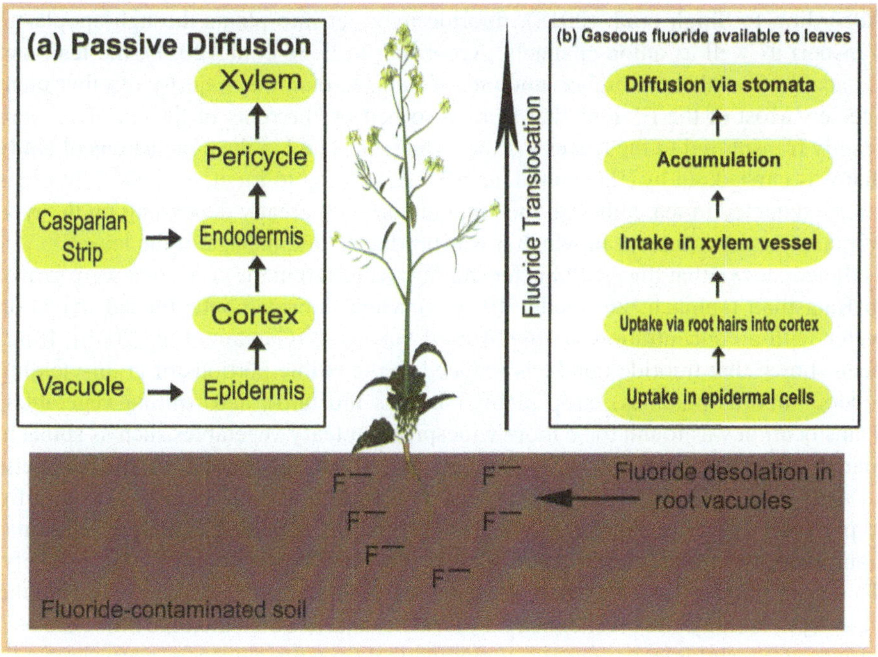

Fig. 15.1 Pathway of F^- movement in case of plants

in a secondary way. According to Mackowiak et al. (2003), the quantity of fluoride that plants are able to absorb is dependent on a number of factors, including the type of plant, the pH of the soil, the fluoride activity and the composition.

It has been demonstrated in a number of studies that fluoride bioconcentration in plants may occur at a variety of various levels (Kalinić et al., 2005; Kozyrenko et al., 2007; Fornasiero, 2001; Saini et al., 2013). According to Gupta and Banerjee (2011), the increased concentration of fluoride in leafy vegetables is due to a faster metabolism and/or photosynthetic rate in green shoots. This is especially true when compared to seeds/grains or other storage components (tubers). In a study by Pal et al. (2012) it was shown that leafy plants, such as spinach, cabbage, radish, and cauliflower, prefer Bioconcentrate Fluoride (BCF > 1), which indicates a faster rate of photosynthesis and higher irrigation water consumption. According to research conducted by Kumpulainen and Kovistoinen (1977), the bulk of fluoride is deposited in the embryo as well as the outer layer of the grain. Because of this, cereals often have a fluoride content of 1 mg per kilogram. According to the findings of several investigations (Haidouti et al., 1993; Saini et al., 2013), spinach has an exceptionally high concentration of fluoride, particularly in regions that are geographically adjacent to industrial centers. Tea is widely consumed as an aperitif in a number of countries, including India, which is one of these countries. Tea was discovered to have the highest fluoride content after it was shown to be one of the traditional beverages responsible for absorbing 67% of the total fluoride found in leaves (Fung et al., 1999).

According to Singh et al. (2018), fluoride may get into plants through apoplastic transport as well as anion channels. According to Peng et al. (2021), the tea plant is capable of withstand higher amounts of fluoride than the majority of other plant species. Most of the F^- ions that were absorbed by the roots of the tea plant were readily transported to the leaves via the xylem, as stated in the conclusions of study that was carried out by Zhang et al. (2013). According to Lu et al. (2004), fluorides can be detected in tea, although the amount can vary greatly depending on the type of plant, the environment in which it was produced, and the age of the leaves.

It was shown that fluoride transfer (mg/kg) occurred in the crops that were grown in Rajasthan (spinach: 26, methi: 19, etc.) when fluoride-contaminated irrigation water with a concentration of 7.4–14 mg/l was used (Gautam et al., 2010). It has been shown that fluoride uptake is reduced in the edible portions of grain-yielding tubers (such as potatoes), crop plants (such as mustard), and fruiting vegetables. Translocation was found to be more widespread in leafy vegetables such as spinach, marsilea, and coriander leaves, as stated by Gupta and Banerjee (2011), respectively.

According to Gautam et al. (2010), high concentrations of fluoride can be harmful to plants in a number of different ways. These methods include inhibiting germination, changing the permeability of membranes, reducing photosynthetic capacity, decreasing productivity and biomass, and creating a range of illnesses related to the plant's physiology and biochemistry. Fluoride can have a considerable influence on a variety of physiological processes, including leaf tip burn, chlorosis, necrosis, modifications in the biochemical ratio of the plant body, and others (McNulty & Newman, 1961; Miller et al., 1999).

Studies (Kundu & Mandal, 2010; McNulty & Newman, 1961) have shown that fluoride can be harmful to the chlorophyll pigment as well as secondary metabolites such as sugar, amino acids, ascorbic acid, and proteins. The levels of fluoride that may be found in a variety of naturally occurring substances, as well as those that are grown, are outlined in Table 15.3.

15.1.3 Detrimental Effects of Fluoride on Plants

The symptoms of fluoride poisoning in plants can be affected by a wide variety of factors, including the amount of fluoride present, the type of fluoride, the length of exposure, the intensity of exposure, temperature, the age of the plant, as well as the other gases present in the environment and the circulation rate of those gases. Fluoride dissolved in the water is absorbed by plants, where it travels through the vascular system and accumulates near the leaf edges (Treshow, 1970). Because of the large amounts of fluoride that are present at the tips of the leaves, the fluoride poisoning symptoms that show up first are often seen on the leaf's perimeter (Treshow, 1970). The accumulation of fluoride may cause a moderate type of necrosis on the upper margin of the leaf, which then may spread to the leaf root. Exposure to HF at low concentrations results in the development of chronic lesions, which can manifest as broad chlorosis or chlorosis along the veins of leaf tissue. Extremely high levels of

Table 15.3 Fluoride level in water, soil, and crops

State	Area	F^- in water	F^- in soil	F^- in crops and vegetables	References
		mg/l			
Gujarat	Vadodra	0.8–3.3	1.3–9.2	–	Parikh et al. (2015)
Rajasthan	Udaipur	0.36–0.71	–	101–189	Bhat et al. (2015)
	Newai Tehsil	0.3–9.8	50–180	8–98	Saini et al. (2013)
	Dausa	5.1–15	–	1.1–14	Yadav et al. (2012)
	Nagaur	0.92–15	–	1.9–26	Gautam et al. (2010)
Uttar Pradesh	Unnao	–	1–4	1.1–55	Jha et al. (2011)
West Bengal	Bankura and Purulia	0.08–1.3	55–399	13–63	Bhattacharya et al. (2017)
	Purulia	0.01–1.7	69–417	–	Bhattacharya (2016)
	Bankura and Purulia	0.08–1.3	55–399	–	Samal et al. (2015)
	West Bengal	0.15–1.8	–	–	Datta et al. (2014)
	Birbhum	0.58–10	–	0.4–4.2	Pal et al. (2012)
	Birbhum	3.2–3.8	–	–	Pal et al. (2012)
	Birbhum	–	–	4–27	Gupta and Banerjee (2011)
	North Parganas	0.01–1.2	–	–	Kundu and Mandal (2010)
	Nadia	0.01–1.2	–	–	Kundu and Mandal (2009)
	Murshidabad	0.02–1.2	–	–	Kundu and Mandal (2009)
	Birbhum	0.62–4.1	140–144	12–13	Gupta et al. (2009)
	Birbhum	0.01–2	–	–	Gupta et al. (2006)

F can cause serious damage, which can manifest as necrosis in the intercostal areas or, if absorbed fast, as necrosis in the leaf tips and margins, which may eventually reach the bases of the leaf. Necrosis can occur anywhere on the leaf, although it most commonly occurs in the intercostal regions. According to Miller et al. (1999), fluoride is absorbed by the plant and then transmitted to the shoots based on the concentration it has in the cell sap. Once there, it has an effect on the physiological, structural, and biochemical components of the cell (Fig. 15.2), which ultimately

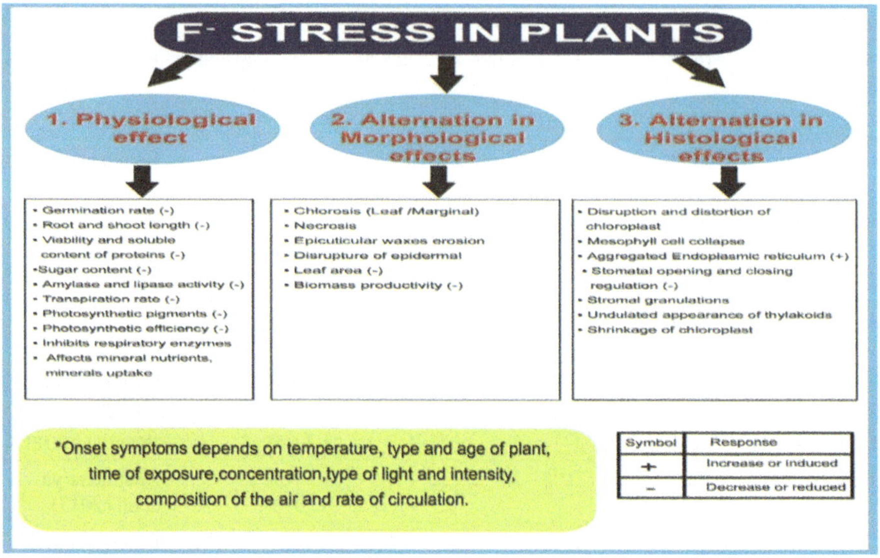

Fig. 15.2 Fluoride stress in the case of plants

leads to cell death. Histochemical experiments in plants that were injured by fluoride revealed that the disintegration and deformation of the chloroplasts of palisade cells followed the initial expression of leaf damage in the lower epidermis and the spongy mesophyll. According to Panda (2015), the upper epidermis is the last layer to show signs of deformation or collapse associated with exposure to fluoride.

Fluoride exposure has been associated with the beginning of the pigmentation process, as well as alterations in the structure of chloroplasts (Trivender et al., 2013). Fluoride has been shown to have a considerable inhibitory effect on a variety of physiological functions, including photosynthesis and other biological activities. According to Landis et al. (2017), necrosis and chlorosis are two obvious symptoms of fluoride poisoning in plants. Necrosis and chlorosis are the two diseases that ultimately result in the death of a plant. Both problems were produced in wheat green leaves when fluoride ions were present at a concentration of 200 mg kg^{-1} (Trivender et al., 2013).

The majority of plant species are susceptible to damage from fluoride (F^-) since it has the ability to disrupt many metabolic pathways (Elloumi et al., 2005). Fluoride has a negative impact on the processes of growth, reproduction, germination, respiration, yield, photosynthesis, and the metabolism of amino acids and proteins (Garrec & Letourneur, 1981). It does this by activating the stromal enzymes and membranes that are involved in the fixing of carbon dioxide. According to Panda (2015), fluoride has the ability to block the action of enzymes that require cofactors such as ions of Mg^{2+}, Mn^{2+}, and Ca^{2+}. According to Elloumi et al. (2005), the levels of chlorophyll, magnesium, calcium, sugar, and starch present in the leaves saw a significant reduction. It would suggest that seeds and seedlings are more vulnerable to the effects of fluoride than fully grown plants. An excessive amount of fluoride deposition to plants leads in damage to the leaves, as well as damage to the fruits and a reduction in production (Ando et al., 1998).

Because farmers have longer cropped seasons, certain crops may be irrigated with fluoridated water for months at a time, increasing the risk of fluoride poisoning (Wollaeger, 2015). In plants that show signs of fluoride poisoning, necrotic patches are more visible around the tips and edges of the leaves. However, despite the severity of the damage, which reduced the amount of photosynthesis that occurs, the green parts of the leaf continued to function normally. After photosynthetic activity was inhibited, there was a very gradual return to normal. There is a lot of mystery around the particular process through which fluorides have an effect on plants.

15.1.4 Changes Observed in Root and Shoot Lengths

Fluoride poisoning shortens the length of seedlings' roots and shoots by causing an imbalance in the amount of nitrogen they take in Sabal et al. (2006). According to Mondal and George (2015), the quantity of fluoride in the soil that included 95 mg NaF kg^{-1} caused the greatest reduction in root biomass, which was 82.5%. The concentration of F^- increased, while the shoot length dramatically reduced. Pant et al. (2008) observed findings that were comparable for wheat, Bengal gram (*Cicer arietinum*), tomato (*Lycopersicon esculentum*), and mustard (*Brassica juncea*). Furthermore, it was found that the growth of shoots and roots in *Prosopis juliflora* was stunted when there was an increase in the amount of NaF (Saini et al., 2012). According to a study, the high concentration of fluoride in the soil caused necrosis and chlorosis in the *Triticum aestivum* plant, which led to a decrease in the plant's productivity. This was also the cause of the decrease in production (Agarwal & Chauhan, 2014).

The presence of dry weight leads to a reduction in metabolic activity, which in turn leads to a drop in fresh weight and a percentage of seedlings (Gupta et al., 2009). This phenomenon occurs linearly. Other crops, such as maize, chili, mustard, mung, and tomato, as well as lady's finger, are not as sensitive to fluoride contamination. Soy yield dropped by as much as 30% when F levels of 375 mg kg^{-1} or more were present (Bustingorri et al., 2015). F^- levels beyond 50 mg L^{-1} led to a reduction in rice production (Singh et al., 1979). Fluoride can be found in vegetables, fruits, and crops in quantities ranging from 0.1 to 0.4 mg kg^{-1}; however, grains can contain up to 2 mg kg^{-1} fluoride (Jolly et al., 1974). Fluoride is generally found in fairly low amounts. The concentrations of total soluble sugar and proline in the leaves were the first to drop. However, when proline progressively accumulated while the plant was going through the germination stage, and the fluoride concentration gradually increased as a result of the synthesis of proline-rich proteins while the plant was stressed, both grew with increasing fluoride concentration. According to research by Greenway and Munns (1980) and Yang and Miller (1963), an increase in the plant's proline and sugar content may make it more able to survive the effects of stress. Yu (1996) discovered that when F^- concentration increased, total soluble sugars, and notably reducing sugars, dropped in mung bean (*Vigna radiata*) seedlings. This was especially true for reducing sugars. The protein content of the seedling leaves progressively decreased as the concentration of fluoride steadily increased as a result of stress (Singh et al., 1985).

Even at low doses, fluoride can have a negative impact on the health of plant species and other types of plants (Sahariya et al., 2021). However, *Lolium multiflorous* and *Trifolium repens* have fluoride absorption values of 50 mg kg^{-1} and 30 mg kg^{-1}, respectively, despite the fact that the indicated lower limit for fluoride accumulation in plants is 10 mg L^{-1} (Arnesen, 1997). This contrasts to the fact that this value is only 10 mg L^{-1}. These figures are significantly higher than the recommended lower limit. As a result of this, it was discovered that the toxicity of fluoride can have a detrimental impact on the enzymatic activity of cells, as well as the germination of seeds and many other processes, as indicated in Table 15.4.

The process through which plants take fluoride from contaminated soil and air, move it to other parts of their bodies, and store it in their cell walls is illustrated in Fig. 15.2. Due to their susceptibility to fluoride buildup, plants may have adverse effects on their growth and development even at low levels of fluoride deposition (Sahariya et al., 2021). A wide variety of plants, such as *Lupinus luteus* and *Zea mays*, have been found to be fluoride-tolerant (Banerjee & Roychoudhury, 2019b). This is likely because these plants are able to manufacture protein and protect themselves against protein breakdown.

Fluoride may have a minor, major, or long-lasting influence on other plant species (Choudhary et al., 2019). This effect may vary depending on the species. Fluoride toxicity may have a negative impact on ROS production, gene expression patterns, enzyme activities, carbohydrate metabolism, accumulation of nucleotide synthesized

Table 15.4 Detrimental impact on crops due to F^- toxicity

Plants/crops	Growth and development	References
Paddy rice (*Oryza sativa*)	Seed germination drops to 96% and 92% for 20 and 30 mg L^{-1} of F^-	Gupta et al. (2009)
Paddy rice (seedlings)	Seed germination drops to 26.14% and 47.47% for 25 and 50 mg L^{-1} of F^-	Singh et al. (2020)
Cajanuscajan L.	Active oxygen species increases	Yadu et al. (2018)
Tomato (*Solanum lycopersicum*)	Metabolic changes increase, and the germination rate decreases	Ahmad et al. (2018)
Paddy rice (*Oryza sativa*)	Inhibits shoot and root length	Banerjee et al. (2020)
Aromatic and nonaromatic indica rice	Tip burning chlorosis, loss of abscisic acid, and inhibition of polyamine biosynthesis and the ascorbate–glutathione cycle	Banerjee and Roychoudhury (2019a)
Rice (*Oryza sativa*)	Oxidative stress, reduction in chlorophyll content	Singh and Roychoudhury (2023)

biomass, protein synthesis, respiration, photosynthesis, fruit and leaf damage, yield and growth, germination, and plant metabolism (Bhardwaj et al., 2021; Choudhary et al., 2019; Kumar et al., 2017).

According to Mondal (2017), the rice germination pattern (*Oryza sativa*) gradually shifts in a different direction when the concentration of fluoride pigment in the soil increases. Three different winter wheat cultivars showed reduced germination, slower root growth, and inhibited catalase activity (Pelc et al., 2020). There is a possibility that several cultivars of a given species will gather varying levels of fluoride in their produce. According to the findings of a study, the aromatic rice variety Gobindobhog (GB) is less susceptible to the adverse effects of rising fluoride concentrations than the non-aromatic rice variety IR-64 (Banerjee & Roychoudhury, 2019a). This is determined by comparing the relative water content, vigor index, dry weight, and fresh weight of the varieties. It is possible for fluoride to enter plants either by its roots or their stomata. The cortex and epidermis of secondary roots are the primary pathways through which fluoride ions reach the xylem and phloem (Fig. 15.1). Stomata are responsible for the release of fluoride from plants (Singh et al., 2018).

It has been discovered that there was a linear association between fluoride concentrations and the physiological measures that were induced in paddy rice (*Oryza sativa*) (Gupta et al., 2009). These physiological measurements included the length of the shoots and roots as well as the total biomass. Increased bioaccumulation of fluoride in paddy rice is caused by high fluoride concentrations (more than 10 mg L^{-1}), which also inhibits seed germination (for example, 75 mg of fluoride is accumulated per kg of dry biomass for every 10 mg L^{-1} of initial fluoride concentration). Singh and Roychoudhary (2020) observed a decrease in biomass, root and shoot length, seed germination, and chlorophyll concentration in rice seedlings that were exposed to two fluoride concentrations of 25 and 50 mg L^{-1}. This finding suggests that oxidative stress may be the source of the observed changes in rice seedlings. In addition, the chlorophyll concentration was lowered by 1.8 and 2.3 times, for 25 and 50 mg of L^{-1}. Yadu et al. (2018) discovered that fluoride poisoning affected membrane stability, protein concentration, genomic template integrity, and fluoride deposition in dry biomass. Additionally, fluoride poisoning caused a drop in the amount of fluoride deposited. The quantities of oxygen species, reactive cell death, DNA polymorphism, protein carbonylation, and lipase activity all increase when there is toxicity.

The ability of plants to store fluoride was investigated by Baunthiyal and Ranghar (2015), who relied on a variety of studies to reach their conclusion. Fluoride has been shown to be predominantly deposited in the root systems of plants, as stated by Rizzu et al. (2021). Fluoride is harmful to plants because it lowers the amount of chlorophyll they have, it causes oxidative stress and it alters the quantities of soluble sugars, betaine, proline, nitrogen, and other macro and micronutrients in their tissues (Rizzu et al., 2021). According to Mondal (2017), fluoride deposition was greater in the roots of four different types of rice types than it was in the shoots. According to Zouari et al. (2017) research, the fluoride concentration in the roots of Olive (*Olea europaea*) plants was much greater than that of the plant's leaves. Olive trees exhibited the greatest decrease in antioxidant enzymes and mineral content, as well as an increase in oxidative stress, at 80 mM sodium fluoride (NaF) concentrations. These results were found in the study conducted by the researchers. Photosynthetic properties, foliar water status, cell membranes and photosynthetic pigments of loquat trees are all negatively impacted by the presence of fluoride in polluted air (Elloumi et al., 2017). In a dose-dependent study, the stress caused by sodium fluoride resulted in a considerable drop in the yield of eggplant or brinjal. The yield decreased the highest when exposed to 600 mg L^{-1} of NaF (Ali et al., 2020).

Fluoride levels in fish and shellfish that are contained in solid meals are comparable to the levels of fluoride found in ocean water. In addition, there is evidence that the amounts of fluoride in bread, baked goods, cereals, and other grain products are higher than they should be (Jackson et al., 2002; Singer et al., 1980).

15.2 Techniques of Removal

Certain fluoride removal techniques could reduce the toxicity of fluoride in water and soil.

15.2.1 Nalgonda Method

Two chemicals, lime (calcium oxide) and alum (aluminum sulfate or potassium and aluminum sulfate), are introduced and promptly combined with fluoride-contaminated water (Dahi et al., 1996). Mild stirring causes flakes (aluminum hydroxides) to develop, which can be filtered out using sedimentation. The flocs and the crucial components of the fluoride by using a combination of ion exchange, sorption, and a few of the hydroxide groups.

15.2.2 Bone Char

The bones of the animals that have only been burned to remove all organics. It is mainly composed of carbon and tricalcium phosphate. It was first documented in defluoridation facilities by Scott et al. (1937) and later by Sorg and Logsdon (1978), and was once less costly than bone because it was used to bleach tube syrup.

15.2.3 Synthetic Tricalcium Phosphate

To create this, phosphoric acid is combined with lime. To refresh the medium, it was washed with a moderately acidic solution (1% NaOH) and dried. Water fluoride levels of as high as 700 mg/L may be reduced.

15.2.4 Florex

It is a commercially available blend of tricalcium phosphate and Hydroxy-apatite, and it can remove fluoride from 600 mg of fluoride/litre when regenerated with a solution of 1.5% sodium hydroxide.

15.2.5 Activated Carbon with Lime

Different types of activated carbon have a high fluoride removal capacity. Following a routine evaluation of raw and treated municipal waste, Scott et al. (1937) observed a drop in fluoride content in effluents from lime softening facilities compared to fluoride in raw water.

15.2.6 Limestone, Peculiar Soils, Clay

Lately, heat-treated soils and limestone have been tested to remove fluoride. Fluoride levels in wastewater have been reduced below the MCL of 4 mg/l using limestone. MCL stands for Maximum Contamination Level. The mechanism of fluoride sorption by clay minerals is postulated on experimental results. The study examined fluoride removal using adsorption on low-cost materials such as bentonite, kaolinite, lignite seeds, and coals.

15.2.7 Natural Minerals and Fly Ash

Red soil, charcoal, bricks, and fly ash are examples of natural materials that have been used to try to remove fluoride. According to the analysis, red soil has the best fluoride removal ability, followed by charcoal, fly ash, and brick.

15.2.8 Electrokinetic Remediation

It is an efficient way to treat polluted organic and inorganic soils, sludges, and sediments. The EK approach provides a direct electrical potential to the contaminated soil through electrodes. Pollutants are mobilized to the electrodes in a direct-current electrical field through a complex series of reactions and transport channels in the diseased soil. An electric field is utilized in this method to stimulate the migration of pollutants toward the electrode. However, other natural chemicals found in the soil can also be mobilized. However, little research has been done on soil fertility after electrokinetic therapy.

However, the efficacy of these fluoride removal procedures is low. Low-cost residential use options include bone charcoal, calcined clay, and the Nalgonda process. The Nalgonda procedure is also a cost-effective solution at the community level. If a high level of fluoride removal is required this method is preferable for electrokinetic remediation.

15.3 Conclusion

This review primarily discussed bioaccumulation in the case of plants and microbial approaches to fluoride removal. Accumulation in plants poses a significant risk, negatively impacting their development and growth, and enters in plants via soil, leaves, or water. Continuous exposure to fluoride can cause physiological, biochemical, and molecular changes in plants. Therefore, fluoride is a common element found in all sections of the environment and comes from natural and artificial sources. By human activities, fluoride is now more readily available to humans, animals, and plants, and this has beneficial and bad consequences for all species. Insufficient evidence suggests that fluoride has positive effects on plant growth. Excessive levels of fluoride in the air or soil can damage plant leaves (albeit at different rates for different species). Although it is critical to recognize the advantages fluoride provides, it is equally crucial to question its industrial emissions and addition to municipal drinking water, as well as its harmful side effects. Fluoride toxicity in people and plants is caused by excessive fluoride consumption by ingestion or inhalation of various sources. Fluoride pollution in groundwater is a major source of concern. Fluoride contamination has the potential to severely damage plant species. Given all these considerations, fluoride toxicity and associated mitigation techniques have become a significant challenge for agricultural scientists trying to achieve consistent production under the effect of fluoride stress. More information is needed before we can make informed recommendations to mitigate the dangers posed by increasing levels of F^- in our pastoral systems and advance sustainable practices for the future.

Acknowledgements The authors acknowledge the financial support of CSIR, New Delhi, and other respective institutions, which are also acknowledged for this work.

References

Agarwal, R., & Chauhan, S. S. (2014). Bioaccumulation of sodium fluoride toxicity in *Triticum aestivum* var. Raj. 3077. *International Journal of Food, Agriculture & Veterinary Sciences, 4*(1), 98–101.

Ahmad, M. A., Bibi, H., Munir, I., Ahmad, M. N., Zia, A., Mustafa, G., Ullah, I., & Khan, I. (2018). Fluoride toxicity and its effect on two varieties of Solanum lycopersicum. *Fluoride, 51*, 267–277.

Ali, S. A. R., Ansari, M., Ahmad, M. N., Haider, A., Jabeen, F., & Zulfiqar, A. (2020). Effect of sodium fluoride stress on the productivity of solanum melongena l., (aubergine, brinjal, eggplant). *Fluoride, 53*(3), 521–530.

Amiri, V., & Berndtsson, R. (2020). Fluoride occurrence and human health risk from groundwater use at the west coast of Urmia Lake, Iran. *Arabian Journal of Geosciences, 13*, 921. https://doi.org/10.1007/s12517-020-05905-7

Amor, Z., Bariou, B., Mameri, N., Taky, M., Nicolas, S., & Elmidaoui, A. (2001). Fluoride removal from brackish water by electrodialysis. *Desalination, 133*(3), 215–223. https://doi.org/10.1016/S0011-9164(01)00102-3

Ando, M., Tadano, M., Asanuma, S., Tamura, K., Matsushima, S., Watanabe, T., Kondo, T., Sakurai, S., Ji, R., & Liang, C. (1998). Health effects of indoor fluoride pollution from coal burning in China. *Environmental Health Perspectives, 106*(5), 239–244. https://doi.org/10.1289/ehp.98106239

Arnesen, A. (1997). Availability of fluoride to plants grown in contaminated soils. *Plant and Soil, 191*, 13–25. https://doi.org/10.1023/A:1004210713596

Banerjee, A., & Roychoudhury, A. (2019a). Differential regulation of defence pathways in aromatic and non-aromatic Indica rice cultivars towards fluoride toxicity. *Plant Cell Reports, 38*, 1217–1233. https://doi.org/10.1007/s00299-019-02438-6

Banerjee, A., & Roychoudhury, A. (2019b). Melatonin application reduces fluoride uptake and toxicity in rice seedlings by altering abscisic acid, gibberellin, auxin and antioxidant homeostasis. *Plant Physiology and Biochemistry, 145*, 164–173.

Banerjee, A., Samanta, S., & Roychoudhury, A. (2020). Spermine ameliorates prolonged fluoride toxicity in soil-grown rice seedlings by activating the antioxidant machinery and glyoxalase system. *Ecotoxicology and Environmental Safety, 189*, 109737. https://doi.org/10.1016/j.ecoenv.2019.109737

Baunthiyal, M., & Ranghar, S. (2015). Accumulation of fluoride by plants: Potential for phytoremediation. *Clean Soil Air Water, 43*, 127–132.

Bera, B., & Ghosh, A. (2019). Fluoride dynamics in hydrogeological diversity and fluoride contamination index mapping: A correlation study of North Singbhum Craton, India. *Arabian Journal of Geosciences, 12*(24), 802. https://doi.org/10.1007/s12517-019-4994-8

Bhat, N., Jain, S., Asawa, K., Tak, M., Shinde, K., Singh, A., Gandhi, N., & Gupta, V. V. (2015). Assessment of fluoride concentration of soil and vegetables in vicinity of zinc smelter, Debari, Udaipur, Rajasthan. *Journal of Clinical and Diagnostic Research: JCDR, 9*(10), ZC63. https://doi.org/10.7860/JCDR/2015/13902.6667

Bhattacharya, P. (2016). Analysis of fluoride distribution and community health risk in Purulia district of West Bengal, India. In *Proceedings of the 9th National Level Science Symposium organized by the Christ college.*

Bhattacharya, P., Samal, A. C., Banerjee, S., Pyne, J., & Santra, S. C. (2017). Assessment of potential health risk of fluoride consumption through rice, pulses, and vegetables in addition to consumption of fluoride-contaminated drinking water of West Bengal, India. *Environmental Science and Pollution Research, 24*(25), 20300–20314. https://doi.org/10.1007/s11356-017-9649-2

Brindha, K., & Elango, L. (2011). Fluoride in groundwater: Causes, implications and mitigation measures. *Fluoride Properties, Applications and Environmental Management, 1*, 111–136.

Buck, B. J., Londono, S. C., McLaurin, B. T., Metcalf, R., Mouri, H., Selinus, O., & Shelembe, R. (2016). The emerging field of medical geology in brief: Some examples. *Environmental Earth Sciences, 75*(6), 449. https://doi.org/10.1007/s12665-016-5362-6

Bustingorri, C., Balestrasse, K. B., & Lavado, R. S. (2015). Effects of high arsenic and fluoride soil concentrations on soybean plants.

Camargo, J. A., & Alonso, Á. (2006). Ecological and toxicological effects of inorganic nitrogen pollution in aquatic ecosystems: A global assessment. *Environment International, 32*(6), 831–849. https://doi.org/10.1016/j.envint.2006.05.002

Chabukdhara, M., Gogoi, M., & Gupta, S. K. (2019). Potential and feasibility of the microalgal system in removal of pharmaceutical compounds from wastewater. In S. K. Gupta, F. Bux (Eds.), *Application of microalgae in wastewater treatment*. Springer, Cham. https://doi.org/10.1007/978-3-030-13913-1_10

Choudhary, S., Rani, M., Devika, O. S., Patra, A., Singh, R. K., & Prasad, S. K. (2019). Impact of fluoride on agriculture: A review on it's sources, toxicity in plants and mitigation strategies. *International Journal of Chemical Studies, 7*(2), 1675–1680.

Dahi, E., Mtalo, F., Njau, B., & Bregnhj, H. (1996). Defluoridation using the Nalgonda technique in Tanzania.

Datta, A., Chakrabortty, A., De Dalal, S., & Lahiri, S. (2014). Fluoride contamination of underground water in West Bengal, India. *Fluoride, 47*(3), 241–248.

Domingos, M., Klumpp, A., Rinaldi, M. C. S., Modesto, I. F., Klumpp, G., & Delitti, W. B. C. (2003). Combined effects of air and soil pollution by fluoride emissions on Tibouchina pulchra Cogn., at Cubatão, SE Brazil, and their relations with aluminium. *Plant and Soil, 249*(2), 297–308. https://doi.org/10.1023/A:1022800225753

Edmunds, W., & Smedley, P. (2005). In B. J. Alloway & O. Selinus (Eds.), *Fluoride in natural waters essentials of medical geology*. Elsevier.

Elloumi, N., Abdallah, F. B., Mezghani, I., Rhouma, A., Boukhris, M., & Tunisia, S. (2005). Effect of fluoride on almond seedlings in culture solution. *Fluoride, 38*(3), 193.

Elloumi, N., Zouari, M., Mezghani, I., Ben Abdallah, F., Woodward, S., & Kallel, M. (2017). Adaptive biochemical and physiological responses of *Eriobotrya japonica* to fluoride air pollution. *Ecotoxicology, 26*(7), 991–1001. https://doi.org/10.1007/s10646-017-1827-y

Fornasiero, R. B. (2001). Phytotoxic effects of fluorides. *Plant Science, 161*(5), 979–985. https://doi.org/10.1016/S0168-9452(01)00499-X

Fung, K., Zhang, Z., Wong, J., & Wong, M. H. (1999). Fluoride contents in tea and soil from tea plantations and the release of fluoride into tea liquor during infusion. *Environmental Pollution, 104*(2), 197–205. https://doi.org/10.1016/S0269-7491(98)00187-0

Garrec, J., & Letourneur, L. (1981). Fluoride absorption by the root and foliar tissues of the horse-bean (calicole) and lupin (calcifuge). *Fluoride (United States), 14*(1).

Gautam, R., Bhardwaj, N., & Saini, Y. (2010). Fluoride accumulation by vegetables and crops grown in Nawa Tehsil of Nagaur district (Rajasthan, India). *Journal of Phytology, 2*(2).

Greenway, H., & Munns, R. (1980). Mechanisms of salt tolerance in nonhalophytes. *Annual Review of Plant Physiology, 31*(1), 149–190.

Gupta, S., & Banerjee, S. (2011). Fluoride accumulation in crops and vegetables and dietary intake in a fluoride-endemic area of West Bengal. *Fluoride, 44*(3), 153.

Gupta, S., Banerjee, S., & Mondal, S. (2009). Phytotoxicity of fluoride in the germination of paddy (*Oryza sativa*) and its effect on the physiology and biochemistry of germinated seedlings. *Fluoride, 42*(2), 142.

Gupta, S., Banerjee, S., Saha, R., Datta, J., & Mondal, N. (2006). Fluoride geochemistry of groundwater in Nalhati-1 block of the Birbhum district, West Bengal, India. *Fluoride, 39*(4), 318.

Haidouti, C., Chronopoulou, A., & Chronopoulos, J. (1993). Effects of fluoride emissions from industry on the fluoride concentration of soils and vegetation. *Biochemical Systematics and Ecology, 21*(2), 195–208. https://doi.org/10.1016/0305-1978(93)90037-R

Hem, J. D. (1985). *Study and interpretation of the chemical characteristics of natural water* (p. 2254). Department of the Interior, US Geological Survey, Washington, DC, USA.

Hossain, M., & Patra, P. K. (2020). Hydrogeochemical characterisation and health hazards of fluoride enriched groundwater in diverse aquifer types. *Environmental Pollution, 258*, 113646. https://doi.org/10.1016/j.envpol.2019.113646

Jackson, R. D., Brizendine, E. J., Kelly, S. A., Hinesley, R., Stookey, G. K., & Dunipace, A. J. (2002). The fluoride content of foods and beverages from negligibly and optimally fluoridated communities. *Community Dentistry and Oral Epidemiology, 30*(5), 382–391. https://doi.org/10.1034/j.1600-0528.2002.00002.x

Jha, S., Nayak, A., & Sharma, Y. (2011). Site specific toxicological risk from fluoride exposure through ingestion of vegetables and cereal crops in Unnao district, Uttar Pradesh, India. *Ecotoxicology and Environmental Safety, 74*(4), 940–946. https://doi.org/10.1016/j.ecoenv.2011.01.002

Jolly, S., Oberoi, D., Sharma, R., & Ralhan, S. (1974). Fluoride balance studies in cases of endemic fluorosis. In *Proceeding of the symposium on fluorosis*, Hyderabad, India.

Kalinić, N., Hršak, J., Vadjić, V. et al. (2005). Fluoride content in soil and vegetation. *Bulletin of Environmental Contamination and Toxicology, 75*, 157–162. https://doi.org/10.1007/s00128-005-0732-x

Kaur, L., Rishi, M. S., & Siddiqui, A. U. (2020). Deterministic and probabilistic health risk assessment techniques to evaluate non-carcinogenic human health risk (NHHR) due to fluoride and nitrate in groundwater of Panipat, Haryana, India. *Environmental Pollution, 259*, 113711. https://doi.org/10.1016/j.envpol.2019.113711

Kozyrenko, M. M., Artyukova, E. V., Shmakov, V. N., & Konstantinov, Y. M. (2007). Effect of fluoride pollution on genetic variability of *Larix gmelinii* (Pinaceae) in East Siberia. *Journal of Forest Research, 12*(5), 388–392. https://doi.org/10.1007/s10310-007-0031-y

Kumar, K., Giri, A., Vivek, P., Kalaiyarasan, T., & Kumar, B. (2017). Effects of fluoride on respiration and photosynthesis in plants: An overview. *Annals of Environmental Science and Toxicology, 2*(1), 043–047.

Kumpulainen, J., & Koivistoinen, P. (1977). Fluorine in foods. In F. A. Gunther (Ed.), *Residues of pesticides and other contaminants in the total environment*, New York, NY.

Kundu, M., & Mandal, B. (2010). Fluoride concentration in groundwater in the North 24-paraganas district of West Bengal, India. *Fluoride, 43*(2), 160–164.

Kundu, M. C., & Mandal, B. (2009). Agricultural activities influence nitrate and fluoride contamination in drinking groundwater of an intensively cultivated district in India. *Water, Air, and Soil Pollution, 198*(1), 243–252. https://doi.org/10.1007/s11270-008-9842-5

Landis, W. G., Sofield, R. M., & Yu, M.-H. (2017). Introduction to environmental toxicology: molecular substructures to ecological landscapes. Landis, Ruth M. Sofield, Ming-Ho Yu.

Lee, Y.-K., Lee, C.-H., Kang, G.-S., Eom, K., Cho, S. Y., Lee, S., & Joh, H.-I. (2021). Understanding an exceptionally fast and stable Li-ion charging of highly fluorinated graphene with fine-controlled C-F configuration. *ACS Applied Materials & Interfaces, 13*(45), 53767–53776. https://doi.org/10.1021/acsami.1c13811

Lu, Y., Guo, W.-F., & Yang, X.-Q. (2004). Fluoride content in tea and its relationship with tea quality. *Journal of Agricultural and Food Chemistry, 52*(14), 4472–4476. https://doi.org/10.1021/jf0308354

Luo, W., Gao, X., & Zhang, X. (2018). Geochemical processes controlling the groundwater chemistry and fluoride contamination in the Yuncheng Basin, China—An area with complex hydrogeochemical conditions. *PLoS ONE, 13*(7), e0199082. https://doi.org/10.1371/journal.pone.0199082

Mackowiak, C., Grossl, P., & Bugbee, B. (2003). Biogeochemistry of fluoride in a plant–solution system. *Journal of Environmental Quality, 32*(6), 2230–2237. https://doi.org/10.2134/jeq2003.2230

Madhnure, P., Sirsikar, D., Tiwari, A., Ranjan, B., & Malpe, D. (2007). Occurrence of fluoride in the groundwaters of Pandharkawada area, Yavatmal district, Maharashtra, India. *Current Science*, 675–679.

Masood, N., Hudson-Edwards, K. A., & Farooqi, A. (2021). Groundwater nitrate and fluoride profiles, sources and health risk assessment in the coal mining areas of Salt Range, Punjab Pakistan. *Environmental Geochemistry and Health*, 1–14.

McNulty, I., & Newman, D. (1961). Mechanism(s) of fluoride induced chlorosis. *Plant Physiology, 36*(4), 385. https://doi.org/10.1104/pp.36.4.385

Mikkonen, H. G., van de Graaff, R., Mikkonen, A. T., Clarke, B. O., Dasika, R., Wallis, C. J., & Reichman, S. M. (2018). Environmental and anthropogenic influences on ambient background concentrations of fluoride in soil. *Environmental Pollution, 242*, 1838–1849. https://doi.org/10.1016/j.envpol.2018.07.083

Miller, G., Shupe, J., & Vedina, O. (1999). Accumulation of fluoride in plants exposed to geothermal and industrial water. *Fluoride, 32*(2), 74–83.

Mondal, N. K. (2017). Effect of fluoride on photosynthesis, growth and accumulation of four widely cultivated rice (*Oryza sativa* L.) varieties in India. *Ecotoxicology and Environmental Safety, 144*, 36–44. https://doi.org/10.1016/j.ecoenv.2017.06.009

Mondal, P., & George, S. (2015). Removal of fluoride from drinking water using novel adsorbent magnesia-hydroxyapatite. *Water, Air, & Soil Pollution, 226*(8), 241. https://doi.org/10.1007/s11270-015-2515-2

Onipe, T., Edokpayi, J. N., & Odiyo, J. O. (2020). A review on the potential sources and health implications of fluoride in groundwater of Sub-Saharan Africa. *Journal of Environmental Science and Health, Part A, 55*(9), 1078–1093. https://doi.org/10.1080/10934529.2020.1770516

Pal, K. C., Mondal, N. K., Bhaumik, R., Banerjee, A., & Datta, J. K. (2012). Incorporation of fluoride in vegetation and associated biochemical changes due to fluoride contamination in water and soil: A comparative field study. *Annals of Environmental Science, 6*.

Panda, D. (2015). Fluoride toxicity stress: Physiological and biochemical consequences on plants. *International Journal of Bioresearch and Environmental Agricultural Science, 1*, 70–84.

Pant, S., Pant, P., & Bhiravamurthy, P. (2008). Effects of fluoride on early root and shoot growth of typical crop plants of India. *Fluoride, 41*(1), 57.

Parikh, P., Parikh, R., & Dhoria, B. (2015). Status of fluoride contamination in soil, water and vegetation at Kadipani fluorspar mine.

Pelc, J., Śnioszek, M., Wróbel, J., & Telesiński, A. (2020). Effect of fluoride on germination, early growth and antioxidant enzymes activity of three winter wheat (Triticum aestivum L.) cultivars. *Applied Sciences, 10*, 6971.

Peng, C. y., Xu, X. f., Ren, Y. f., Niu, H. l., Yang, Y. q., Hou, R. y., Wan, X. c., & Cai, H. m. (2021). Fluoride absorption, transportation and tolerance mechanism in *Camellia sinensis*, and its bioavailability and health risk assessment: a systematic review. *Journal of the Science of Food and Agriculture, 101*(2), 379–387.https://doi.org/10.1002/jsfa.10640

Pickering, W. (1985). The mobility of soluble fluoride in soils. *Environmental Pollution Series b, Chemical and Physical, 9*(4), 281–308. https://doi.org/10.1016/0143-148X(85)90004-7

Rizzu, M., Tanda, A., Cappai, C., Roggero, P. P., & Seddaiu, G. (2021). Impacts of soil and water fluoride contamination on the safety and productivity of food and feed crops: A systematic review. *Science of the Total Environment, 787*, 147650. https://doi.org/10.1016/j.scitotenv.2021.147650

Sabal, D., Khan, T., & Saxena, R. (2006). Effect of sodium fluoride on cluster bean (*Cyamopsis tetragonoloba*) seed germination and seedling growth. *Fluoride, 39*(3), 228.

Sahariya, A., Bharadwaj, C., Emmanuel, I., & Alam, A. (2021). Fluoride toxicity in soil and plants: An overview. *Asian Journal of Advances in Research*, 26–34.

Saini, P., Khan, S., Baunthiyal, M., & Sharma, V. (2012). Organ-wise accumulation of fluoride in *Prosopis juliflora* and its potential for phytoremediation of fluoride contaminated soil. *Chemosphere, 89*(5), 633–635. https://doi.org/10.1016/j.chemosphere.2012.05.034

Saini, P., Khan, S., Baunthiyal, M., & Sharma, V. (2013). Mapping of fluoride endemic area and assessment of F−1 accumulation in soil and vegetation. *Environmental Monitoring and Assessment, 185*(2), 2001–2008. https://doi.org/10.1007/s10661-012-2683-0

Samal, A. C., Bhattacharya, P., Mallick, A., Ali, M. M., Pyne, J., & Santra, S. C. (2015). A study to investigate fluoride contamination and fluoride exposure dose assessment in lateritic zones of West Bengal, India. *Environmental Science and Pollution Research, 22*(8), 6220–6229. https://doi.org/10.1007/s11356-014-3817-4

Scott, R., Kimberly, A., Van Horn, A., Ey, L., & Waring, F. (1937). Fluoride in Ohio water supplies—Its effect, occurrence and reduction. *Journal (American Water Works Association), 29*(1), 9–25.

Singer, L., Ophaug, R., & Harland, B. (1980). Fluoride intake of young male adults in the United States. *The American Journal of Clinical Nutrition, 33*(2), 328–332. https://doi.org/10.1093/ajcn/33.2.328

Singh, A., Banerjee, A., & Roychoudhury, A. (2020). Seed priming with calcium compounds abrogate fluoride-induced oxidative stress by upregulating defence pathways in an indica rice variety. *Protoplasma, 257*, 767–782.

Singh, A., Chhabra, R., & Abrol, I. (1979). Effect of fluorine and phosphorus on the yield and chemical composition of rice (*Oryza sativa*) grown in soils of two sodicities. *Soil Science, 127*(2), 86–93.

Singh, A., & Roychoudhury, A. (2020). Silicon-regulated antioxidant and osmolyte defense and methylglyoxal detoxification functions co-ordinately in attenuating fluoride toxicity and conferring protection to rice seedlings. *Plant Physiology and Biochemistry, 154*, 758–769. https://doi.org/10.1016/j.plaphy.2020.06.023

Singh, A., & Roychoudhury, A. (2023). Salicylic acid–mediated alleviation of fluoride toxicity in rice by restricting fluoride bioaccumulation and strengthening the osmolyte, antioxidant and glyoxalase systems. *Environmental Science and Pollution Research, 30*(10), 25024–25036. https://doi.org/10.1007/s11356-021-14624-9

Singh, G., Kaur, P., & Sharma, R. (1985). Effect of CCC and kinetin on certain biochemical parameters in wheat under different salinity levels. *Plant Physiology & Biochemistry.*

Singh, G., Kumari, B., Sinam, G., Kumar, N., & Mallick, S. (2018). Fluoride distribution and contamination in the water, soil and plants continuum and its remedial technologies, an Indian perspective—A review. *Environmental Pollution, 239*, 95–108. https://doi.org/10.1016/j.envpol.2018.04.002

Sorg, T. J., & Logsdon, G. S. (1978). Treatment technology to meet the interim primary drinking water regulations for inorganics: Part 2. *Journal-American Water Works Association, 70*(7), 379–393. https://doi.org/10.1002/j.1551-8833.1978.tb04198.x

Treshow, M. (1970). Environment and plant response. *Environment and Plant Response.*

Trivender, K., Dhaka, T., & Arya, K. (2013). Effect of fluoride toxicity on biochemical parameters (chlorophyll, nitrogen, protein and phosphorus) of wheat (*Triticum aestivum* L.). *International Journal of Forestry and Crop Improvement, 4*(2), 80–83.

Wollaeger, H. (2015). Fluoride toxicity in plants irrigated with city water. Assessed in http://www.msue.anr.msu.edu/news/fluoride_toxicity_in_plants_irrigated_with_city_water.

Yadav, K. K., Gupta, N., Kumar, V., Khan, S. A., & Kumar, A. (2018). A review of emerging adsorbents and current demand for defluoridation of water: Bright future in water sustainability. *Environment International, 111*, 80–108. https://doi.org/10.1016/j.envint.2017.11.014

Yadav, R. K., Shipra, S., Megha, B., Ajay, S., Vivek, P., & Raaz, M. (2012). Effects of fluoride accumulation on growth of vegetables and crops in Dausa district, Rajasthan, India. *Advances in Bio Research, 3*(4), 14–16.

Yadu, B., Chandrakar, V., Meena, R. K., Poddar, A., & Keshavkant, S. (2018). Spermidine and melatonin attenuate fluoride toxicity by regulating gene expression of antioxidants in *Cajanus cajan* L. *Journal of Plant Growth Regulation, 37*(4), 1113–1126. https://doi.org/10.1007/s00344-018-9786-y

Yang, S., & Miller, G. (1963). Biochemical studies on the effect of fluoride on higher plants. 1. Metabolism of carbohydrates, organic acids and amino acids. *Biochemical Journal, 88*(3), 505. https://doi.org/10.1042/bj0880505

Yidana, S. M., Ophori, D., Banoeng-Yakubo, B., & Samed, A. A. (2012). A factor model to explain the hydrochemistry and causes of fluoride enrichment in groundwater from the middle Voltaian sedimentary aquifers in the northern region, Ghana. *ARPN Journal of Engineering and Applied Sciences, 7*(1), 50–68.

Yu, M.-H. (1996). Effect of Fluoride on growth and soluble sugars in germinating Mung Bean (*Vigna radiata*) seeds. *Society for, 3.*

Yu, Y.-Q., Cui, S.-F., Fan, R.-J., Fu, Y.-Z., Liao, Y.-L., & Yang, J.-Y. (2020). Distribution and superposed health risk assessment of fluorine co-effect in phosphorous chemical industrial and agricultural sources. *Environmental Pollution, 262*, 114249. https://doi.org/10.1016/j.envpol.2020.114249

Zhang, L., Li, Q., Ma, L., & Ruan, J. (2013). Characterization of fluoride uptake by roots of tea plants (*Camellia sinensis* (L.) O. Kuntze). *Plant and Soil, 366*(1), 659–669. https://doi.org/10.1007/s11104-012-1466-2

Zouari, M., Elloumi, N., Bellassoued, K., Ahmed, C. B., Krayem, M., Delmail, D., Elfeki, A., Rouina, B. B., Abdallah, F. B., & Labrousse, P. (2017). Enzymatic antioxidant respones and mineral status in roots and leaves of olive plants subjected to fluoride stress. *South African Journal of Botany, 111*, 44–49. https://doi.org/10.1016/j.sajb.2017.03.039

Chapter 16
Performance Analysis of Passive Solar Still for De-fluoridation of Water

Krishn Pratap Singh, Abhishek Dixit, Bhanu Pratap Singh, and Deepesh Singh

Abstract Due to pollution and rapid resource depletion, the availability of clean water is decreasing at an alarming rate. Different techniques have been devised to utilize available polluted water to produce potable water. Most desalination systems use high-grade energy. Solar energy is a very large and inexhaustible source of energy. Solar desalination reduces the amount of energy used. The success of Solar still is dependent not only on the enhancement of its yield, but also the quality of distilled output. The objective of the present study is to examine the performance of solar energy with the addition of phase change material (PCM) to the bottom of it and affects energy storage capacity. The presence of fluoride may cause several health issues such as dental and skeleton fluorosis, bone rupture, muscle damage, neurological and thyroid damage, etc. Today, the removal and reduction from drinking water is a global concern. Solar energy is still one of the most suitable and economical techniques for reducing the level of fluoride in the water. A total of seven synthetic samples of fluoride-contaminated water were prepared as a feed for solar distillation. The concentration of fluoride was assessed in the distillate samples and it was found that the removal efficiency of passive solar still was greater than 90%.

Keywords Contaminated water · Fluoride · Phase-changing material · Solar still

K. P. Singh · D. Singh
Harcourt Butler Technical University, Kanpur 208002, India
e-mail: pratapkrishn23@gmail.com

D. Singh
e-mail: dr.deepeshsingh@gmail.com

A. Dixit (✉)
Pranveer Singh Institute of Technology, Kanpur 209305, India
e-mail: dixit.abhishek87@gmail.com

B. P. Singh
Department of Biotechnology, National Institute of Technology, Warangal 506004, India
e-mail: bhanupratapsingh.bce19@itbhu.ac.in

A. K. Yadav et al. (eds.), *Advanced Treatment Technologies for Fluoride Removal in Water*, Water Science and Technology Library 125,
https://doi.org/10.1007/978-3-031-38845-3_16

16.1 Introduction

In many developing countries, most cities are facing a scarcity of water due to the exponential growth of the population, unplanned water supply schemes, and the growing sanitation pressure. Pure water availability is rapidly dwindling as a result of fast resource depletion and contamination. The different sources of clean water are arctic ice zones, lakes, rivers, and underground water tables (Shirin & Yadav, 2014; Yadav et al., 2014). As a result of the presence of salt and undesirable materials and organisms, direct use of water from existing water reserves should be avoided. Earth has one part of land and three parts of water. Only 2% of the water on Earth is salt-free water, and 98% of the three parts of water available are salty (Alqsair et al., 2022; Jahanpanah et al., 2021). Drinking this kind of brackish water can cause a variety of contagious diseases in people. Therefore, it is crucial to clean up the contaminated water. In the arid area, the surplus availability of solar energy plays a vital role in the desalination of water. Solar desalination is important for cleaning the water (Moreno et al., 2022; Thakur et al., 2022). The sun's energy is also used in the solar desalination process to purify the water. Desalination is one of the many techniques used to address the issue of water purification. Solar stills are very easy, economical, and driven by solar energy. The preferred technique for daily use is solar desalination, and it can be used particularly in remote and small villages in many arid places (Kumar et al., 2020). When comparing traditional thermal and membrane filtration technologies, solar energy is considered the most affordable choice for desalinating water (Al-Harahsheh et al., 2022; Al-Karaghouli et al., 2009). Solar stills are a practical choice for small to medium-scale systems (Maridurai et al., 2022). They do not need high-intensity of energy and they do not emit any dangerous gases that could harm the environment. Solar stills are also simple to construct and maintain (Yadav, 2017). The simple design and structure of solar stills produce 4–6 l/h (m^2 day) (Al-Karaghouli et al., 2009). Phase-changing material (PCM) can store solar energy and emit in the absence of sunlight., they can be used to reduce heat losses during the peak period of solar irradiation (Thakur et al., 2022). Recent studies on solar stills have revealed that PCM can drastically improve their performance (Kumar et al., 2022; Nandhakumar et al., 2020). PCM acts as a source of heat for the evaporation of water from the basin in the absence of sun irradiation (Alqsair et al., 2022). Fluoride in drinking water is a concern that affects millions of people's health on a global scale (Mittal et al., 2020). Most fluoride ions found in drinking water are geogenic. This indicates that at certain places groundwater has significant levels of fluoride ions. (Kashyap et al., 2021). Fluoride consumption in excess results in dental effects, bone disorders, and, when consumed in large amounts over an extended period, potentially serious possible skeletal problems (Yadav, 2017). It is widely known that excessive fluoride consumption can cause serious dental and skeletal fluorosis at values greater than 1 mg/L (Kumar et al., 2019). Around 20 developed and growing economic nations, including India, have found that fluorosis as a result of unacceptable fluoride in groundwater (Yadav, 2017). To meet the demand for fluoride-free clean water, the main issue is to improve the daily distillation yield per unit area while minimizing energy losses (Kalita et al., 2016). One of the oldest ways of defluoridation, the Nalgonda procedure was developed in an Indian village in the Andhra Pradesh region (Kumar et al., 2019). India has 18,000 remote villages where residents face interrupted electricity supply (Yadav, 2017). India is fortunate to have an abundance of solar energy. It gets sunlight almost all year, and nearly all of its regions receive enough solar radiation. In rural and tribal regions, it is not possible to purify the water using costly and electricity-based equipment. Therefore, this method can be useful for such regions. The objective of this chapter is to determine the fluoride removal efficiency of passive solar still from water.

16.2 Materials and Methods

The set-up of the experiment consists of one distillation unit: passive solar with a glass cover with an inclination angle of 26°50′. Figures 16.1 and 16.2 shows the working model of passive solar still with thermoplastic polymer and covered with glass respectively. The city of Kanpur is situated between the latitudes of 25°26′ and 26°58′ north and the longitudes of 79° and 80°34′ east (Bhadauria & Dixit, 2022; Dixit et al., 2022). Therefore, it was appropriate to select a glass cover that would receive the most sunlight. This 3 mm thick tilting glass cover functioned as a condensing surface for the water vapour produced in the still basin as well as a solar energy transmitter. The basin has a 1 m^2 effective area constructed of galvanized iron (GI), and it was set within a frame built of wood. The distiller basin was also blackened to increase solar energy absorption (Kalita et al., 2016). PCM was placed below the basin in the still. The addition of PCM increased nocturnal output while reducing daylight production (Moreno et al., 2022). A 5 kg quantity of calcium chloride hexahydrate was used as PCM. At the end of the basin, a distillate output channel was built for the purpose of collecting the distillate output, and a plastic pipe was attached through the channel's hole using adhesive (Araldite). Thermocol, a thermoplastic polymer, was provided as a heat insulation material between the GI sheet water basin and wooden frame (Yadav, 2017). The edges were covered in a rubber gasket. These preparations are made to ensure that the still is airtight. On the inside of the glass cover, water condenses and evaporates. It descends on the lower edge of the glass cover. The thermocouples were attached to a digital temperature indicator that shows the temperatures (Jahanpanah et al., 2021). The thermocouple is a device made of two different metal wires of semiconducting rods that are joined at the ends.

Experiments were conducted from 20 to 27 May 2019 for defluoridation. A known amount of NaF was dissolved in tap water to make a synthetic water sample of varying fluoride content. Fluoridated water was fed to the still in the batches in still. The inlet was then closed with a plug and let the water was suspended for some time to remove the fluoride content from the water through the adsorbing characteristics of the material. Most of the fluoride is removed through the adsorbing property of activated charcoal and most of the remaining fluoride was removed through the

Fig. 16.1 Working model of passive solar still with thermoplastic polymer inside the surface

Fig. 16.2 Working model of passive solar still covered with glass plate

evaporation and condensation process of solar still. The distillate was collected in a bottle. The concentration of fluoride in the distillate (C_{fd}) and in feed water (C_{ff}) was estimated using a spectrophotometric method. The defluoridation efficiency of solar still is determined by Eq. 16.1.

$$\text{Passive solar still fluoride removal efficiency} = \left(\frac{C_{ff} - C_{fd}}{C_{ff}}\right) \times 100 \quad (16.1)$$

16.3 Results and Discussion

The desired limit of fluoride is 1 mg/L as per BIS 10500:2012. It can be relaxed up to 1.5 mg/l depending on the unviability of any alternate source. A known amount of sodium fluoride (NaF) solution concentration was mixed every day for every 7 days to form a synthetic raw sample (mentioned as A, B, C, D, E, F, G) of fluoridated water. The concentration of fluoride in the raw sample is denoted as C_{ff} and the concentration of fluoride in the distillate is denoted by C_{fd}. The variation in fluoride concentration in raw feed and corresponding distillate samples is presented in Fig. 16.3.

The efficiency of passive solar still the removal of fluoride-contaminated water with different concentrations was calculated by Eq. (16.1). The variation in fluoride removal efficiency is presented in Fig. 16.4.

From Fig. 16.4, it can be observed that the fluoride concentration of feed (C_{ff}) < 5 mg/l (sample A), fluoride was removed with 100% efficiency. In other samples (samples B, C, D, E, F, and G) with a fluoride concentration of more than 5 mg/l, the fluoride with more than 90% efficiency.

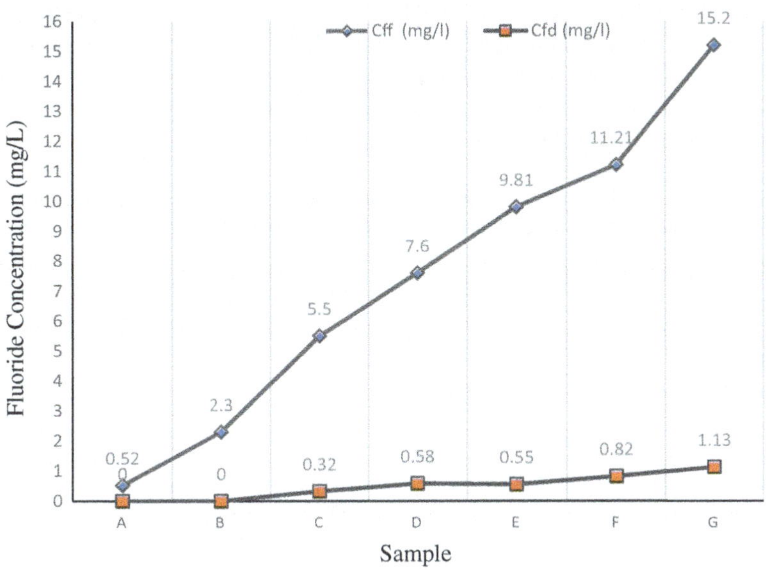

Fig. 16.3 Fluoride in raw and distillate

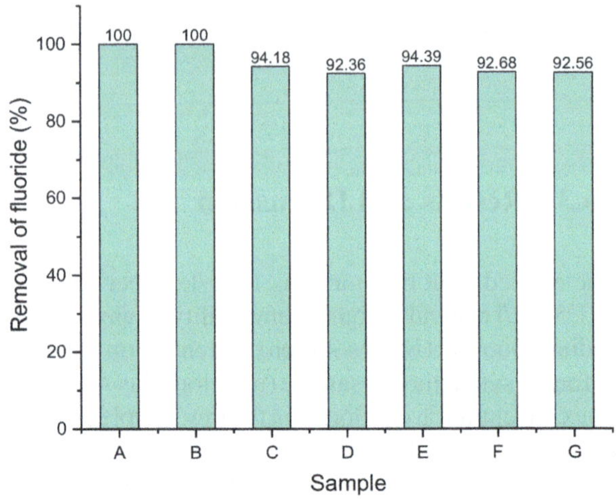

Fig. 16.4 Percentage removal of fluoride using passive solar still

16.4 Conclusions

On the basis of the results, it can be concluded that passive solar stills are also very effective in the removal of fluoride from the water. The removal efficiency of the passive solar stills is also high and ranges from 92 to 100% depending upon the fluoride content present in the water. The fluoride present in the distillate water lies within the allowed range recommended by IS 10500:2012. Passive solar still does not involve the use of any mechanical and electrical devices. It does not rely on pumps and sensors. The saline water poured inside the still will only begin evaporating when a certain amount of internal energy has built up in the system. Therefore, it is economical to activate solar stills and other technologies.

Acknowledgements The authors acknowledge the unconditional support of the laboratory staff of Harcourt Butler Technical University, Kanpur (Uttar Pradesh), India.

References

Al-Harahsheh, M., Abu-Arabi, M., Ahmad, M., & Mousa, H. (2022). Self-powered solar desalination using solar still enhanced by external solar collector and phase change material. *Applied Thermal Engineering, 206*, 118118. https://doi.org/10.1016/j.applthermaleng.2022.118118

Al-Karaghouli, A., Renne, D., & Kazmerski, L. L. (2009). Solar and wind opportunities for water desalination in the Arab regions. *Renewable and Sustainable Energy Reviews, 13*(9), 2397–2407. https://doi.org/10.1016/j.rser.2008.05.007

Alqsair, U. F., Abdullah, A. S., & Omara, Z. M. (2022). Enhancement the productivity of drum solar still utilizing parabolic solar concentrator, phase change material and nanoparticles' coating. *Journal of Energy Storage, 55*(PA), 105477. https://doi.org/10.1016/j.est.2022.105477

Bhadauria, S., & Dixit, A. (2022). Estimation of air pollution tolerance and anticipated performance index of roadside plants along the national highway in a tropical urban city. *Environmental Monitoring and Assessment*, 1–14.https://doi.org/10.1007/s10661-022-10483-0

BIS 10500:2012. Indian Standard Drinking Water Specification (Second Revision). *Bureau of Indian Standards, IS 10500*(May), 1–11. http://cgwb.gov.in/Documents/WQ-standards.pdf

Dixit, A., Singh, D., & Kumar, S. (2022). Changing scenario of municipal solid waste management in Kanpur city, India. *Journal of Material Cycles and Waste Management, 2020*.https://doi.org/10.1007/s10163-022-01427-4

Jahanpanah, M., Sadatinejad, S. J., Kasaeian, A., Jahangir, M. H., & Sarrafha, H. (2021). Experimental investigation of the effects of low-temperature phase change material on single-slope solar still. *Desalination, 499*(May 2020), 114799. https://doi.org/10.1016/j.desal.2020.114799

Kalita, P., Dewan, A., & Borah, S. (2016). A review on recent developments in solar distillation units. *Indian Academy of Sciences, 41*(2), 203–223.

Kashyap, S. J., Sankannavar, R., & Madhu, G. M. (2021). Journal of hazardous materials letters fluoride sources, toxicity and fluorosis management techniques—A brief review. *Journal of Hazardous Materials Letters, 2*, 100033. https://doi.org/10.1016/j.hazl.2021.100033

Kumar, P. S., Suganya, S., Srinivas, S., Priyadharshini, S., Karthika, M., Karishma Sri, R., Swetha, V., Naushad, M., & Lichtfouse, E. (2019). Treatment of fluoride-contaminated water. A review. *Environmental Chemistry Letters, 17*(4), 1707–1726. https://doi.org/10.1007/s10311-019-00906-9

Kumar, P. S., Singh, D., Singh, D., Kumar, P., & Singh, D. (2020). CHEM-CONFLUX 20 Special Issue Physicochemical Parametricand Water Quality Index (WQI) Analysis of Gomti River, Lucknow using MDSSS. *Journal of Indian Chemical Society, 97*.

Kumar, M. P., Chauhan, P., Sharma, A. K., Rinawa, M. L., Rahul, A. J., Srinivas, M., & Tamilarasan, A. (2022). Performance study on solar still using nano disbanded phase change material (NDPCM). *Materials Today: Proceedings, 62*, 1894–1897. https://doi.org/10.1016/j.matpr.2022.01.050

Maridurai, T., Rajkumar, S., Arunkumar, M., Mohanavel, V., Arul, K., Madhesh, D., & Subbiah, R. (2022). Performance study on phase change material integrated solar still coupled with solar collector. *Materials Today: Proceedings, 59*, 1319–1323. https://doi.org/10.1016/j.matpr.2021.11.539

Mittal, Y., Srivastava, P., Kumar, N., & Yadav, A. K. (2020). Remediation of fluoride contaminated water using encapsulated active growing blue-green algae, Phormidium Sp. *Environmental Technology and Innovation, 19*, 100855. https://doi.org/10.1016/j.eti.2020.100855

Moreno, S., Álvarez, C., Hinojosa, J. F., & Maytorena, V. M. (2022). Numerical analysis of a solar still with phase change material under the basin. *Journal of Energy Storage, 55*(April). https://doi.org/10.1016/j.est.2022.105427

Nandhakumar, S., Thirumalai, R., Viswaaswaran, J., Senthil, T. A., & Vishnuvardhan, V. T. (2020). Investigation of production costs in manufacturing environment using innovative tools. *Materials Today: Proceedings, 37*(Part 2), 1235–1238. https://doi.org/10.1016/j.matpr.2020.06.433

Shirin, S., & Yadav, A. K. (2014). Physico chemical analysis of municipal wastewater discharge in ganga river, Haridwar district of Uttarakhand, India. *Current World Environment, 9*, 536–543. https://doi.org/10.12944/CWE.9.2.39

Thakur, V., Kumar, N., Kumar, S., & Kumar, N. (2022). A brief review to improve the efficiency of solar still using efficient phase change materials. *Materials Today: Proceedings, 64*, 1295–1299. https://doi.org/10.1016/j.matpr.2022.04.119

Yadav, A. (2017). Water desalination system using solar heat: A review. *Renewable and Sustainable Energy Reviews, 67*, 1308–1330.https://doi.org/10.1016/j.rser.2016.08.058

Yadav, A. K., Sahoo, S. K., Mahapatra, S., Kumar, A. V., Pandey, G., Lenka, P., & Tripathi, R. M. (2014). Concentrations of uranium in drinking water and cumulative, age-dependent radiation doses in four districts of Uttar Pradesh, India. *Toxicological and Environmental Chemistry, 96*, 192–200. https://doi.org/10.1080/02772248.2014.934247

Chapter 17
A Statistical Approach to the Prediction of Fluoride in River Water Using the Best Subset Method

Madhusudana Rao Chintalacheruvu and Prakhar Modi

Abstract Water quality parameters were predicted by many methods in the past, but the application of the best subset method is yet to be explored much. This method is unique in its application as it adopts the statistical regression modeling procedure to fit a separate model for every combination of any number of predictors while selecting the best subset for making the optimal decision. The present chapter demonstrates the application of the best subset model for the prediction of fluoride in river water by developing the regression equations for various combinations of other physicochemical parameters of the water quality of the Godavari River, the second largest river in India. The desired fluoride concentrations were evaluated using the concentrations of 16 other water quality parameters data collected for a period of 13 years (1993–2005) from 13 monitoring sites during the Monsoon (June-October) and Post-Monsoon (November-February) periods in the study region. Several best subsets of independent variables are chosen based on the percentage variation in the dependent variable, with a gradual increase in the number of predictors using various combinations of water quality parameters, and the best models are selected in agreement with the highest R^2 value and the lowest RSS value. The final best subset regression model is selected among the various best models using *F*-value statistics criteria. The best subset regression model is successful in explaining 90% variation in fluoride concentrations for both seasons in the study area. It would be interesting to try this method for the study of both surface water and groundwater quality parameters.

Keywords Fluoride · Prediction · Best subset method · Regression model · Godavari River

M. R. Chintalacheruvu (✉) · P. Modi
Department of Civil Engineering, National Institute of Technology, Jamshedpur 831014, India
e-mail: cmrao.civil@nitjsr.ac.in

P. Modi
e-mail: prkhr.08@gmail.com

A. K. Yadav et al. (eds.), *Advanced Treatment Technologies for Fluoride Removal in Water*, Water Science and Technology Library 125,
https://doi.org/10.1007/978-3-031-38845-3_17

17.1 Introduction

Water is characterized as a natural, valuable but scarce resource, having a finite and comparatively stable amount of available supply. The global availability of freshwater is progressively decreasing, leading to a deficient resource. River water is considered to be one of the most significant and versatile components with respect to its multiple uses. The principal applications of river water include its utilization as a potable water source, for irrigation, fisheries, and hydroelectric power generation. Therefore, water must be devoid of any potentially hazardous chemical substances or living microorganisms to ensure its safety (WHO, 2017). The determination of the quality of the water resource depends on various physiochemical parameters and the source and magnitude of any pollution load. To evaluate this, monitoring of the parameters is crucial. Surface water quality within a specified geographical region is influenced by both natural factors, including precipitation rates, soil degradation, and weathering mechanism, and anthropogenic activities, such as urbanization, industrialization, and agricultural methods, as well as water resource utilization (Shirin et al., 2021, 2022; Yadav, 2021). Over time, these factors have intensified, resulting in the degradation of water resources and rendering them unsuitable for primary and secondary uses.

In India, the population predominantly experiences two major health conditions as a result of the prevalence of arsenic (As) along with fluoride (F) in potable water. However, it should be noted that the health consequences resulting from F contamination are significantly more pervasive compared to those arising from As contamination within the nation (Subba, 2011). According to Susheela (1999), in India, an estimate of 62 million individuals, which comprises 6 million minors, are affected by fluorosis, due to the ingestion of fluoride contaminated water. The prevalence of endemic fluorosis is extensive in multiple states in India, including Uttar Pradesh, Delhi, Karnataka, Tamil Nadu, Gujrat, Andhra Pradesh, Bihar, Haryana, Punjab, Odhisha, Rajasthan, Telangana, Maharashtra, and Madhya Pradesh (Subba et al., 2015). According to Ali et al. (2016), fluorosis, a fatal disease, affects around 200 million individuals in 25 nations. Reports of regions affected by fluorosis have surfaced from all over the world. These include Nigeria (Gbadebo, 2012), China (Li et al., 2018), Korea (Kim & Jeong, 2005), Africa (Gizaw, 1996) and India (Adimalla, 2018; Adimalla & Rajitha, 2018; Adimalla & Li, 2018; Adimalla et al., 2018a, 2018b, 2018c; Subba et al., 2015).

The predominant sources of fluoride (F) are largely natural origin, emanating from mineral deposits such as apatite, cryolite, fluorspar, mica, and rock phosphate. Geographically extensive belts characterized by many high concentrations of F are commonly associated with three primary geological factors: (i) marine sediment deposits located in mountainous regions, (ii) volcanic rock formations, and (iii) granitic and gneissic rock structures (Murray, 1986). The process of weathering primary minerals, such as fluorite, results in the release of F into the water and soil. Additionally, fluoride is released into the environment through hydro geothermal vents and volcanic activity, resulting in contamination. The emission of gases containing hydrogen fluoride also contributes to the contamination of air with fluoride, as noted by Vithanage and Bhattacharya (2015). However, the hot and arid pre-monsoon season is distinguished by an elevated rate of evapotranspiration, which results in the accumulation of various salts, including fluoride, in the upper soil layers, as observed by Singh et al. (2018).

Numerous research investigations have been conducted worldwide on the geochemistry of groundwater that contains fluoride. Numerous researchers, including Das et al. (2003), Gupta et al., (1999, 2005), Kundu and Mandal (2009a, 2009b), Misra and Mishra (2006), and Vikas et al. (2009), conducted an extensive examination of the geochemistry and origin of groundwater containing F in diverse geological formations throughout India. Research studies on regional-scale fluoride contamination in India were conducted by Saxena and Ahmed (2003) and Jacks et al. (2005). Mukherjee and Singh (2018) conducted a thorough review of numerous studies on the presence of fluoride in groundwater, examining the various mechanisms of contamination involved. Several researchers, including Adimalla et al. (2019), Allam et al. (2020), Davne and Pradhan (2021), Magesh et al. (2016), Sadat (2012), Solunke (2021) and Subba et al. (2017), have conducted a study to assess the extent of fluoride contamination in the vicinity of South India. A significant observation is that most studies are centred on the occurrence and distribution of fluoride, as well as its remedial measures. However, Amini et al. (2008), Podgorski et al. (2018), Reddy et al. (2019), and Satish et al. (2022) conducted research studies that focused on predicting the fluoride concentrations using GIS and other statistical modeling techniques. Forecasting or predicting fluoride concentration can be highly advantageous as it enables appropriate measures to be implemented before the onset of severe health problems in the vicinity.

Numerous techniques are available to predict the concentration of water quality parameters, each with its own merits and demerits. However, the majority of these techniques necessitate the collection of data, which is prone to man-made errors. Additionally, there is a significant possibility of encountering missing data for various parameters. As the lack of concentration data holds notable implications for water quality analysis, numerous researchers have suggested various methodologies to gauge these missing concentrations values. As an example, Abyaneh (2014) employed multivariate linear regression in conjunction with ANN to anticipate the concentrations of water quality parameters. Najafzadeh et al. (2018) and Shirin and Yadav (2014) approximated the concentrations of BOD, DO, and COD by gene expression programming (GEP), model tree (MT) and evolutionary polynomial regression (EPR) techniques. Abdelmalik (2016) applied remote sensing techniques, statistical approaches, and ASTER data sets to estimate various parameters of water quality in Qaroun Lake. Reddy et al. (2019) endeavored to provide a potential fluoride contamination zone for an entire state at a regional scale, utilizing a GIS based prediction model in Telangana, India.

In addition, statistical regression techniques can facilitate the identification of correlations among various parameters. Monteiro and Costa (2018) evaluated the efficacy of time series statistical models for predicting and forecasting dissolved oxygen levels at multiple monitoring sites. Hirtle and Rencz (2003) employed spectral reflectance measurements obtained from both ground-based and satellite-based platforms to construct regression models for forecasting dissolved organic carbon. Singanan et al. (1995) demonstrated the ability of correlation analysis to assess the efficacy of association between independent and dependent variables. The use of correlation analysis in conjunction with the routine measurement of water quality parameters can explain the variance in river water quality, as evidenced by the work of Mulla et al. (2007). Sattari et al. (2016) used the k-nearest neighbor algorithm and support vector regression to estimate conductivity and TDS in Iran. Gedefaw et al. (2018) and Haque et al. (2018) used multiple regression analysis to determine the optimal predictor variable and develop an urban water demand forecasting model. Satish et al. (2022) attempted to predict the water quality parameters in the Godavari river basin by using statistical and ANN models. There is a scarcity of literature on the prediction of fluoride concentration. Some researchers have employed statistical methods and machine learning algorithms such as ANN, Random Forest, and Logistic Regression to predict fluoride contamination (Barzegar et al., 2016; Nafouanti et al., 2021).

The preceding discussion highlights the abundance of research studies on river water quality modeling. However, the crucial process of optimal decision making for sustainable river water quality management remains unexplored. In particular, the best subset method, which is an accurate statistical regression technique capable of fitting separate models for every feasible combination of predictors and selecting the optimal subset to make informed decisions in water quality analysis, has not gained widespread adoption in the water quality modelling. Limited research is available on the use of best subset methods in other applications, such as groundwater conductivity modeling using the best subset procedure (Akoteyon et al., 2013; Bhatia et al., 1997), river water quality (Chintalacheruvu & Modi, 2023), and urban lakes (Biswas et al., 2017). However, the literature does not feature any work utilizing the best subset method to assess river water quality parameters. The scarcity of research studies using the best subset method in various applications highlights the need for further exploration and investigation of its potential utility. Therefore, this chapter aims to present the utilization of the best subset method to predict the concentration of Fluoride. Specifically, this will be demonstrated through the regression equations using physicochemical water quality parameters in the Godavari River in Monsoon (Jun-Oct) and Post-Monsoon (Nov-Feb) seasons.

17.2 Study Area

The Godavari river is positioned within 16°16′00″ to 22°36′00’ North and 73°26′00″ to 83°07′00″ East of the Deccan region. With its drainage basin that covers nearly 10% of India's total geographical area, this river is the largest peninsular and the third largest river in India. The river basin area spreads around 312,812 km^2 (CWC Report, 2021) and runs through 7 states namely Maharashtra (48.8%), Telangana (20%), Chhattisgarh (12.4%), Madhya Pradesh (7.9%), Odisha (5.7%), Andhra Pradesh (3.7%), and Karnataka (1.5%). The river rises in the Western Ghats with 1067 m elevation near the Triyambak Hills in Nashik, Maharashtra. After covering a total of almost 1465 km in the southeast direction, it descends in the Bay of Bengal. Of several tributaries the Maner, Manjira and Pravara are right bank major tributaries whereas the Indravati, Pranhita, the Purna, and Sabari are the main tributaries of the left bank of Godavari River (CWC Report, 2016). The Godavari river basin experiences its maximum precipitation throughout the southwest monsoon from July to September with an annual average rainfall of 1110 mm ranging from 881 to 1395 mm (IMD, 2021). The Godavari basin is divided into eight subbasins of which five subbasins are chosen in the present investigation namely Godavari lower, Indravati, Godavari middle, Pranhita, and Manjra basin, which collectively accounts for around 14.73, 12.68, 12.01, 11.96, and 9.76% of the entire area of Godavari basin. The location map of the Godavari River Basin is presented in Fig. 17.1. The observed data are collected using 13 water quality gauging stations scattered across the five chosen sub-basins in the study area. The 13 stations are Dhalegaon, GR Bridge, Yelli, Degloore, Saigaon, Mancherial, Pathagudem, Purna, Jagdalpur, Nowrangpur, Perur, Konta, and Polavaram. In recent years, it is observed that the quality of river water is harshly contaminated by agricultural, industrial and man-made activities. This causes many harmful pollutants to adversely affect the surrounding of river basin. Of these, fluoride is one of most harmful pollutant whose permissible limits in water shall be strictly controlled. In this sense, regular monitoring of fluoride concentrations should be considered. This can also be made possible by using significant correlations between fluoride and other water quality parameters.

17.3 Materials and Methodology

17.3.1 Field Observations

In the present investigation, the observed data of 17 water quality parameters recorded in the monsoon (June-October) and Post-Monsoon (November-February) periods are used. The observed data were recorded at 13 monitoring stations for a duration of 13 years (1993–2005) in the Godavari River. The 17 water quality parameters are (1) Total Alkalinity (Alk), (2) Calcium (Ca), (3) Chlorine (Cl), (4) Carbonate (CO_3), (5) Electrical Conductivity (EC), (6) Fluoride (F), (7) Total Hardness (Hard), (8) Bicarbonate (HCO_3), (9) Potassium (K), (10) Magnesium (Mg), (11) Sodium (Na), (12) Total Nitrogen (N), (13) pH, (14) Residual Sodium Carbonate (RSC), (15) Sodium Adsorption Ratio (SAR), (16) Silica (SiO_2), and (17) Sulphate (SO_4).

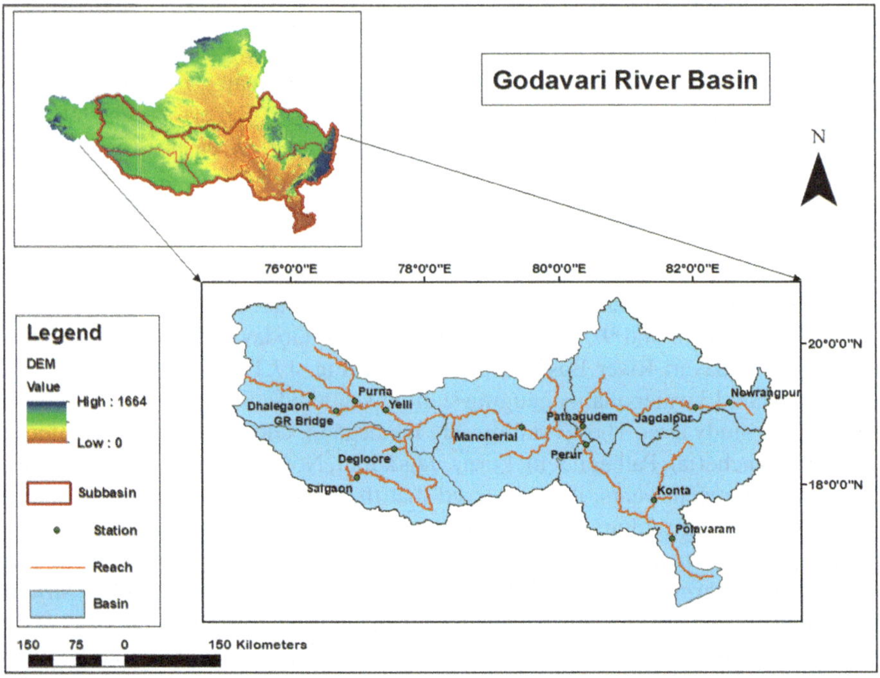

Fig. 17.1 Location map of Godavari River

17.3.2 Model Formulation

The best subset method is used to assess fluoride concentrations in the Godavari river at 13 monitoring stations using 16 water quality parameters. The general equation of a statistical model is given as:

$$Y_i = \sum B_j x_{ij} + e \tag{17.1}$$

with $x_{i0} = 1$ and j = 0 to k.

Here x_{ij} and Y_i are the independent variables and dependent variables respectively for ith observation. The model aims to estimate the unknown coefficient, B_j, and the number of coefficients, k. Furthermore, e represents the error in evaluating Y_i that is assumed to have 0 as mean and standard deviation as constant. The widely used least squares method is applied to determine the value of B_j as it is simple and does not have any constraint regarding statistical distribution. This method will determine the coefficients of independent water quality parameters (x_{ij}) to predict the concentration of the dependent water quality parameter, i.e. fluoride (Y_i).

17.3.3 Statistical Analysis

Prior to performing a statistical regression analysis on the observed data, it is of utmost importance to identify and remove the input error associated with the datasets for which skewness and kurtosis tests, outlier detection, and scatter plots are used. Upon elimination of the input error, a regression analysis is performed with fluoride as the dependent variable and the 16 water quality parameters as the independent variable. Using the corrected data, the correlation of all the water quality parameters with fluoride is calculated for both seasons for all 13 years of observed data in the Godavari river.

The correlation matrix between independent water quality parameters is calculated for the monsoon season in year 1993 as shown in Table 17.1. The correlation matrix can be used to determine the independent parameters, which will highly influence the prediction of the dependent water quality parameters. Also, it can be used to remove the parameters that are very less influencing which will help to increase the efficiency of the prediction model.

To increase the visualization of the correlation matrix, the correlation of fluoride with other independent parameters is shown in bold in Table 17.1. This will help as a preliminary analysis to explain the variations in the dependent variable fluoride. The parameters with very less correlation such as Nitrate in the postmonsoon season can be removed from further analysis because its capability of explaining the variations in the concentration of fluoride is very low and hence it will not be considered further during model formulations.

17.3.4 Selection of Variables for Regression Analysis

To allow the model to be implemented for prediction, it is favourable to include a larger number of independent variables as possible to achieve reliable fitted values. However, the use of a lesser number of independent variables will make the prediction process much easier. It is necessary that the R^2 value be close to 1, this indicates that a substantial portion of the variation in the dependent variable is explained by the regression model. It is evident that the value of R^2 increases as more variables are added to the fitted regression. However, the addition of too many independent variables can make the model difficult to manage. Therefore, a decision is made between large and small numbers of independent variables along with an acceptable value of R^2, and this is known as the selection of optimal regression variables leading to the selection of the best model (Draper & Smith, 1981). Although there exist some statistical approaches, the present investigation selects the best independent variables using the best subset method.

Table 17.1 Correlation matrix for Monsoon season water quality parameters

	Alk	Ca	CO_3	Cl	EC	K	HCO_3	Hard	Mg	Na	pH	Nitrate	RSC	SAR	SiO_2	SO_4	F
Alk	1.00																
Ca	0.96	1.00															
CO_3	0.77	0.62	1.00														
Cl	0.87	0.88	0.54	1.00													
EC	0.90	0.90	0.54	0.94	1.00												
K	0.83	0.89	0.50	0.89	0.88	1.00											
HCO_3	1.00	0.97	0.73	0.89	0.92	0.84	1.00										
Hard	0.99	0.99	0.67	0.88	0.90	0.84	0.99	1.00									
Mg	0.92	0.84	0.72	0.79	0.82	0.60	0.92	0.91	1.00								
Na	0.92	0.92	0.61	0.98	0.96	0.89	0.94	0.92	0.83	1.00							
pH	0.85	0.82	0.80	0.71	0.65	0.69	0.84	0.83	0.78	0.74	1.00						
Nitrate	0.19	0.26	0.06	− 0.02	0.02	0.19	0.19	0.22	0.06	0.05	0.14	1.00					
RSC	0.92	0.81	0.86	0.78	0.86	0.74	0.90	0.85	0.85	0.85	0.75	0.04	1.00				
SAR	0.92	0.92	0.61	0.98	0.94	0.91	0.93	0.92	0.80	1.00	0.77	0.08	0.83	1.00			
SiO_2	0.61	0.55	0.37	0.67	0.74	0.61	0.62	0.57	0.59	0.68	0.39	− 0.07	0.68	0.67	1.00		
SO_4	0.68	0.73	0.24	0.91	0.90	0.83	0.71	0.71	0.57	0.89	0.41	− 0.08	0.62	0.88	0.68	1.00	
F	**0.76**	**0.69**	**0.75**	**0.51**	**0.50**	**0.36**	**0.74**	**0.75**	**0.83**	**0.56**	**0.85**	**0.04**	**0.64**	**0.55**	**0.18**	**0.19**	**1.00**

17.3.5 Best Subset Regression Model

The best subset model involves a comparison amongst a full model that includes all independent variables and a reduced or subset model that contains fewer independent variables. The reduced model is a subset of the full model and is chosen if it meets the required statistical criteria.

Opting for a subset model over a full model is advantageous, as the subset model can determine the regression coefficients and future outcomes of lower variability. It seems to be an efficient way of determining the models that can accomplish the goal with a lesser number of independent variables (Hocking, 1976). The correlation matrix enables the selection of various subsets of independent variables based on the amount of variation that can be considered in predicting the dependent variables. Consequently, the regression of each subset is assessed using the R^2 value attained. The model obtained with a higher value of R^2 and F value will be considered as the best model for a particular year. Furthermore, the final best subset model is selected among all the best models that are developed for each year's data set using widely used statistical criteria such as R^2, F value, adjusted R^2 (R^2a), and RSS (Residual Sum of Squares). The criteria used in the selection of the best subset model are described below:

(i) **R^2 Criterion:**

A computing formula for R^2 is given as

$$R^2 = 1 - \frac{SSE}{SS_y} = \frac{SSR}{SS_y} \tag{17.2}$$

With $SS_y = \sum (Y_i - \overline{Y})^2$; $SSE = \sum \left(\hat{Y}_i - Y_i\right)^2$; $SSR = \sum \left(\hat{Y}_i - \overline{Y}\right)^2$.

where SSR is the sum of squared residuals and SSE is the sum of the acquired estimation of errors. Here Y is the average value of the dependent variable, and $\hat{Y}_i$ are the model-computed or simulated values of the dependent variable. The strong association between Y_i and $\hat{Y}_i$ yields a large value of R^2, while their weak association gives a small value of R^2. Relying solely on the R^2 criterion is incompetent for selecting a subset model, as the subset model will invariably have a lower R^2 value than the full model. Consequently, the full model will always show the maximum possible R^2 value. However, for a fixed number of independent variables (equal to k), R^2 can be used to compare different models with a large value of R^2 indicating the preferred model.

(ii) **F-value Criterion:**

The equation for the F value is described (Draper & Smith, 1981).

$$F = \frac{(N - k - 1)R^2}{k(1 - R^2)} \tag{17.3}$$

The ratio of explained and unexplained variation in Y_i is known as the F value statistic. The term $[(N - k - 1)/k]$ is the constant multiple. If the large amount of variation is explained, only then is the regression considered to be significant. This is only possible when the F value is large. Moreover, the residual sum of squares SSE_r reflects the variation in the dependent variable that is not accounted for by the model. If the independent variables excluded from the subset model are significant, then there shall be a significant increase in unexplained variation Y_i because of removing them. Therefore, SSE_r should become larger than SSE_f (for the full model). A convenient test statistic (Weisberg, 1980) using this idea is:

$$F_{k-m,N-k-1} = \frac{(SSE_r - SSE_f)/(k-m)}{SSE_f/(N-k-1)} \tag{17.4}$$

where SSE_f and SSE_r are the residual sums of squares of the full model with k number of independent variables, and the subset model with *the k-m* number of independent variables (where m is the number of independent variables dropped from the full model), respectively. The full or large model shall be chosen when $F_{k-m,\,N-k-1} > F^*$ (i.e., the $F_{k-m,\,N-k-1}$ statistic is significantly large), where F^* is $\alpha \times 100\%$ point of the $F_{k-m,\,N-k-1}$ distribution. The significance level $\alpha = 0.05$ is typical (i.e., at 5%) (Weisberg, 1980).

(iii) ***Adjusted R^2 (R_a^2)***:

The equation of R_a^2 is described (Heumann et al., 2016)

$$R_a^2 = 1 - \left[\frac{(1-R) \times n - 1}{n-k-1}\right] \tag{17.5}$$

In linear models, adjusted R^2 is the measure of model precision that evaluates the proportion of variance in the target variable that is accounted for by the input variables. The fit of linear regression is frequently overestimated by R^2. Adjusted R^2 (R^2_a) is a revised version of R^2 that considers insignificant predictors in a regression model. A reduced value of R^2_a indicates that the additional input variables do not enhancing the model compared to the previous model. However, a higher R^2_a shows that additional variables are improving the model and reducing the test error.

(iv) **Residual Sum of Squares (RSS)**:

RSS, also called the sum of the squared estimate of errors (SSE) or the sum of squared residuals (SSR), is a metric that quantifies the discrepancy between the observed and simulated model data. A lower value of RSS indicates the best fit for a model.

17.4 Results and Discussions

The present investigation aims to arrive at regression models to estimate the fluoride concentration in the Godavari River. An attempt is made to predict the concentration of fluoride using other substitute water quality parameters, and several combinations are developed and tested in both seasons. The simulation of the best subset model to estimate fluoride concentration as an independent variable is carried off with 16 water quality parameters as independent variables for a duration of 13 years. The model that meets the test statistic criteria is chosen and declared as the best subset model. The *F* value statistics are the test statistics criteria used to choose the best subset model. The final best subset model out of 13 models, one for each year, is selected in the sense of largest R^2, smallest RSS value and number of independent variables.

As is clearly evident from Table 17.1 for the Monsoon season in 1993, pH is the best lone variable capable of explaining around 85% variation in fluoride concentrations. While using remaining parameters such as Ca, Cl, K, Na, EC etc. as a single independent variable are capable of explaining roughly 69%, 51%, 36%, 56% and 50% variations in the fluoride, respectively.

Likewise, using the correlation matrix for post-monsoon in 1996 shows pH is capable of explaining almost 74% variations in fluoride concentrations if only a single parameter is considered as independent variable. Now, the ability to explain the variations in fluoride or simply the R^2 value can be increased using the various parameters with pH in both seasons. It is quite noticeable from Table 17.2 that the pair of pH and K shows larger values of R^2 and *F* among the other pairs perhaps these are the best set for the second iteration. Similarly, the third parameter is added and the pair that shows the highest R^2 and *F* values is chosen as the best set in the 3rd iteration. The process continues until the eighth iteration in which the eighth parameter is added to the final model. The final sets of Independent variables for all 8 iterations in Monsoon are presented in Table 17.3. Using the similar process for the post-monsoon season and similarly determining the final selected sets of independent variables.

Now to select the best model among the sets of independent variables as shown in Table 17.3, one shall always make sure that the chosen model has a maximum R^2 and minimum number of independent variables. For this reason, the decision is made using *F* distribution statistics, which makes a trade-off between the above two criteria. It is evident from Table 17.4 for Monsoon that when the condition $F_{k-m, N-k-1} > F^*$ is not fulfilled, the full model gets reduced to a subset model and when the condition is met, the full model is accepted as the best subset model. Also, since 13 models are chosen as the best model one of each year, a post-optimality analysis shall be conducted to determine the final best subset model. For which, the behavior of RSS and R^2 with increase in predictor variable i.e., independent variable for Monsoon and Post Monsoon season are observed. It is observed that with an increase in the number of independent variables the value of R^2 increases and that of RSS decreases. Perhaps, it is not only the highest R^2 and lowest RSS that shall

Table 17.2 Several sets of models and their statistics for monsoon

No. of independent parameters	Parameters	R^2	RSS	F Value
2	pH, Alk	0.72	0.12	12.98
	pH, Ca	0.72	0.12	12.63
	pH, Cl	0.73	0.11	13.69
	pH, CO_3	0.73	0.11	13.73
	pH, EC	0.72	0.12	12.90
	pH, Hard	0.72	0.12	13.03
	pH, HCO_3	0.72	0.12	12.88
	pH, K*	**0.81**	**0.08**	**21.96**
	pH, Mg	0.79	0.09	19.03
	pH, Na	0.73	0.11	13.30
	pH, Nitrate	0.72	0.12	12.94
	pH, RSC	0.72	0.12	12.63
	pH, SAR	0.74	0.11	14.12
	pH, SiO_2	0.74	0.11	14.23
	pH, SO_4	0.75	0.11	14.70
3	pH, K, Alk	0.94	0.03	45.15
	pH, K, Ca	0.95	0.02	55.18
	pH, K, Cl	0.85	0.06	17.32
	pH, K, CO_3	0.82	0.07	14.02
	pH, K, EC	0.90	0.04	25.87
	pH, K, Hard*	**0.96**	**0.02**	**65.61**
	pH, K, HCO_3	0.94	0.02	48.69
	pH, K, Mg	0.92	0.03	33.71
	pH, K, Na	0.87	0.06	19.76
	pH, K, Nitrate	0.82	0.08	13.25
	pH, K, RSC	0.85	0.06	16.53
	pH, K, SAR	0.85	0.06	17.25
	pH, K, SiO_2	0.81	0.08	13.18
	pH, K, SO_4	0.84	0.07	15.41
4	pH, K, Hard, Alk	0.96	0.02	43.89
	pH, K, Hard, Ca	0.96	0.02	44.02
	pH, K, Hard, Cl	0.96	0.02	43.79
	pH, K, Hard, CO_3	0.96	0.02	45.95
	pH, K, Hard, EC	0.96	0.02	43.82
	pH, K, Hard, HCO_3	0.96	0.02	43.75
	pH, K, Hard, Mg	0.96	0.02	44.36
	pH, K, Hard, Na	0.96	0.02	44.13

(continued)

Table 17.2 (continued)

No. of independent parameters	Parameters	R^2	RSS	F Value
	pH, K, Hard, Nitrate	0.96	0.02	52.61
	pH, K, Hard, RSC	0.96	0.02	43.81
	pH, K, Hard, SAR	0.96	0.02	44.79
	pH, K, Hard, SiO_2*	**0.96**	**0.02**	**53.52**
	pH, K, Hard, SO_4	0.96	0.02	43.82

*Indicates the selected model

Table 17.3 Selected sets of independent variables in Monsoon

Independent variable	K	RSS	R^2	F-value	R_a^2
pH	1	0.119	0.716	27.768	0.690
pH + K	2	0.077	0.815	21.964	0.777
pH + K + Hard	3	0.018	0.956	65.615	0.942
pH + K + Hard + SiO_2	4	0.015	0.964	53.523	0.946
pH + K + Hard + SiO_2 + Nitrate	5	0.010	0.977	59.435	0.961
pH + K + Hard + SiO_2 + Nitrate + CO_3	6	0.008	0.980	49.272	0.960
pH + K + Hard + SiO_2 + Nitrate + CO_3 + Na	7	0.007	0.982	39.414	0.957
pH + K + Hard + SiO_2 + Nitrate + CO_3 + Na + SO_4	8	0.007	0.984	31.499	0.953

be considered in selection of the final best subset model. In this regard, as shown in Table 17.2 the prime goal is on selection of the best final subset model with the minimum number of independent variables along with the highest R^2 and lowest RSS value.

The selection of the final best subset model using post-optimality analysis is shown in Table 17.5 for the monsoon season. It can be visualized from Table 17.5, that the RSS value of 1997 model, is lower than the 1993 model but still the 1993 model is selected as the final best subset model because of a lesser number of independent variables along with a satisfactorily lower RSS value. Similar analysis is conducted for Post-Monsoon season, through which it is evident that the 1996 model is selected over 1997 model although the number of parameters is same but the R^2 of 1996 is higher.

The final best subset model equations for the Monsoon and Post-Monsoon season are presented by Eq. (17.6) and Eq. (17.7) respectively as:

Final Model for Monsoon Season (1993):

$$F = -3.562 + 0.545 * pH - 0.198 * K + 0.004 * Hard \tag{17.6}$$

Table 17.4 Selection of best subset model using *F* distribution Statistics in Monsoon

Full model with k parameters			The reduced model with k-m Parameters		k − m	N − k − 1	$F_{k-m, N-k-1}$	F^*	Preferred model
Model	N	SSE_f	Model	SSE_r					
pH + K + Hard + SiO_2 + Nitrate + CO_3 + Na + SO_4	13	0.007	pH + K + Hard + SiO_2 + Nitrate + CO_3 + Na	0.007	7	4	− 0.142	6.09	Reduce the model
pH + K + Hard + SiO_2 + Nitrate + CO_3 + Na	13	0.007	pH + K + Hard + SiO_2 + Nitrate + CO_3	0.008	6	5	− 0.143	4.95	Reduce the model
pH + K + Hard + SiO_2 + Nitrate + CO_3	13	0.008	pH + K + Hard + SiO_2 + Nitrate	0.010	5	6	0.188	4.39	Reduce the model
pH + K + Hard + SiO_2 + Nitrate	13	0.010	pH + K + Hard + SiO_2	0.015	4	7	0.989	4.12	Reduce the model
pH + K + Hard + SiO_2	13	0.015	pH + K + Hard	0.018	3	8	0.570	4.07	Reduce the model
pH + K + Hard*	13	0.018	pH + K	0.077	2	9	14.585	4.26	Use full model
pH + K	13	0.077	pH	0.119	1	10	5.301	4.96	Use full model

*Indicates best subset model

($R^2 = 0.956$, RSS = 0.018, F-value = 65.615).

Final Model for Post-Monsoon Season (1996):

$$F = -4.12 + 0.59 * pH + 0.32 * RSC - 0.03 * SiO_2 + 0.04 * Mg \quad (17.7)$$

Table 17.5 Selection of the final best model using post optimal analysis in Monsoon

Year	K	RSS	R^2	F-value	Final equation
1993*	**3**	**0.018**	**0.956**	**65.615**	**$F = -3.562 + 0.545 * pH - 0.198 * K + 0.004 * Hard$**
1994	1	0.227	0.604	16.809	$F = 0.143 + 0.0079 * Na$
1995	2	0.043	0.941	79.935	$F = 0.158 + 0.031 * CO_3 + 0.328 * RSC$
1996	1	0.163	0.810	46.930	$F = 0.238 + 0.015 * Na$
1997	5	0.005	0.991	161.898	$F = 0.0829 + 0.424 * RSC + 0.011 * CO_3 + 0.002 * SO_4 - 0.058 * K + 0.029 * Mg$
1998	1	0.038	0.858	66.683	$F = 0.213 + 0.849 * RSC$
1999	3	0.034	0.919	34.239	$F = 0.157 + 0.847 * RSC + 0.143 * SAR - 0.018 * Mg$
2000	1	0.030	0.932	151.795	$F = 0.195 + 0.076 * CO_3$
2001	2	0.132	0.774	17.158	$F = -3.335 + 0.527 * RSC + 0.449 * pH$
2002	1	0.607	0.546	13.209	$F = 0.534 + 0.066 * CO_3$
2003	3	0.071	0.817	13.400	$F = -0.373 + 0.028 * CO_3 + 0.041 * SiO_2 + 0.003 * SO_4$
2004	1	0.079	0.772	37.227	$F = 0.233 + 0.014 * SO_4$
2005	1	0.049	0.601	16.598	$F = 0.190 + 0.499 * RSC$

*Final best subset model

($R^2 = 0.943$, RSS = 0.060, F-value = 32.914).

It is evident that the subset models developed in the present investigation are capable of predicting more than 90% of variations in fluoride concentration for both seasons. The *F* value of both the models depicts a statistically significant regression model. Using the final best subset model equation, Fig. 17.2 shows a plot between observed and simulated fluoride concentrations in Godavari River for the monsoon and post monsoon period. The plot also shows an agreement between the observed and model computed fluoride concentrations. It is established from the final model equations that the first parameter appearing is by far, of highest significance. The efficiency of the final model to estimate fluoride concentrations largely depends on the accuracy of the concentrations of parameters that occur first. The pH of the water quality parameter is found to be the most useful variable in the prediction of Fluoride concentration in both seasons. But the second variable differs in both seasons, which suggests that no single general model can be developed for all the seasons.

This can also be explained by the fact that the nearby activities during monsoon and after monsoon differ due to which the final developed model for both seasons are different. Furthermore, the R^2 values of these models explain the variation of fluoride

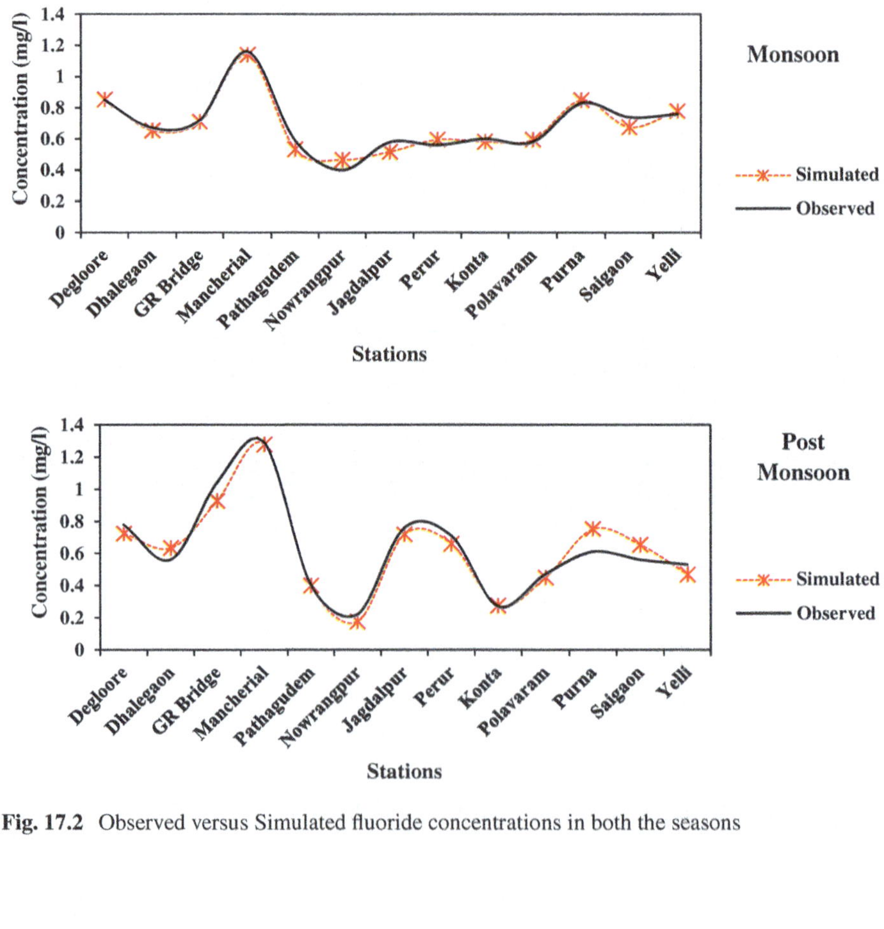

Fig. 17.2 Observed versus Simulated fluoride concentrations in both the seasons

concentrations. Thus, if the concentrations of independent variables are available for the required period, the final developed model can be effectively utilized to predict the missing value of fluoride in Godavari River.

17.5 Conclusions

The Godavari river is currently considered as one of the important river basins that is affected by Fluoride from various pollutant sources. As a result, regression models are developed for both seasons that help predict the concentration of Fluoride using other water quality parameters concentration. In this regard, several regression models are developed using the best subset method. In all the models developed, the addition of independent variables results in an increase in the R^2 value and a decrease in the RSS value. Therefore, it is crucial to determine the final model that has the least number of independent variables, while also minimizing errors that may arise during data collection. To accomplish this, it is worth emphasizing that the highest R^2 value alone cannot be used to determine the final subset model, and hence, smallest RSS and *F* distribution statistics are also considered. Since all the final developed models are capable of explaining over 90% of the variation in fluoride concentration, they can effectively be used in estimating missing observed values. It can also be inferred that the use of combinations with multiple parameters is more effective in comparison to a single-parameter model. However, no universal model can be developed to predict the variation in both seasons, despite observing some similar parameters highly influential for both seasons, such as pH. A change in season affects the proportion of independent parameters used to predict the concentration of fluoride. Since these regression models are created through correlation analysis, it is expected that the final subset models will be able to accurately predict the missing values within acceptable levels, barring any significant changes in agricultural and industrial activities or the introduction of pollutant point sources that can lead to undesired spikes in the concentrations of water quality parameters. This study provides information on key predictors of water quality parameters and the relationship between them, which can inform decision-making processes aimed at improving water quality in the Godavari River.

Acknowledgements The authors are thankful to the reviewers and editor for their valuable suggestions in improving the quality of the manuscript.

References

Abdelmalik, K. W. (2016). Role of statistical remote sensing for Inland water quality parameters prediction. *The Egyptian Journal of RS and Space Sciences, 21*,193–200.

Abyaneh, H. M. (2014). Evaluation of multivariate linear regression and artificial neural networks in prediction of water quality parameters. *Journal of Environmental Health Science & Engineering.*

Adimalla, N. (2018). Elevated fluoride concentration levels in rural villages of Siddipet Telangana State South India. *Data in Brief, 16*, 693–769.

Adimalla, N., & Li, P. (2018). Occurrence, health risks, and geochemical mechanisms of fluoride and nitrate in groundwater of the rock-dominant semi-arid region, Telangana State, India. *Human and Ecological Risk Assessment: An International Journal, 24*(8), 2143–2161.

Adimalla, N., & Rajitha, S. (2018). Spatial distribution and seasonal variation in fluoride enrichment in groundwater and its associated human health risk assessment in Telangana State, South India. *Human and Ecological Risk Assessment: An International Journal, 24*(8), 2119–2132.

Adimalla, N., Li, P., & Qian, H. (2018c). Evaluation of groundwater contamination for fluoride and nitrate in semi-arid region of Nirmal Province, South India: A special emphasis on human health risk assessment (HHRA). *Human and Ecological Risk Assessment: An International Journal, 24*(7), 1949–1971.

Adimalla, N., Li, P., & Venkatayogi, S. (2018b). Hydrogeochemical evaluation of groundwater quality for drinking and irrigation purposes and integrated interpretation with water quality index studies. *Environmental Processes, 5*(4), 363–383.

Adimalla, N., Vasa, S. K., & Li, P. (2018a). Evaluation of groundwater quality, Peddavagu in Central Telangana (PCT), South India: An insight of controlling factors of fluoride enrichment model. *Earth Systems and Environment, 4*(3), 841–852.

Adimalla, N., Venkatayogi, S., & Das, S. V. G. (2019). Assessment of fluoride contamination and distribution: A case study from a rural part of Andhra Pradesh, India. *Applied Water Science, 9*(94), 1–10. https://doi.org/10.1007/s13201-019-1023-4

Akoteyon, I. S., Akintuyi, A. O., & Mbata, U. A. (2013). Regression model for predicting water quality parameters: Study of Igando, Lagos-Nigeria. *American Academic & Scholarly Research Journal, 5*(4).

Ali, S., Thakur, S. K., Sarkar, A., & Shekhar, S. (2016). Worldwide contamination of water by fluoride. *Environmental Chemistry Letters, 14*(3), 291–315.

Allam, E., Gangula, H. K., Arukonda, R., & Muralidhar, M. (2020). Groundwater quality with special reference to fluoride concentration in the granitic and basaltic contact zone of Southern India. *Data in Brief, 33*, 106462.

Amini, M., Mueller, K., Abbaspour, K. C., Rosenberg, T., Afyuni, M., Moller, K. N., Sarr, M., & Johnson, C. A. (2008). Statistical modeling of global geogenic fluoride contamination in groundwaters. *Environmental Science & Technology, 42*, 3662–3668.

Barzegar, R., Moghaddam, A. A., Adamowski, J., & Fijani, E. (2016). Comparison of machine learning models for predicting fluoride contamination in groundwater. *Stochastic Environmental Research and Risk Assessment, 30*, 1207–1225.

Bhatia, K. K. S., Tyagi, A., & Sharma, M. K. (1997). Groundwater conductivity modelling using best subset procedure. In *Proceeding of International Symposium on Emerging Trends in Hydrology* (pp. 41–53), Department of Hydrology, University of Roorkee, Roorkee, 25–27.

Biswas, J. K., Bera, B., Chanda, R., Sarkar, S. K., Majumdar, J., Majumdar, S., & Hazra, S. (2017) Nutrient Modelling of an Urban Lake using Best Subset Method. Int. J. of Environ. Sci. Technology.

Central Water Commission (2016) Godavari Basin Water Quality Year Book, Volume III

Central Water Commission (2021) Reports on Water Quality Scenario of Rivers, Volume-1

Chintalacheruvu, M. R., & Modi, P. (2023). Application of best subset method for river water quality modeling: A study on Godavari River, India. *Environmental Quality Management, 33*(1), 393–409.

Das, B., Talukdar, J., Sarma, S., Gohain, B., Dutta, R. K., Das, H. B., & Das, S. C. (2003). Fluoride and other inorganic constituents in groundwater of Guwahati, Assam, India. *Current Science, 85*, 657–661.

Davne MP, Pradhan V (2021) The analytical study of heavy metals present in water samples of Godavari River. International Journal of Scientific Development and Research, 6(11), 138–141. https://doi.org/10.31695/IJSDR.2021.34824

Draper, N. R., & Smith, H. (1981). *Applied Regression Analysis* (2nd ed.). John Wiley and Sons.

Gbadebo, A. M. (2012). Groundwater fluoride and dental fluorosis in southwestern Nigeria. *Environmental Geochemistry and Health, 34*(6), 597–604.

Gedefaw, M., Hao, W., Denghua, Y., & Girma, A. (2018). Variable selection methods for water demand forecasting in Ethiopia: Case study Gondar town. *Cogent Environmental Science, 4*(1), 1537067.

Gizaw, B. (1996). The origin of high bicarbonate and fluoride concentrations in waters of the main Ethiopian Rift Valley. *Journal of African Earth Sciences, 22*(4), 391–402. https://doi.org/10.1016/S0899-5362(96)00097-6

Gupta, M. K., Singh, V., Rajwanshi, P., Agarwal, M., Rai, K., Srivastava, S., Shrivastav, R., & Dass, S. (1999). Groundwater quality assessment of Tehsil Kheragarh Agra (India) with special reference to fluoride. *Environmental Monitoring and Assessment, 59*, 272–285.

Gupta, S. K., Deshpande, R. D., Agarwal, M., & Raval, B. R. (2005). Origin of high fluoride in groundwater in the North Gujarat-Cambay region India. *Hydrogeology Journal, 13*, 596–605.

Haque, M. M., Rahman, A., Hagare, D., & Chowdhury, R. K. (2018). A comparative assessment of variable selection methods in urban water demand forecasting. *Water, 10*, 419.

Heumann, C., Schomaker, M., & Shalabh (2016). *Introduction to statistics and data analysis.* Springer International Publishing Switzerland.

Hirtle, H., & Rencz, A. (2003). The relation between spectral reflectance and dissolved organic carbon in lake water: Kejimkujik National Park, Nova Scotia, Canada. *International Journal of Remote Sensing, 24*(5), 953–967.

Hocking, R. R. (1976). The analysis and selection of variables in linear regression. *Biometrics, 32*, 1–49.

India Meteorological Department (IMD). (2021). Annual Report.

Jacks, G., Bhattacharya, P., Chaudhary, V., & Singh, K. (2005). Controls on the genesis of some high-fluoride groundwaters in India. *Applied Geochemistry, 20*(2), 221–228.

Kim, K., & Jeong, Y. (2005). Factors influencing natural occurrence of fluoride-rich ground waters: A case study in the southeastern part of the Korean Peninsula. *Chemosphere, 58*(10), 1399–1408.

Kundu, M. C., & Mandal, B. (2009a). Agricultural activities influence nitrate and fluoride contamination in drinking groundwater of an intensively cultivated district in India. *Water, Air, and Soil Pollution, 198*, 243–252.

Kundu, M. C., & Mandal, B. (2009b). Assessment of potential hazards of fluoride contamination in drinking groundwater of an intensively cultivated district in West Bengal, India. *Environmental Monitoring and Assessment, 152*, 97–103.

Li, P., He, X., Li, Y., et al. (2019). Occurrence and health implication of fluoride in groundwater of loess aquifer in the Chinese loess plateau: A case study of Tongchuan, Northwest China. *Exposure and Health, 11*, 95–107.

Magesh, N. S., Chandrasekar, N., & Elango, L. (2016). Occurrence and distribution of fluoride in the groundwater of the Tamiraparani River basin, South India: A geostatistical modeling approach. *Environmental Earth Sciences, 75*, 1483.

Misra, A. K., & Mishra, A. (2006). Study of quaternary aquifers in Ganga Plain, India: Focus on groundwater salinity, fluoride and fluorosis. *Journal of Hazardous Materials, 144*, 438–448.

Monteiro, M., & Costa, M. (2018). A time series model comparison for monitoring and forecasting water quality variables. *Hydrology, 5*, 37.

Mukherjee, I., & Singh, U. K. (2018). Groundwater fluoride contamination, probable release, and containment mechanisms: A review on Indian context. *Environmental Geochemistry and Health, 40*(6), 2259–2301.

Mulla, J. G., Farooqui, M., & Zaheer, A. (2007). A correlation and regression equations among water quality parameters. *International Journal of Chemical Sciences, 5*(2), 943–952.

Murray, J. J., & World Health Organization. (1986). *Appropriate use of fluorides for human health*. World Health Organization.

Nafouanti, M. B., Li, J., Mustapha, N. A., Uwamungu, P., & AL-Alimi, D. (2021). Prediction on the fluoride contamination in groundwater at the Datong Basin, Northern China: Comparison of random forest, logistic regression and artificial neural network. *Applied Geochemistry, 132*, 105054.

Najafzadeh, M., Ghaemi, A., & Emamgholizadeh, S. (2018). Prediction of water quality parameters using evolutionary computing-based formulations. *International Journal of Environmental Science and Technology, 16*, 6377–6396.

Podgorski, J. E., Labhasetwar, P., Saha, D., & Berg, M. (2018). Prediction modeling and mapping of groundwater fluoride contamination throughout India. *Environmental Science & Technology, 52*, 9889–9898. https://doi.org/10.1021/acs.est.8b01404

Rao, N. S., Rao, P. S., Dinakar, A., Rao, P. V. N., & Marghade, D. (2017). Fluoride occurrence in the groundwater in a coastal region of Andhra Pradesh, India. *Applied Water Science, 7*(4), 1467–1478. https://doi.org/10.1007/s13201-016-0511-y

Reddy, S. K. K., Sahadevan, D. K., Gupta, H., & Reddy, D. V. (2019). GIS-based prediction of groundwater fluoride contamination zones in Telangana, India. *Journal of Earth System Science, 128*(5), 132. https://doi.org/10.1007/s12040-019-1167-8

Sadat, N. (2012). Study of fluoride concentration in the river (Godavari) and groundwater of Nanded City. *International Journal of Engineering Inventions, 1*(1), 11–15.

Satish, N., Jagadeesh, A., Rajitha, K., & Varma, R. R. M. (2022). Prediction of stream water quality in Godavari River Basin, India using statistical and artificial neural network models. *H2Open Journal, 1*(1), 1–14. https://doi.org/10.2166/h2oj.2021.066

Sattari, M. T., Joudi, A. R., & Kusiak, A. (2016). Estimation of water quality parameters with data-driven model. *Journal-American Water Works Association*.

Saxena, V., & Ahmed, S. (2003). Inferring the chemical parameters for the dissolution of fluoride in groundwater. *Environmental Geology, 43*(6), 731–736.

Shirin, S., Jamal, A., Emmanouil, C., Singh, V. P., & Yadav, A. K. (2022). Assessment and characterization of waste material used as backfilling in an abandoned mine. *International Journal of Coal Preparation and Utilization*, 1–9. https://doi.org/10.1080/19392699.2022.2118259

Shirin, S., Jamal, A., Emmanouil, C., & Yadav, A. K. (2021). Assessment of characteristics of acid mine drainage treated with fly ash. *Applied Sciences, 11*, 3910. https://doi.org/10.3390/app11093910

Shirin, S., & Yadav, A. K. (2014). Physico chemical analysis of municipal wastewater discharge in ganga river, Haridwar district of Uttarakhand, India. *Current World Environment, 9*, 536–543. https://doi.org/10.12944/CWE.9.2.39

Singh, G., Kumari, B., Sinam, G., Kriti, N. K., & Mallick, S. (2018). Fluoride distribution and contamination in the water, soil and plants continuum and its remedial technologies, an Indian perspective: A review. *Environmental Pollution, 239*, 95–108.

Singanan, M., Rao, K. S., & Rambabu, C. (1995). A correlation study on Physico-chemical characteristics of groundwater in Rameswaram Island. *Indian Journal of Environmental Protection, 15*(3), 212–217.

Solunke, K. R. (2021). Study of fluoride concentration in the Godavari River near Shahagad. *Journal of Emerging Technologies and Innovative Research, 8*(12), 111–116.

Subba, R. N. (2011). High-fluoride groundwater. *Environmental Monitoring and Assessment, 176*, 637–645.

Subba, R. N., Vidyasagar, G., Surya, R. P., & Bhanumurthy, P. (2015). Assessment of hydrogeochemical processes in a coastal region: Application of multivariate statistical model. *Journal of the Geological Society of India, 86*(1), 89–96.

Susheela, A. K. (1999). Fluorosis management programme in India. *Current Science, 77*, 1250–1256.

Vikas, C., Kushwaha, R. K., & Pandit, M. K. (2009). Hydrochemical status of groundwater in district Ajmer (NW India) with reference to fluoride distribution. *Journal of the Geological Society of India, 73*, 773–784.

Vithanage, M., & Bhattacharya, P. (2015). Fluoride in drinking water: Health effects and remediation. *CO_2 Sequestration, Biofuels and Depollution*, 105–151.

Weisberg, S. (1980). *Applied linear regression*. Wiley and Sons.

World Health Organization. (2017). *Guidelines for drinking water quality: First addendum to the fourth edition*.

Yadav, M. K. (2021). *Seasonal variation in water quality of Krishna river*. CWC Reports on Water Quality Scenario of Rivers.

Part IV
Future Framework and Advanced Technologies

Chapter 18
Environmental and Health Effects of Fluoride Contamination and Treatment of Wastewater Using Various Technologies

Ankit Kumar, Ramakrishna Chava, Sonam Gupta, Saba Shirin, Aarif Jamal, and Akhilesh Kumar Yadav

Abstract The growth of the economy is highly dependent on industrial development, and most industries discharge their effluent into streams or lakes. Due to industries growing at a very fast rate, the rate of pollution increased several times. Various industries discharge their effluent into the mainstream of water. As a result, the water

A. Kumar · R. Chava
Department of Chemical Engineering, Indian Institute of Technology Gandhinagar, Gandhinagar 382355, India
e-mail: ankit.rs.che14@iitbhu.ac.in

R. Chava
e-mail: ramakrishna.chava@iitgn.ac.in

S. Gupta
Department of Environmental Science, Babasaheb Bhimrao Ambedkar University, Lucknow 226025, India
e-mail: guptasonam3333@gmail.com

S. Shirin (✉)
Department of Environmental Science, School of Vocational Studies and Applied Sciences, Gautam Buddha University, Greater Noida 201312, India
e-mail: sabashirin83@gmail.com

S. Shirin · A. Jamal · A. K. Yadav (✉)
Department of Mining Engineering, Indian Institute of Technology (Banaras Hindu University), Varanasi 221005, India
e-mail: yadavbasti@gmail.com

A. Jamal
e-mail: ajamal.min@itbhu.ac.in

A. K. Yadav
Department of Environmental Engineering and Management, Chaoyang University of Technology, Taichung 413310, Taiwan

Environmental Science and Engineering Department, Indian Institute of Technology Bombay, Mumbai 400076, India

A. K. Yadav et al. (eds.), *Advanced Treatment Technologies for Fluoride Removal in Water*, Water Science and Technology Library 125,
https://doi.org/10.1007/978-3-031-38845-3_18

becomes polluted and humans, aquatic animals, and even plants are affected. Fluoride is a chemical contaminant. It is toxic to plants and animals, including humans. This chapter explores the impact of fluoride contamination on the environment and contaminations in drinking water that can harm human health and various species at high levels. We also examine strategies to prevent and treat fluorosis, both globally and in India, and recent advances in fluoride remediation, as well as review various methods for fluoride removal from drinking water, such as coagulation-precipitation, membrane separation process, ion exchange, adsorption techniques, hybrid technology, and others. However, membrane and ion exchange processes are not widely adopted due to high installation and maintenance costs. Coagulation-precipitation and adsorption techniques are more prevalent in India. Among the different methods for water defluoridation, the adsorption technique is widely used and provides satisfactory results. It is also a more attractive method in terms of cost, design, and operation simplicity. The literature survey reveals that various methods have demonstrated novel potential for fluoride removal. However, it is still a need to evaluate the feasibility of such methods on a commercial scale, leading to improved pollution control.

18.1 Introduction

Drinking water scarcity is a global challenge that affects both urban and rural areas, and more than 1.5 billion people rely on groundwater for their drinking needs (Nakayama et al., 2022). Access to potable water is an indispensable prerequisite for maintaining good health. However, several regions worldwide encounter challenges in obtaining safe drinking water due to geographical, economic and financial constraints. Deficient water quality in these areas poses severe health and social predicaments. Consequently, developing and underdeveloped countries struggle to adhere to prescribed potable water quality standards established by regulatory authorities. As a result, ensuring access to safe drinking water has been considered a significant priority in numerous countries (Ali et al., 2019; Kashyap et al., 2021b; Onipe et al., 2020; Poonia et al., 2021). According to the World Health Organization (WHO), an estimated 783 million people lack access to fundamental drinking water services. Furthermore, projections suggest that more than 50% of the global population will encounter a dearth of potable drinking water by 2025 (Lacson et al., 2021). Groundwater constitutes a significant portion of freshwater that is suitable for human and animal consumption as a result of its higher microbial quality than surface water. However, various geochemical processes and anthropogenic activities have resulted in the accumulation and contamination of groundwater by different chemical elements and compounds. Furthermore, improper disposal of wastewater from urban, industrial, and agricultural sources has adversely affected the quality of groundwater in many regions. These water quality issues pose serious threats to human health and aquatic ecosystems (Lacson et al., 2021; Pearcy et al., 2015). Waterborne diseases, which result from the ingestion of contaminated water, have a considerable impact on human health and entail substantial economic costs for society. Therefore, improving drinking water quality would produce significant health benefits (WHO, 2011).

Among the various chemical contaminants, high concentrations of nitrate, arsenic, and fluoride ions pose serious health risks to living organisms. The presence of excessive F in drinking water (above 1.5 mg F^-/L) has been associated with several negative health outcomes in humans, which have attracted considerable attention from the research community (Li et al., 2021; Skórka-Majewicz et al., 2020; Vandana et al., 2021; Wimalawansa, 2020). Fluorosis is a collective term for a variety of diseases caused by chronic ingestion of excessive fluoride. The major adverse health effects on human beings due to the ingestion of excess fluoride through drinking water are presented in Fig. 18.1. The severity of fluorosis depends on the concentration of fluoride in drinking water, which can range from dental fluorosis (1.5–4.0 mg F^-/L) to crippling fluorosis (>10 mg F/L) (Ali et al., 2019; Lacson et al., 2021; Mohapatra et al., 2009) The main source of fluoride in drinking water is geogenic, that is, groundwater in some areas contains high levels of fluoride due to the natural occurrence of fluoride-bearing minerals such as fluorapatite ($Ca_5(PO_4)_3F$), sellaite (MgF_2), fluorite (CaF_2) and cryolite (Na_2AlF_6) in sedimentary and igneous rocks (Chowdhury et al., 2019; Jha et al., 2011). From the facts presented above, it is clear that the demerits of F consumption outweigh the merits. The detrimental health impacts of fluoride are predominantly observed in developing and underdeveloped nations, with India being the most affected country characterized by numerous areas of endemic fluorosis. India also has significant fluorite deposits, resulting in high levels of fluoride contamination in its groundwater.

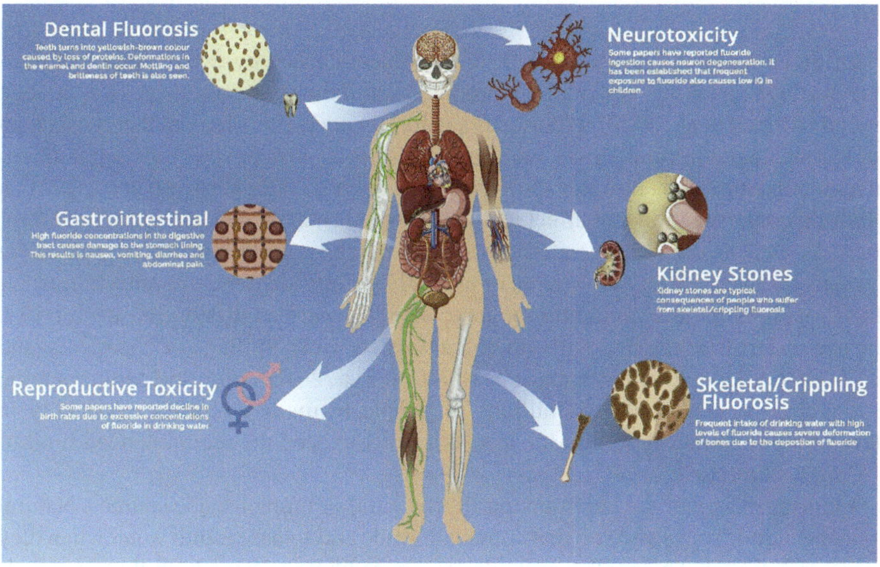

Fig. 18.1 Adverse health effects on humans due to ingestion of excess fluoride from drinking water

Several regions in Asia and Africa are susceptible to diseases associated with fluorosis. Specifically, India and China exhibit the majority of fluorosis cases in Asia, while Tanzania stands out as a prominent region in Africa with elevated fluoride concentrations in groundwater, which serves as a major source of drinking water (Kashyap et al., 2021a). To prevent excessive fluoride exposure through drinking water, some countries have established fluoride standards for drinking water according to WHO guidelines which are generally between 1–1.5 mg/L (Pillai et al., 2021). However, these standards are primarily based on the prevention of fluorosis and do not account for other health effects of fluoride. The exact number of people affected by fluoride contamination is unknown, but it is estimated that about 200 million people worldwide are at high risk of crippling fluorosis (Kabir et al., 2020). Endemic fluorosis is prevalent in some regions of India, China, South Africa, and Bangladesh (Chaudhary & Prasad, 2015). It is important to understand the extent and impact of fluoride toxicity on human health to address this global problem (Johnston & Strobel, 2020; Maheshwari et al., 2021). There are various methods for defluoridation of water, such as precipitation-coagulation, membrane-based processes, ion exchange and adsorption processes, etc. However, each method has its own advantages and disadvantages in terms of cost, efficiency, environmental impact, and health effects. Some of the methods are not suitable for certain regions or water sources due to the presence of other ions or contaminants. Therefore, it is important to understand and evaluate the different methods based on their performance and feasibility.

18.1.1 Sources of Fluoride Contamination and Its Causes

Fluoride contamination is a public health problem caused by excess intake of fluoride through drinking water, food products, or industrial pollutants over a long period (Ahmad et al., 2022). Fluoride is a naturally occurring element that is essential for human health in small doses, but can cause serious problems when it accumulates in high concentrations in water sources (Ahmad et al., 2022; Kashyap et al., 2021b). Fluoride contamination can affect teeth and bones, causing dental and skeletal fluorosis, as well as other adverse effects on the nervous, endocrine and reproductive systems (Dawes, 2003; Guissouma et al., 2017). Fluoride contamination can occur due to various natural and anthropogenic sources, such as dissolution from rocks, the use of fertilizers, irrigation with contaminated water, spillage of chemicals, and atmospheric deposition (Vithanage & Bhattacharya, 2015). Fluoride contamination is a global issue that affects millions of people in many countries, especially in developing regions where fluoride levels in water exceed the recommended limit of 1.5 mg/L (Ahmad et al., 2022).

Fluoride can enter the environment from natural or human-made sources. Natural sources include rocks and soil that contain fluoride and release it into water when they erode or dissolve. This makes groundwater more likely to have high fluoride levels than surface water. Human-made sources include industrial activities that produce fluoride as a pollutant and discharge it into water (Ahmad et al., 2022). Fluoride contamination is the presence of an excessive amount of fluoride in water, soil or food, which can cause health problems for humans and animals. Fluoride is a naturally occurring mineral found in rocks, soil and water sources, but high levels of fluoride can be harmful to human health. Sources of fluoride contamination include natural sources, industrial activities, and human activities (Rizzu et al., 2021). Natural sources of fluoride contamination include rocks and soil that contain high levels of fluoride. When water passes through these rocks and soils, it dissolves fluoride and becomes contaminated (Kashyap et al., 2021a; Raj & Shaji, 2017). In some areas, groundwater sources can contain high levels of fluoride due to geological factors. According to a study by Tiwari et al. (2020), fluoride contamination in groundwater is a widespread problem in the Indian state of Rajasthan, affecting approximately 60% of the population. In some cases, natural fluoride contamination can be severe, leading to dental and skeletal fluorosis, a condition that affects bones and teeth (Kashyap et al., 2021a).

Industrial activities such as mining, fertilizer production, and the use of fluoride-containing chemicals can also contribute to fluoride contamination. For example, aluminum smelting plants can release large amounts of fluoride into the environment through wastewater discharge. According to a study by Wang et al. (2019), fluoride contamination in soil and water sources near an aluminum smelting plant in China was found to be a significant problem, with some areas showing fluoride concentrations up to 19 times the safe limit. Additionally, industries that use fluoride-containing chemicals can release fluoride into the environment through wastewater discharge. The contamination of groundwater with fluoride is geogenic, that is, the dissolution of fluoride-bearing minerals in groundwater. However, anthropogenic activities such as industrial wastewater discharge can also contribute to fluoride pollution. Industrial wastewater containing high levels of fluoride can originate from various sectors such as the metallurgy, smelting, battery, cement, and fertilizer industries. The discharge of such wastewater into the environment poses a threat to human health and aquatic ecosystems (Chatterjee et al., 2017). Human activities, such as the use of fluoride-containing toothpaste and mouthwash, and the improper disposal of fluoride-containing products, can also contribute to fluoride contamination. The use of fluoride-containing toothpaste and mouthwash can contribute to the fluoride contamination of the water supply if they are not properly disposed of. In addition, the use of fluoride-containing pesticides can lead to contamination of soil and water sources. According to a study established by Mukherjee and Singh (2018), in arid and semi-arid regions of India, which are mostly affected by the prevalence of fluoride, the use of fluoride-containing pesticides is a significant contributor to the fluoride contamination of agricultural soils in India. Overall, fluoride contamination is a significant public health concern, and its sources are diverse. Natural sources, industrial activities, and human activities can all contribute to fluoride contamination and the effects can be severe, leading to dental and skeletal fluorosis. Therefore, it is important to monitor and regulate fluoride levels in drinking water, soil, and food sources to ensure public health and safety.

18.1.2 Environmental Aspects of Fluoride

Fluoride is a naturally occurring element found in soil, rocks, and water. It is also added to toothpaste and drinking water to prevent tooth decay. Although fluoride is beneficial for dental health, high concentrations of fluoride in the environment can have negative effects on both human health and the environment. Fluoride can enter the environment through natural processes such as weathering and erosion of rocks and soil. However, anthropogenic activities such as mining, industrial processes, and agricultural practices can also contribute to high concentrations of fluoride in the environment (Yadav et al., 2019). For example, phosphate fertilizers used in agriculture often contain high levels of fluoride, which can leach into groundwater and surface water (He et al., 2021). Exposure to high levels of fluoride can have negative effects on human health. Long-term exposure to high levels of fluoride can cause dental fluorosis, a condition that affects tooth enamel and can lead to discoloration and pitting of the teeth (Johnston & Strobel, 2020). High levels of fluoride can also cause skeletal fluorosis, a condition that affects bones and can lead to joint pain and stiffness (Susheela, 1999). Although dental fluorosis is mainly caused by ingestion of fluoride, skeletal fluorosis is mainly caused by inhalation of fluoride dust in occupational settings (Miranda et al., 2021).

In addition to human health effects, high levels of fluoride in the environment can also have negative effects on ecosystems. Aquatic organisms, such as fish and amphibians, are particularly susceptible to the effects of fluoride. High levels of fluoride can cause skeletal abnormalities in fish, reduce fish growth rates, and affect fish behavior (Camargo, 2003). Amphibians exposed to high levels of fluoride can develop deformities such as extra limbs or missing limbs (Collins, 2010). High levels of fluoride can also affect plant growth and soil microbial activity, leading to decreased soil fertility (Singh et al., 2018).

To prevent high levels of fluoride in the environment, it is important to regulate anthropogenic sources of fluoride. For example, regulations on phosphate fertilizers can help reduce fluoride levels in groundwater and surface water. Furthermore, water treatment plants can use techniques such as reverse osmosis or activated alumina filtration to remove fluoride from drinking water (Organization, 2017). In general, fluoride is beneficial for dental health, high concentrations of fluoride in the environment can have negative effects on both human health and the environment. It is important to regulate anthropogenic sources of fluoride to prevent high levels of fluoride in the environment and its negative effects on ecosystems. Further research is also needed to better understand the long-term effects of exposure to low levels of fluoride in the environment.

18.1.3 Brief Overview of the Importance of Treating Fluoride-Contaminated Wastewater

It is important to treat industrial wastewater contaminated with fluoride before it into the environment. Waste water treatment is a crucial process that aims to remove contaminants and pollutants from water sources to ensure safe reuse or discharge into the environment. Several methods have been proposed for fluoride removal from industrial wastewater, such as precipitation, membrane technology, ion exchange, adsorption, electrochemical method, and hybrid technology (Wan et al., 2021; Yadav et al., 2019). These techniques have shown varying degrees of success in removing fluoride from wastewater, depending on factors such as initial fluoride concentration, pH, temperature, and presence of other contaminants. For example, adsorption using activated carbon has been shown to be effective in removing fluoride from wastewater, with removal efficiencies ranging from 81 to 96% depending on the pH and initial fluoride concentration of the wastewater (Wan et al., 2021) Another technique that has been extensively studied for the removal of fluoride is ion exchange using various types of resins. For example, a study by Waghmare et al. (2015) demonstrated the high fluoride removal efficiency using a synthetic zeolite resin, with a maximum removal efficiency of 98.8%.

18.2 Technologies for Fluoride Treatment in Wastewater

Fluoride is a common pollutant in industrial wastewater, especially from the semiconductor, solar cell, and metal plating industries. Fluoride can cause adverse effects on human health and the environment, such as dental and skeletal fluorosis. Therefore, various technologies have been developed to remove fluoride from wastewater to meet discharge standards. One of the most widely used methods is chemical precipitation and coagulation, which involves the addition of calcium and aluminum salts to form insoluble calcium fluoride and aluminum hydroxide complexes. This method can reduce fluoride concentrations to less than 5 mg/L, as well as remove other contaminants such as acid, silica, and heavy metals. However, this method also has some drawbacks, such as high chemical consumption, large sludge production, and sensitivity to pH and temperature. Other methods that have been reported for fluoride removal include adsorption, ion exchange, membrane filtration, electrochemistry, and induced crystallization. These methods have different advantages and disadvantages in terms of efficiency, cost, selectivity, and applicability. Therefore, the selection of the best technology for fluoride treatment depends on the characteristics of the wastewater and the specific requirements of each case.

18.2.1 The Coagulation and Precipitation

This method is widely used and is suitable for the large-scale treatment of wastewater. The principle is to add a coagulant with a coagulation effect to the fluorine-containing wastewater and then adjust the pH to an appropriate value, so that the fluoride in the wastewater is adsorbed by the formed colloid or precipitate, thereby achieving the purpose of removing fluoride ions (Waghmare & Arfin, 2015). Coagulation is a process that uses a specific chemical, known as a coagulant, to destabilize colloidal particles in water. Various substances can act as coagulants for the removal of fluoride from water, such as aluminum, iron, alum, lime, zeolites, sodium aluminate, ferric chloride, and silica gel (Waghmare & Arfin, 2015). However, alum and lime are the most widely used. The Nalgonda technique is a prominent example of fluoride removal by coagulation/precipitation. This technique consists of six sequential steps: rapid mixing, coagulation/flocculation, sedimentation, filtration, disinfection, and sludge concentration. In this technique, aluminum salts, bleaching powder, and lime are added to the fluoride-contaminated water (Waghmare & Arfin, 2015).

The following steps constitute this process: (a) Formation of insoluble aggregates of aluminum hydroxide, (b) sedimentation of the aggregates at the bottom, and (c) co-precipitation of fluoride and bleaching powder (Bhatnagar et al., 2011). However, this technique, known as the Nalgonda technique, has a limited efficiency of approximately 70% for water defluoridation and is not suitable for treating water with high fluoride concentrations. Moreover, this method is costly due to the high price of alum (Ahmad et al., 2022). In the precipitation method, the fluoride in the calcium fluoride is removed from the water (Ahmad et al., 2022). Phosphate and calcium are used to induce the precipitation of fluoride, followed by filtration using bone char presaturated with fluoride ions. A medium containing saturated bone charcoal acts as a catalyst for the precipitation of fluoride in the form of fluorapatite and CaF_2 (Aljerf & Choukaife, 2017).

18.2.2 Electrocoagulation

Electrocoagulation is an effective method to remove flocculating agents that are produced by electro-oxidizing a sacrificial anode, usually made of aluminum or iron. This process does not require any chemical flocculants or coagulants and reduces the disposal of the sludge. This method involves three fundamental processes: electrochemistry, hydrodynamics, and coagulation (Emamjomeh et al., 2011). The electrocoagulation reactor has an electrolytic cell with a cathode and an anode. The main chemical reactions that occur at the electrodes (aluminum and iron electrodes) (Mollah et al., 2001).

$$Al \leftrightarrow Al^{3+} + 3e^-$$

$$3H_2O + 3e^- \leftrightarrow 3/2H_2 + 3OH^-$$

Moreover, Al^{3+} and OH^- ions produced at electrode surfaces react in the bulk wastewater to form aluminum hydroxide, which is a common coagulant in water treatment.

$$Al^{3+} + 3OH^- \leftrightarrow Al(OH)_3$$

Similar chemical reactions also occurred in electrocoagulation over iron catalysts.

$$Fe(s) \leftrightarrow Fe^{3+}(aq) + 3e^-$$
$$3H_2O + 3e^- \leftrightarrow 3H_2(g) + 3OH^-(aq)$$
$$Fe^{3+}(aq) + 3OH^-(aq) \leftrightarrow Fe(OH)_3$$

The aluminum and iron hydroxide flocs formed during the process mainly act as adsorbents and/or traps for metal ions, thereby removing them from the solution. The main objective of this study was to evaluate the efficiency of the electrocoagulation process for the removal of fluoride from aqueous media using iron and aluminum electrodes and to determine the effects of voltage, pH, initial fluoride concentration, and reaction time on the removal efficiency (Bazrafshan et al., 2012).

Ghosh et al. (2008)studied the removal of fluoride from drinking water with 2 to 10 mg L^{-1} fluoride concentration using mono and bipolar connections. They found that the bipolar connection was more effective for the removal of fluoride. Under bipolar connection, the fluoride level reached the recommended limit of 1 mg L^{-1} at 625 A m^{-2} in 30 min. The operational costs for the monopolar and bipolar connection were 0.38 and 0.62 US$ m^{-3}, respectively. Vasudevan et al. (2011) compared the effect of direct and alternating current on fluoride removal from water using aluminum alloy electrodes. They reported that direct current formed an impermeable oxide layer on the cathode surface, while anode corrosion occurred due to oxidation. This reduced the method efficiency because the current between the electrodes could not be controlled. This problem was solved by using an alternating current. They observed that at pH 7.0 and current density of 1.0 A dm^{-2}, the removal efficiencies of alternating and direct current were 93% and 91.5%, respectively. The energy consumption was 1.883 and 2.541 kW h kL^{-1}, respectively. They also found that the electrocoagulation process was spontaneous and exothermic.

18.2.3 Nanofltration

Nanofiltration is recognized as a highly selective membrane process for effectively removing fluoride due to its specific membrane characteristics (Tahaikt et al., 2007). Nanofiltration membranes possess larger pore sizes compared to reverse osmosis membranes, facilitating the movement of both solvent and solute. As a result, nanofiltration operates under low pressure conditions, enabling fast flow rates (Hu & Dickson, 2006). In the context of fluoride removal and brackish water desalination, nanofiltration showed superior performance compared to other membrane techniques (Diawara et al., 2011). Specifically, in a comparative study between two commercially available nanofiltration membranes, NF-270 and NF-90, NF-270 reduced the fluoride concentration from 10 to 1.5 mg L^{-1}, while NF-90 achieved a reduction from 20 to 0.5 mg L^{-1}. Furthermore, they also reported that the combined use of BW30 and NF-90 membranes resulted in a decrease in fluoride concentration from 417.9 mg L^{-1} to less than 1.5 mg L^{-1} (Shen & Schäfer, 2014).

18.2.4 Reverse Osmosis (RO) Technology

It employs a semi-permeable membrane to remove fluoride ions from water, is another commonly used method (Ahmad et al., 2022). Reverse osmosis (RO) is a common form of water purification that uses a semipermeable membrane to filter out contaminants from water. RO can remove fluoride from water because fluoride molecules are larger than water molecules and cannot pass through the membrane. RO filters can remove up to 98% fluoride from tap water, making it safe for drinking (Mastropietro et al., 2021). The RO process works by applying pressure to the water source to separate it from dissolved solids and other impurities, including fluoride. High pressure forces water through a semi-permeable membrane, which allows only pure water molecules to pass through while trapping fluoride and other contaminants (Waghmare & Arfin, 2015). Several studies have shown that RO technology is effective in removing fluoride from water sources. For example, a study by Yang et al. (2022) demonstrated that RO systems with a cellulose acetate membrane were capable of removing more than 98% fluoride from a synthetic water source. Similarly, a study by Shan et al. (Shen & Schäfer, 2014) showed that an RO membrane with a 98% rejection rate could remove fluoride from groundwater sources with concentrations ranging from 0.5 to 5.8 mg/L. Additionally, RO systems have been found to remove other impurities such as bacteria, viruses, and heavy metals (Waghmare & Arfin, 2015).

Despite its effectiveness, RO technology does have some limitations. The process can be energy-intensive and require high-pressure pumps to operate effectively. Additionally, the RO process can remove essential minerals from the water, which can affect the taste and quality of the water. However, these limitations can be mitigated through the use of energy-efficient pumps and remineralization techniques to add essential minerals back into the water (Waghmare & Arfin, 2015). Some factors that affect the efficiency of RO technology are:

- The quality and pore size of the membrane
- The pressure and flow rate of the water
- The concentration and type of contaminants in the water
- The temperature and pH of the water
- The maintenance and replacement of the filters and membrane

Some advantages and limitations of RO technology are:

- Advantages: RO can remove a wide range of contaminants, including bacteria, lead, arsenic, nitrates, pesticides, and fluoride. RO can improve the taste, odor, and appearance of water. RO can save money and reduce plastic waste by replacing bottled water (Min et al., 1984).
- Limitations: RO can be wasteful because it purifies only a fraction of the water that passes through the system. RO can remove some beneficial minerals from water, such as calcium and magnesium. RO can require high pressure and electricity to operate. RO can be expensive to install and maintain (Min et al., 1984).

Overall the reverse osmosis (RO) technology is an effective method for removing fluoride from drinking water sources. Several studies have shown that RO systems can remove fluoride with rejection rates of over 98%. Although the process can be energy-intensive and remove essential minerals, these limitations can be addressed through the use of energy-efficient pumps and remineralization techniques.

18.2.5 Adsorption Method

The adsorption method is the most widely used fluorine removal method, which can be used directly for low fluorine content wastewater treatment and can also be used as an advanced treatment after chemical precipitation and precipitation of coagulation.

Adsorption is a boundary phenomenon that involves the accumulation of particles from the bulk phase to the solid or liquid phase (Obulapuram et al., 2021b). The formation of a substance layer on the surface of adsorbents is termed adsorption. The adsorption process can be described by the following steps. First, a layer of fluoride is formed on the surface of adsorbent particles from the heterogeneous solution. Second, fluoride ions are adsorbed on the surface of the adsorbent particles. Third, intraparticle diffusion occurs, where the adsorbed fluoride moves to the inner surfaces of the porous adsorbent materials (Obulapuram et al., 2021a). The amount of fluoride adsorbed per unit mass of the adsorbent indicates the degree of water purification from fluoride contamination (Mallakpour & Hussain, 2021). In the adsorption method, which utilizes various adsorbents such as activated carbon, zeolites, and alumina to remove fluoride ions from wastewater (Kumar et al., 2019). According to the different raw materials used, the adsorbents are divided into conventional adsorbents and new high-efficiency adsorbents. Table 18.1 shows the various types of adsorbents. The efficiency of an adsorbent in removing fluoride depends on factors such as the initial fluoride concentration, the type of adsorbent used, the pH of the water, the presence of interfering ions, and the contact time (Waghmare & Arfin, 2015). Adsorption is considered the most effective method for defluorination for a small population due to its simpler operation, simple design, availability of various adsorbents, and relatively low setup cost (Waghmare & Arfin, 2015). However, some adsorbents are very costly, while some are not technically suitable for defluoridation in rural areas (Ghorai & Pant, 2005).

Table 18.1 Various types of adsorbents

Type of adsorbent	Examples
Conventional adsorbents	Activated carbon, activated alumina, zeolite, bentonite, etc.
New high-efficiency adsorbents	Polymer-based adsorbents, metal-based adsorbents, rare earth-based adsorbents, resin-based adsorbents, etc.

18.2.5.1 Conventional Adsorbents

Most of the polymer-based adsorbent materials are derived from natural biomass and its derivatives, such as chitosan and diatomaceous earth. Although natural chitosan can remove fluoride ions in water through surface adsorption, complexation, and ion exchange, it usually needs to be loaded and modified to improve fluoride removal performance (Singh et al., 2016; Vinati et al., 2015). The main chemical composition of diatomite is SiO_2 and contains a small amount of impurities such as Al_2O_3, CaO, MgO, etc. (Peter, 2009). It has a huge specific surface area and silicon and fluoride ions in water form stable fluorosilicic acid, which will enhance the defluorination effect. Natural minerals such as zeolite, bentonite, etc. They have low cost and good development prospects as fluoride removal adsorbents. However, both original zeolite and original bentonite need to be modified properly to further improve the fluorine removal performance (Peter, 2009). Take, for example, fluorine-containing wastewater produced by a chemical company. The hydroxy complex $Al(OH)_2^+$ rich in positively charged aluminum is the key to improving the fluorine removal performance (Mohapatra et al., 2009). Metal-based adsorbents are mainly oxides or hydroxides of metals such as aluminum, iron, and magnesium. Activated alumina is the first metal oxide used for fluorine removal. It has a large specific surface area, high mechanical strength, good high-temperature resistance, and corrosion resistance. It has an ideal fluorine removal effect in acidic solutions and its adsorption capacity is generally 0.8–2.0 mg/g. However, activated alumina also has low adsorption capacity, difficult separation, and rapid decline in adsorption capacity after repeated regeneration is the main disadvantage (Ahmad et al., 2022; Mohapatra et al., 2009). Iron-based adsorbents have properties similar to aluminum-based adsorbents but are more stable, while MgO needs to be pretreated due to its certain defluorination ability (Devi et al., 2014).

18.2.5.2 New High-Efficiency Adsorbents

To improve the shortcomings of traditional adsorbents, such as low adsorption capacity and low mechanical strength when absorbing fluorine in wastewater, rare earth elements with high affinity for F, large adsorption capacity and high adsorption rate can be added to fluorine removal materials. Therefore, new high-efficiency adsorbents for fluoride removal have been explored and developed in recent years. Some examples of such new adsorbents are summarized below:

Polypyrrole-based adsorbents: These are adsorbents that are composed of polypyrrole (PPy), a conducting polymer, and various supports or modifiers, such as biosorbents, clay minerals, metal oxides, and magnetic nanoparticles. PPy-based adsorbents can exhibit high adsorption capacity and selectivity for fluoride due to the electrostatic attraction between the positively charged PPy backbone and negatively charged fluoride ions, as well as the π–π interactions between the aromatic rings of PPy and the fluoride ions. Furthermore, PPy-based adsorbents can be easily synthesized by chemical or electrochemical polymerization methods and can be modified by doping or functionalization to enhance their performance (Ghosh et al., 2022).

Fe–Al–Ce nanoadsorbent: This is an adsorbent that is composed of nanosized particles of iron (Fe), aluminum (Al), and cerium (Ce) oxides or hydroxides. Fe–Al–Ce nanoadsorbent can effectively remove fluoride from water by multiple mechanisms, such as surface complexation, electrostatic attraction, ion exchange, and precipitation. The presence of Ce in the adsorbent can increase the surface area, pore volume, and surface charge density of the adsorbent, as well as inhibit the formation of $Al(OH)_3$ gel that can reduce the adsorption sites. The Fe–Al–Ce nanoadsorbent can be prepared by co-precipitation method and can be granulated or coated with glass beads to improve its stability and separation (Wang et al., 2022).

Hu Jiapeng et al. (Razzaq et al., 2020) applied the prepared hydroxylanthanum modified resin to simulated fluorine-containing wastewater in the laboratory and found that the hydroxylanthanum modified resin had better selective adsorption performance for F, which not only enabled the treated wastewater to meet the national discharge standard and can improve the pH of discharged wastewater.

18.2.6 Ion-Exchange Process

The ion-exchange process involves the use of an ion exchange material, which allows water to flow through its bed and remove unwanted ions according to the following chemical equation:

$$\text{Matrix} - NR_3 + Cl^- + F^- \rightarrow \text{Matrix} - NR_3 + F^- + Cl^-$$

During the ion exchange process, fluoride ions displace the chloride ions in the resin, gradually filling up all available binding sites. To regenerate the resin, a backwashing procedure is employed using a sodium chloride salt solution dissolved in water. This facilitates the replacement of fluoride ions with chloride ions, allowing the resin to resume its ion exchange functionality (Ojekunle et al., 2016). The replacement of chloride ions in the resin is driven by the higher electronegativity of fluoride ions. Chikuma et al. (1987) conducted a modification to the anion exchange method of fluoride removal using lanthanum. In their study, Chikuma and Nishimura (1990) utilized Amberlite IRA-400 resin in an aqueous solution to remove fluoride and observed that fluoride ions displaced the chloride ions originally present in the resin. Ho et al. (2004) studied the fluoride removal by the ion exchange method and enhanced the capacity of the ion exchange method by using titanium oxyhydroxide. They investigated the doping of zirconia and silica with small particle sizes, over iron oxyhydroxide (a mesoporous material) and reported an increased ion exchange capacity using this material. However, this approach proved to be costly and raised concerns about membrane integrity. Meenakshi and Viswanathan (2007) conducted a study to assess the fluoride removal capability of chelating resins, namely Ceralite IRA 400 (CER) and Resin FR 10 (IND). They found that the chelating resin exhibited a high selectivity for fluoride removal compared to that of anion exchange resins. The ion exchange technique demonstrated significant potential for eliminating fluoride from aqueous solutions, achieving removal rates of up to 95%. However, the high cost of resins renders the treatment economically unfavorable, despite the possibility of resin regeneration. Furthermore, the regeneration process generates a substantial amount of fluoride-loaded waste, which poses a drawback to the overall technique (Jadhav et al., 2015).

18.2.7 Hybrid Methods

Hybrid technology for fluoride removal involves the integration of different treatment technologies to efficiently remove fluoride from contaminated water sources. For example, the hybrid adsorption-electrocoagulation method combines the advantages of both adsorption and electrocoagulation methods to remove fluoride from wastewater (Gai & Deng, 2021). Kapenja et al. (2017)investigated the impact of submerged Amaranthus root sprouts in F-contaminated water with varying concentrations of $FeCl_3$ (0.1, 1, and 10 mM) on F coagulation. The researchers observed that within a span of 12 h, the removal efficiency increased to 20%, 37% and 40% respectively, in the presence of the plant, compared to a mere 10% removal without the plant. This improvement in effectiveness is attributed to the formation of a Fe (III) oxide coating on the root surface, which potentially encapsulates organic molecules and CO_2. Furthermore, Khusboo and Khan (Chaudhary & Khan, 2016) also reported that the introduction of different microbial consortiums to *P. juliflora* resulted in a bioaccumulation factor ranging from 0.12 to 3.3. They have investigated a sequential adsorption bioaccumulation approach (SABA) using leaves and *Citrus limetta* peels in conjunction with bacteria, the F concentration was successfully reduced below 1.5 mg L^{-1} from an initial concentration of 20 mg L^{-1} (Dwivedi et al., 2017).

18.3 Conclusion and Future Perspective

The contamination of water with fluoride presents a significant challenge to human health, necessitating effective environmental protection measures and water management strategies. Although anthropogenic activities contribute to the release of fluoride into the environment, the primary concern arises from the ingestion of fluoride from natural sources. The United Nations Sustainable Development Goals (UNSDG) guide for clean drinking water provides a mandatory, cost-effective, efficient, and technically feasible approach to fluoride removal, addressing the inherent limitations associated with other methods. This comprehensive review aims to encompass a wide range of techniques used so far for the removal of fluoride from drinking water and industrial wastewater. The conventional method for the removal of fluoride from drinking water involves liming and subsequent precipitation of fluoride. However, these methods often suffer from drawbacks such as high operational and maintenance costs, potential secondary contamination through the generation of toxic sludge, and complex treatment processes. Various technological systems utilized for defluoridation were critically evaluated. Several technologies are available for the treatment of fluoride in wastewater, each with its advantages and disadvantages. The choice of technology depends on various factors and a combination of technologies may be necessary to achieve effective fluoride removal. Further research is needed to develop more efficient and cost-effective treatment methods for fluoride-contaminated wastewater. This study adopts a comprehensive approach to address the intricate complexities associated with F pollution, establishing apparently the importance of considering various exposure pathways and associated human health risks. Consequently, future researchers employ a dedicated and holistic methodology when investigating fluoride concentrations.

Acknowledgements The authors thank our prestigious institute and other people who are involved in this study for their direct and indirect contributions.

References

Ahmad, S., Singh, R., Arfin, T., & Neeti, K. (2022). Fluoride contamination, consequences and removal techniques in water: A review. *Environmental Science: Advances.*

Ali, S., Fakhri, Y., Golbini, M., Thakur, S. K., Alinejad, A., Parseh, I., Shekhar, S., & Bhattacharya, P. (2019). Concentration of fluoride in groundwater of india: A systematic review, meta-analysis and risk assessment. *Groundwater for Sustainable Development, 9*, 100224

Aljerf, L., & Choukaife, A. E. (2017). Hydroxyapatite and fluoroapatite behavior with ph change. *Internal Medicine Journal, 24*, 407–410.

Bazrafshan, E., Ownagh, K. A., & Mahvi, A. H. (2012). Application of electrocoagulation process using iron and aluminum electrodes for fluoride removal from aqueous environment. *E-Journal of Chemistry, 9*, 2297–2308.

Bhatnagar, A., Kumar, E., & Sillanpää, M. (2011). Fluoride removal from water by adsorption—A review. *Chemical Engineering Journal, 171*, 811–840.

Camargo, J. A. (2003). Fluoride toxicity to aquatic organisms: A review. *Chemosphere, 50*, 251–264.

Chatterjee, A., Sarah, S., Sreedevi, P. D., Selles, A., & Ahmed, S. (2017). Demarcation of fluoride vulnerability zones in granitic aquifer, semi-arid region, Telengana, India. *Arabian Journal of Geosciences, 10*, 558.

Chaudhary, K., & Khan, S. (2016). Physiochemical characterization of fluoride (f) contaminated soil and its microbe-assisted bioremediation by prosopis juliflora. *Plant Biol. Soil Health, 3.*

Chaudhary, V., & Prasad, S. (2015). Rapid removal of fluoride from aqueous media using activated dolomite. *Analytical Methods, 7*, 8304–8314.

Chikuma, M., & Nishimura, M. (1990). Selective sorption of fluoride ions by anion-exchange resin modified with alizarin fluorine blue-praseodymium(iii) complex. *Reactive Polymers, 13*, 131–138.

Chikuma, M., Okabayashi, Y., Nakagawa, T., Inoue, A., & Tanaka, H. (1987). Separation and determination of fluoride ion by using ion exchange resin loaded with alizarin fluorine blue. *Chemical & Pharmaceutical Bulletin, 35*, 3734–3739.

Chowdhury, A., Adak, M. K., Mukherjee, A., Dhak, P., Khatun, J., & Dhak, D. (2019). A critical review on geochemical and geological aspects of fluoride belts, fluorosis and natural materials and other sources for alternatives to fluoride exposure. *Journal of Hydrology, 574*, 333–359.

Collins, J. P. (2010). Amphibian decline and extinction: What we know and what we need to learn. *Diseases of Aquatic Organisms, 92*, 93–99.

Dawes, C. (2003). What is the critical ph and why does a tooth dissolve in acid? *Journal (Canadian Dental Association), 69*, 722–724.

Devi, R. R., Umlong, I. M., Raul, P. K., Das, B., Banerjee, S., & Singh, L. (2014). Defluoridation of water using nano-magnesium oxide. *Journal of Experimental Nanoscience, 9*, 512–524.

Diawara, C. K., Diop, S. N., Diallo, M. A., Farcy, M., & Deratani, A. (2011). Performance of nanofiltration (nf) and low pressure reverse osmosis (lpro) membranes in the removal of fluorine and salinity from brackish drinking water. *Journal of Water Resource and Protection, 3*, 912.

Dwivedi, S., Mondal, P., & Balomajumder, C. (2017). Bioremoval of fluoride from synthetic water using gram-negative bacteria *shewanella putrefaciens*. *Journal of Hazardous, Toxic, and Radioactive Waste, 21*, 04016023.

Emamjomeh, M. M., Sivakumar, M., & Varyani, A. S. (2011). Analysis and the understanding of fluoride removal mechanisms by an electrocoagulation/flotation (ecf) process. *Desalination, 275*, 102–106.

Gai, W.-Z., & Deng, Z.-Y. (2021). A comprehensive review of adsorbents for fluoride removal from water: Performance, water quality assessment and mechanism. *Environmental Science: Water Research & Technology, 7*, 1362–1386.

Ghorai, S., & Pant, K. K. (2005). Equilibrium, kinetics and breakthrough studies for adsorption of fluoride on activated alumina. *Separation and Purification Technology, 42*, 265–271.

Ghosh, D., Medhi, C. R., & Purkait, M. K. (2008). Treatment of fluoride containing drinking water by electrocoagulation using monopolar and bipolar electrode connections. *Chemosphere, 73*, 1393–1400.

Ghosh, S., Malloum, A., Igwegbe, C. A., Ighalo, J. O., Ahmadi, S., Dehghani, M. H., Othmani, A., Gökkuş, Ö., & Mubarak, N. M. (2022). New generation adsorbents for the removal of fluoride from water and wastewater: A review. *Journal of Molecular Liquids, 346*, 118257.

Guissouma, W., Hakami, O., Al-Rajab, A. J., & Tarhouni, J. (2017). Risk assessment of fluoride exposure in drinking water of Tunisia. *Chemosphere, 177*, 102–108.

He, L., Tu, C., He, S., Long, J., Sun, Y., Sun, Y., & Lin, C. (2021). Fluorine enrichment of vegetables and soil around an abandoned aluminium plant and its risk to human health. *Environmental Geochemistry and Health, 43*, 1137–1154.

Ho, L. N., Ishihara, T., Ueshima, S., Nishiguchi, H., & Takita, Y. (2004). Removal of fluoride from water through ion exchange by mesoporous ti oxohydroxide. *Journal of Colloid and Interface Science, 272*, 399–403.

Hu, K., & Dickson, J. M. (2006). Nanofiltration membrane performance on fluoride removal from water. *Journal of Membrane Science, 279*, 529–538.

Jadhav, S. V., Bringas, E., Yadav, G. D., Rathod, V. K., Ortiz, I., & Marathe, K. V. (2015). Arsenic and fluoride contaminated groundwaters: A review of current technologies for contaminants removal. *Journal of Environmental Management, 162*, 306–325.

Jha, S. K., Mishra, V. K., Sharma, D. K., & Damodaran, T. (2011). Fluoride in the environment and its metabolism in humans. *Reviews of Environmental Contamination and Toxicology, 211*, 121–142.

Johnston, N. R., & Strobel, S. A. (2020). Principles of fluoride toxicity and the cellular response: A review. *Archives of Toxicology, 94*, 1051–1069.

Kabir, H., Gupta, A. K., & Tripathy, S. (2020). Fluoride and human health: Systematic appraisal of sources, exposures, metabolism, and toxicity. *Critical Reviews in Environmental Science and Technology, 50*, 1116–1193.

Kapenja, A. J., Msigala, L. C., & Mwakabona, H. T. (2017). The effect of amaranthus hybridus on fluoride removal by iron (iii) salts as fluoride coagulants. *African Journal of Environmental Science and Technology, 11*, 207–212.

Kashyap, S. J., Sankannavar, R., & Madhu, G. (2021). Fluoride sources, toxicity and fluorosis management techniques—A brief review. *Journal of Hazardous Materials Letters, 2*, 100033.

Kumar, P. S., Suganya, S., Srinivas, S., Priyadharshini, S., Karthika, M., Karishma Sri, R., Swetha, V., Naushad, M., & Lichtfouse, E. (2019). Treatment of fluoride-contaminated water. A review. *Environmental Chemistry Letters, 17*, 1707–1726.

Lacson, C. F. Z., Lu, M.-C., & Huang, Y.-H. (2021). Fluoride-containing water: A global perspective and a pursuit to sustainable water defluoridation management-an overview. *Journal of Cleaner Production, 280*, 124236.

Li, M., Zhao, Y., Tian, X., Liu, P., Xie, J., Dong, N., Feng, J., Gao, Y., Fan, Y., & Qiu, Y. (2021). Fluoride exposure and blood pressure: A systematic review and meta-analysis. *Biological Trace Element Research, 199*, 925–934.

Maheshwari, N., Qasim, N., Anjum, R., & Mahmood, R. (2021). Fluoride enhances generation of reactive oxygen and nitrogen species, oxidizes hemoglobin, lowers antioxidant power and inhibits transmembrane electron transport in isolated human red blood cells. *Ecotoxicology and Environmental Safety, 208*, 111611.

Mallakpour, S., & Hussain, C. M. (2021). *Environmental applications of carbon nanomaterials-based devices*: Wiley Online Library.

Mastropietro, T. F., Bruno, R., Pardo, E., & Armentano, D. (2021). Reverse osmosis and nanofiltration membranes for highly efficient pfass removal: Overview, challenges and future perspectives. *Dalton Transactions, 50*, 5398–5410.

Meenakshi, S., & Viswanathan, N. (2007). Identification of selective ion-exchange resin for fluoride sorption. *Journal of Colloid and Interface Science, 308*, 438–450.

Min, B. R., Gill, A. L., & Gill, W. N. (1984). A note on fluoride removal by reverse osmosis. *Desalination, 49*, 89–93.

Miranda, G. H. N., Alvarenga, M. O. P., Ferreira, M. K. M., Puty, B., Bittencourt, L. O., Fagundes, N. C. F., Pessan, J. P., Buzalaf, M. A. R., & Lima, R. R. (2021). A systematic review and meta-analysis of the association between fluoride exposure and neurological disorders. *Scientific Reports, 11*, 22659.

Mohapatra, M., Anand, S., Mishra, B. K., Giles, D. E., & Singh, P. (2009). Review of fluoride removal from drinking water. *Journal of Environmental Management, 91*, 67–77.

Mollah, M. Y. A., Schennach, R., Parga, J. R., & Cocke, D. L. (2001). Electrocoagulation (ec)—Science and applications. *Journal of Hazardous Materials, 84*, 29–41.

Mukherjee, I., & Singh, U. K. (2018). Groundwater fluoride contamination, probable release, and containment mechanisms: A review on Indian context. *Environmental Geochemistry and Health, 40*, 2259–2301.

Nakayama, H., Yamasaki, Y., & Nakaya, S. (2022). Effect of hydrogeological structure on geogenic fluoride contamination of groundwater in granitic rock belt in Tanzania. *Journal of Hydrology, 612*, 128026.

Obulapuram, P. K., Arfin, T., Mohammad, F., Khiste, S. K., Chavali, M., Albalawi, A. N., & Al-Lohedan, H. A. (2021a). Adsorption, equilibrium isotherm, and thermodynamic studies towards the removal of reactive orange 16 dye using cu(i)-polyaninile composite. *Polymers, 13*, 3490.

Obulapuram, P. K., Arfin, T., Mohammad, F., Kumari, K., Khiste, S. K., Al-Lohedan, H. A., & Chavali, M. (2021b). Surface-enhanced biocompatibility and adsorption capacity of a zirconium phosphate-coated polyaniline composite. *ACS Omega, 6*, 33614–33626.

Ojekunle, O. Z., Ojekunle, O. V., Adeyemi, A. A., Taiwo, A. G., Sangowusi, O. R., Taiwo, A. M., & Adekitan, A. A. (2016). Evaluation of surface water quality indices and ecological risk assessment for heavy metals in scrap yard neighbourhood. *Springerplus, 5*, 560.

Onipe, T., Edokpayi, J. N., & Odiyo, J. O. (2020). A review on the potential sources and health implications of fluoride in groundwater of Sub-Saharan Africa. *Journal of Environmental Science and Health, Part A, 55*, 1078–1093.

Organization, W. H. (2017). *Guidelines for drinking-water quality: First addendum to the fourth edition.*

Pearcy, K., Elphick, J., & Burnett-Seidel, C. (2015). Toxicity of fluoride to aquatic species and evaluation of toxicity modifying factors. *Environmental Toxicology and Chemistry, 34*, 1642–1648.

Peter, K. (2009). Defluoridation of high fluoride waters from natural water sources by using soils rich in bauxite and kaolinite. *JEAS-Journal of Engineering & Applied Sciences, 4*, 240–246.

Pillai, P., Dharaskar, S., Pandian, S., & Panchal, H. (2021). Overview of fluoride removal from water using separation techniques. *Environmental Technology & Innovation, 21*, 101246.

Poonia, T., Singh, N., & Garg, M. (2021). Contamination of arsenic, chromium and fluoride in the Indian groundwater: A review, meta-analysis and cancer risk assessment. *International Journal of Environmental Science and Technology, 18*, 2891–2902.

Raj, D., & Shaji, E. (2017). Fluoride contamination in groundwater resources of Alleppey, Southern India. *Geoscience Frontiers, 8*, 117–124.

Razzaq, A. A., Yuan, X., Chen, Y., Hu, J., Mu, Q., Ma, Y., Zhao, X., Miao, L., Ahn, J.-H., & Peng, Y. (2020). Anchoring mof-derived cos 2 on sulfurized polyacrylonitrile nanofibers for high areal capacity lithium–sulfur batteries. *Journal of Materials Chemistry A, 8*, 1298–1306.

Rizzu, M., Tanda, A., Cappai, C., Roggero, P. P., & Seddaiu, G. (2021). Impacts of soil and water fluoride contamination on the safety and productivity of food and feed crops: A systematic review. *Science of the Total Environment, 787*, 147650.

Shen, J., & Schäfer, A. (2014). Removal of fluoride and uranium by nanofiltration and reverse osmosis: A review. *Chemosphere, 117*, 679–691.

Singh, G., Kumari, B., Sinam, G., Kumar, N., & Mallick, S. (2018). Fluoride distribution and contamination in the water, soil and plants continuum and its remedial technologies, an Indian perspective—A review. *Environmental Pollution, 239*, 95–108.

Singh, K., Lataye, D. H., & Wasewar, K. L. (2016). Removal of fluoride from aqueous solution by using low-cost sugarcane bagasse: Kinetic study and equilibrium isotherm analyses. *Journal of Hazardous, Toxic, and Radioactive Waste, 20*, 04015024.

Skórka-Majewicz, M., Goschorska, M., Żwierełło, W., Baranowska-Bosiacka, I., Styburski, D., Kapczuk, P., & Gutowska, I. (2020). Effect of fluoride on endocrine tissues and their secretory functions—review. *Chemosphere, 260*, 127565.

Susheela, A. K. (1999). Fluorosis management programme in India. *Current Science, 77*, 1250–1256.

Tahaikt, M., El Habbani, R., Ait Haddou, A., Achary, I., Amor, Z., Taky, M., Alami, A., Boughriba, A., Hafsi, M., & Elmidaoui, A. (2007). Fluoride removal from groundwater by nanofiltration. *Desalination, 212*, 46–53.

Tiwari, K., Krishan, G., Prasad, G., Mondal, N., & Bhardwaj, V. (2020). Evaluation of fluoride contamination in groundwater in a semi-arid region, Dausa district, Rajasthan, India. *Groundwater for Sustainable Development, 11*, 100465.

Vandana, K., Srishti Raj, B., & Desai, R. (2021). Dental fluorosis and periodontium: An original research report of in vitro and in vivo institutional studies. *Biological Trace Element Research, 199*, 3579–3592.

Vasudevan, S., Kannan, B. S., Lakshmi, J., Mohanraj, S., & Sozhan, G. (2011). Effects of alternating and direct current in electrocoagulation process on the removal of fluoride from water. *Journal of Chemical Technology & Biotechnology, 86*, 428–436.

Vinati, A., Mahanty, B., & Behera, S. K. (2015). Clay and clay minerals for fluoride removal from water: A state-of-the-art review. *Applied Clay Science, 114*, 340–348.

Vithanage, M., & Bhattacharya, P. (2015). Fluoride in the environment: Sources, distribution and defluoridation. *Environmental Chemistry Letters, 13*, 131–147.

Waghmare, S., Arfin, T., Rayalu, S., Lataye, D., Dubey, S., & Tiwari, S. (2015). Adsorption behavior of modified zeolite as novel adsorbents for fluoride removal from drinking water: Surface phenomena, kinetics and thermodynamics studies. *International Journal of Science, Engineering and Technology Research, 4*, 4114–4124.

Waghmare, S. S., & Arfin, T. (2015). Fluoride removal from water by various techniques. *International Journal of Innovative Studies in Sciences and Engineering Technology, 2*, 560–571.

Wan, K., Huang, L., Yan, J., Ma, B., Huang, X., Luo, Z., Zhang, H., & Xiao, T. (2021). Removal of fluoride from industrial wastewater by using different adsorbents: A review. *Science of the Total Environment, 773*, 145535.

Wang, M., Li, X., He, W.-Y., Li, J.-X., Zhu, Y.-Y., Liao, Y.-L., Yang, J.-Y., & Yang, X.-E. (2019). Distribution, health risk assessment, and anthropogenic sources of fluoride in farmland soils in phosphate industrial area, Southwest China. *Environmental Pollution, 249*, 423–433.

Wang, T., Yan, L., He, Y., Alhassan, S. I., Gang, H., Wu, B., Jin, L., & Wang, H. (2022). Application of polypyrrole-based adsorbents in the removal of fluoride: A review. *RSC Advances, 12*, 3505–3517.

WHO, G. (2011). Guidelines for drinking-water quality. *World Health Organization, 216*, 303–304.

Wimalawansa, S. J. (2020). Does fluoride cause the mysterious chronic kidney disease of multifactorial origin? *Environmental Geochemistry and Health, 42*, 3035–3057.

Yadav, K. K., Kumar, S., Pham, Q. B., Gupta, N., Rezania, S., Kamyab, H., Yadav, S., Vymazal, J., Kumar, V., & Tri, D. Q. (2019). Fluoride contamination, health problems and remediation methods in Asian groundwater: A comprehensive review. *Ecotoxicology and Environmental Safety, 182*, 109362.

Yang, W., Shi, F., Jiang, W., Chen, Y., Zhang, K., Jian, S., Jiang, S., Zhang, C., & Hu, J. (2022). Outstanding fluoride removal from aqueous solution by a la-based adsorbent. *RSC Advances, 12*, 30522–30528.

Chapter 19
Future Frameworks for Fluoride and Algorithms for Environmental System

Mridu Kulwant, Divya Patel, Saba Shirin, Shiv Nath Sharma, and Akhilesh Kumar Yadav

Abstract In recent years, there has been intensification in community discernment of the damaging possessions of fluoride to human health (because of its effects on teeth and bones), as well as effects on the environment. Artificial activities are one of the biggest issues in the world today. It is crucial to develop efficient and reliable solutions to remove excess fluoride from water environments. The fluoride

M. Kulwant (✉) · D. Patel
Department of Environmental Studies, The Maharaja Sayajirao University of Baroda, Vadodara 390002, India
e-mail: mridu.k-envphd@msubaroda.ac.in

D. Patel
e-mail: pateldivya9030@gmail.com

D. Patel
National Sugar Institute, Kalyanpur, Kanpur 208017, India

S. Shirin
Department of Environmental Science, School of Vocational Studies and Applied Sciences, Gautam Buddha University, Greater Noida 201312, India
e-mail: sabashirin83@gmail.com

S. Shirin · A. K. Yadav
Department of Mining Engineering, Indian Institute of Technology (Banaras Hindu University), Varanasi 221005, India
e-mail: yadavbasti@gmail.com

S. N. Sharma
Central Soil and Materials Research Station, Department of Water Resources, River Development and Ganga Rejuvenation, Ministry of Jal Shakti, Government of India, Olof Palme Marg, Hauz Khas, New Delhi 110016, India
e-mail: shiv9sharma@gmail.com

A. K. Yadav
Department of Environmental Engineering and Management, Chaoyang University of Technology, Taichung 413310, Taiwan

Environmental Science and Engineering Department, Indian Institute of Technology Bombay, Mumbai 400076, India

A. K. Yadav et al. (eds.), *Advanced Treatment Technologies for Fluoride Removal in Water*, Water Science and Technology Library 125,
https://doi.org/10.1007/978-3-031-38845-3_19

parameter stands out among the others. The goals included describing the framework for monitoring water quality, defining a proposed set of indicators, and informing the public about fluoride content. It is still essential to effectively clean fluoride-contaminated water. Metal–organic framework (MOF) materials are thought to be one of the several attractive adsorption materials for the various technologies used to remove fluoride from water. This article reviews current developments in the synthesis of MOFs and their use in aquatic defluoridation. MOFs are classified as the core metal ions. The mechanism of adsorption and an assessment of potential real-world uses of fluoride removal by various MOFs are explained. The practice of artificial water fluoridation should be reviewed globally as part of efforts to prevent harmful fluoride consumption, and industrial safety regulations should be improved to reduce the unethical release of fluoride compounds into the environment. The practice of artificial water fluoridation should be reviewed globally as part of efforts to prevent harmful fluoride consumption, and industrial safety regulations should be improved in order to reduce the unethical release of fluoride compounds into the environment. Public health strategies to reduce dental caries without relying on systemic fluoride intake.

Keywords Metal–organic framework · Fluoridation · Environment · Human health · Artificial water · Water quality · Monitoring

19.1 Introduction

Fluoride in consumable water poses an essential health danger. Hexafluoro-silicic acid, a frequently used fluoride molecules additive, is added to aqueous community sources, known as community or artificial water fluoridation, and has proved contentious since its 1950 debut in the United States (Aoun et al., 2018; Peckham & Awofeso, 2014; USPHS, 2015). Fluoride ion (F) is a necessary micronutrient for human health, excessive ingestion of F can have negative consequences such as bone fluorosis or DNA impairment (Hou et al., 2021).

Fluoride is a minor component that the human system requires. In terms of clinical prevention and osteoporosis treatment, fluoride is commonly used in dental care. Enhances the resistance of something like the outermost layer to acid attack and aids in preventing tooth decay by preventing cavities. Fluorine pollution has increased as the fluorine industry expands and more fluorine is released into the environment (Sun et al., 2020). According to Podgorski et al. (2018) reports, the northwest states and regions of different parts of India (such as—Delhi, Gujarat, Haryana, Punjab, and Rajasthan, as well as the south Indian states (Andhra Pradesh, Karnataka, Tamil Nadu, and Telangana), have been the hardest hit. Around 120 million individuals, or 9% of the population, are estimated to have fluoride in groundwater, which poses a general risk of fluorosis. They used the random forest machine learning technique along with geology, climate, and soil factors, 12,600 underground concentrations located throughout the country were modelled (Podgorski et al., 2018). There have been reports of widespread geographical variations in fluoride pollution in the Sindhudurg area. Between 2009 and 2016, 225 samples of water were taken from various wells dug, including wells, including wells and wells in that province (latitude: 15.37° and 16.40°, longitude: 73.19° and 74.18°). 80% of the overall data was divided into two subsets: training and testing. The fluoride (F) simulation employs nine distinct biogeochemical parameters: A variety of parameters were measured and expressed pH, EC, TDS, Ca^{2+}, Mg^{2+}, Na^{+}, Cl, HCO^{3-}, and $SO_4{}^{2+}$ by Gupta and Maiti (2022). Fluoride is a waste product from aluminium, fertilizers, and iron ore, but is used in fluorine compounds, ceramics, insecticides, aerosol propellants, refrigerants, and glassware, including teflon-made cookware (Peckham & Awofeso, 2014). The strong mediator in water, ion fluoride produces a variety of soluble complexes with polyvalent metal ions, such as Mg ions, Fe ions, Al ions, and Ca ions that vary with the acidity that exists in the substrate (Sivasankar, 2016). Water pollutants come from anthropogenic and geogenic sources (Haldar et al., 2020).

A ground-breaking architecture is called "BDA-GLSS" that helps companies integrate Big Data Analytics (BDA) into GLSS to enhance sustainability impact. This BDA-GLSS improves process capability and environmental capability, as well as providing both academics and practitioners with a new perspective to assist GLSS initiatives in achieving higher levels of sustainability technologies (Belhadi et al., 2021). According to the report by Mikulčić et al. (2021), the two main industries that are connected to water use are agriculture and electricity generation. Feeding water with too much fluoride has been studied, as has the viability of treating it with zirconium-impregnated hybrid anion exchange resin. The study's findings demonstrated that the suggested approach may successfully eliminate fluoride from groundwater (Mikulčić et al., 2021).

Several investigators agree that eating fluoride rich foods in tiny amounts can help prevent cavities in the mouth, as well as reinforce cartilage; however, fluorine consumption in high doses over time can contribute to thyroid dysfunction, dental fluorosis, skeletal fluorosis, a higher probability of bone cracks, decreased fertility proportions, amplified amounts of urolithiasis (stones in the kidney), along with decreased intellect within young children (Chen et al., 2013; Ozsvath, 2009). Fluoride promotes remineralization of tooth enamel, but its antagonistic consequence is highly subject to the proportion of both minerals in enamel on the teeth (Ozsvath, 2009). Fluoride ingestion and tooth contact cause hypocalcification and/or hypoplasia in young people with low levels of minerals (calcium and magnesium) within the tooth enamel (because of malnutrition), making them more vulnerable to dental caries (Peckham & Awofeso, 2014). Fluoridation of water supply is regularly promoted as reliable and efficient, including advocates claiming credible evidence from scientists or the encouragement of influential community well-being, as well as dental organizations (Peckham & Awofeso, 2014). Skeletal fluorosis has been documented in cases of high fluoride levels or sustained intake at 2 ppm. Fluoride poisoning induces physiological orthopedic disorders, including skeletal system (Ozsvath, 2009). Skeletal fluorosis, which causes joint discomfort in the upper and lower extremities, numbness and tingling, back pain, and knock-knees, is common (> 20%) in areas with a water

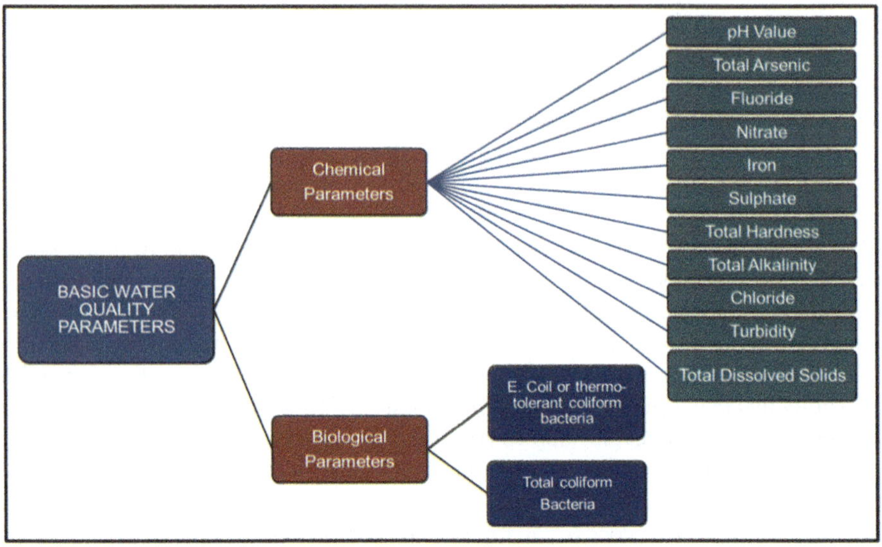

Fig. 19.1 Framework for water quality testing and monitoring parameters

fluoride content greater than 2 ppm or among employees in agricultural fertilizers or metal industries (Peckham & Awofeso, 2014).

Figure 19.1 shows the framework for water quality testing and monitoring parameters (Nayla Hassan, 2019). Ali Akbar Mohammadi created an artificial neural network (ANN) through the fluoride within groundwater resources in Khaf and adjacent districts based on the physicochemical frameworks of the water itself. ANN modeling was used with respect to a variety of inputs. The simulation results from the MLP1 testing phase, it was also proposed that this model, with its high confidence coefficient, may be used to estimate the fluoride amount within underground water provisions in line for significant level of consistency with practical versus projected information (Ozsvath, 2009).

19.2 Framework for Monitoring Water Quality

New research has challenged whether or not adding fluoride to water is beneficial since they discovered no differences among young people who consume fluoridated water versus others (Armfield, 2010; Peckham & Awofeso, 2014). Benedict et al. (2018) presented an instantaneous fashion water quality assessment methodology, the water quality analytical framework (WQAF) which evaluates numerous measurements of water purity has been published in response to new research at predetermined water quality monitoring stations, including pH, temperature, turbidity, and others (and predicts otherwise). If the monitoring station is not accessible at the specified location, WQAF either models and forecasts the factors relating to water quality standards or gathers certain predetermined water quality monitoring station characteristics (if a prompt selection is available). The WQAF categorizes the chosen location as depending on the values of the water quality criteria, a location for drinking water, or a location for agriculture (Benedict et al., 2018). A Remote Sensing (RS)-based framework for water natural water and monitoring for the coming era. The authors evaluate and analyze the major current approaches that used RS to monitor to improve quality water parameters are evaluated and analyzed by the authors in terms of their advantages and disadvantages. The following five guidelines were looked at, comprising three optical ones (turbidity, total suspended sediments, and chlorophyll-a) and two non-optical ones (total phosphorus and total nitrogen) (Hassan et al., 2020). Because their structures may be altered and contain more holes, certain organic materials, such as covalent organic frameworks (COFs) and metal–organic frameworks (MOFs), used as conductors in lithium-ion batteries (LIBs) additional sources of energy preservation (Yabushita et al., 2021).

Although synthetic water, in conjunction with additional fluoride-using systems, most disinfection research has shown that the overall decrease in tooth cavities worldwide has remained uneven (Peckham & Awofeso, 2014). Therefore, caries prevention must be implemented within a robust oral treatment network that includes diet habits, effective community-based oral health promotion activities, and readily available dental services (Peckham & Awofeso, 2014). Advances in environmental science and engineering (ESE) research that can be accommodated by statistical platforms are described in this article (Palanichamy et al., 2021). The safe disposal of sludge and the clean water produced by the effective and environmentally friendly adsorbent aluminum/olivine composite (AOC) was demonstrated by the low desorption potential, which increased the viability of the viability of the green technology at the field scale (Bishayee et al., 2021). Since it began, artificial fluoridation of water has been a controversial wellness initiative. However, researchers, including well-known scientists and academics, have had a hard time getting in dental publications, and thoughtful pieces on municipal water fluoridation have been published (Peckham & Awofeso, 2014). Iheozor-Ejiofor et al. (2015) conducted their study on artificial water to assess the impact of water fluoridation on the prevention of dental caries. Consequently, fluoridation in the water system is successful by lowering emerging stages in children's permanent and deciduous teeth (Iheozor-Ejiofor et al. 2015). Its development of a framework for controlling or dispensing water that utilizes the Internet of Things (IoT) platform, also using drives and ultrasonic detection devices, with flow gauges, was presented as either a solution or a method for implementation (Maroli et al., 2021). IoT, cloud computing data, and mobile sensor networks are just a few of the new technologies that have emerged in the modern era. These technologies are being used to collect data and analyze it in more detail for monitoring (Rathore et al., 2021).

Future intercomparison campaigns should focus on organic compounds that appear to be the most problematic, according to the evaluation by Coquery et al. from four years of interlaboratory testing (1999–2002) for the assessments of chemical pollutants in water (Hu et al., 2016). Recognizing satellite observation as an evaluation technique about the Water Framework Directive amendment, a clear signal to the Member States that its use is supported will be sent by explicitly encouraging the use of satellite-based monitoring to supplement national and statutory monitoring and reporting, as is already being done by a small number of countries (examples in this paper). This monitoring is already possible due to existing academic, governmental, and private sector capabilities (Lakshmikantha et al., 2021). Halliday et al. (2015) discussed among the initial research using onsite analyzing equipment in an arrangement severely impacted by urban waste. In light of this, on-site chemical analyzers including Phosphor are able to provide high-frequency hadrochemical data of excellent quality under these circumstances. The significant seasonal and annual fluctuation found demonstrated that the timing and season of sampling affect water quality assessments done in accordance with the Water Framework Directive (WFD). It may be deduced that it was constructed around an analysis of high-frequency signals and conventional data from surveillance benthic algae, primarily diatoms, accounted for a major portion of the primary production (Halliday et al., 2015).

Shrestha et al. (2008) anticipated a scheme for calculating contaminant discharge ratios using readily available data from in-stream water quality monitoring. It is suggested to reduce the ratios of contaminant discharge using a mixture of easily accessible and cost-free statistical, geographical, and hydrological methods employing a multiple deterioration methodology (Shrestha et al., 2008). Decisions about the development of a long-term surveillance system for a long time have long-term effects, and water quality monitoring generally has a long-term nature. The continual flow of knowledge, as is explained in this chapter, is a way of thinking about how the requirements for information and thc collcction mcthod information work together. It also gives you the opportunity to control the quality of the process (Timmerman et al., 2000).

19.3 Practices of Artificial Water Fluoridation

The comprehensive assessment of drinking water fluoridation conducted by the NHS Division of Evaluation and Outreach research studies of UK, prior to the late 1990s resulted in this lower projection for synthetic water fluoridation's effects on tooth decay today, which discovered revealed the share of children lacking cavities in their teeth increased by 14.6% overall among children who live in places where water is artificially fluoridated, according to a study decay alongside an overall reduction of 2.2 dmft in the average number of young ones, the investigations indicated dmft values between 0.5 particularly 4.4, and in terms of the size or individuals holds dental difference, amongst a 5% rise as well as 64% rise (Peckham & Awofeso, 2014).

Researchers are using machine learning (ML) approaches to examine complicated environmental systems and the data associated with them (Sarker, 2021). Current analytical chemistry trends and approaches for determining fluoride levels in drinking water are also discussed (Eugenio Hernan et al., 2022). Suraj Gupta et al. suggested algorithms for machine learning implementations in the ESE sector using three case studies as examples: (1) sequencing data evaluation for environmental antimicrobial susceptibility identification along with tracking; (2) nontarget analysis for environmental contaminant characterization; and (3) finding abnormalities in ongoing information provided by designed water supply networks (Gupta et al., 2021).

Chen et al. (2007) briefly described the tools that make use of the river basin districts (RBD) framework, demonstrating the value of coordinated conservation of water resources and stream drainage governance throughout the WFD, as well as relevance of remote sensing-based tracking the water condition within Finland's execution of WFD's water ordinance (Chen et al., 2007). Alilou et al. (2018) provide crucial information on how to create a highly accurate water quality monitoring network that can withstand modifications from sources other than points. All the findings of their reading would also be helpful for organizations that monitor water quality and seek a practical method to decide where to conduct sampling. The analytical network method (ANP) then suggested an innovative nonpoint source potential pollution score (NPPS) to categorize the importance of each sampling point before choosing the best sites for a river arrangement. The river mixing length technique (RML) was used and suggest prospective test group locations were suggested to achieve this goal (Alilou et al., 2018).

19.4 Future Fluoride Algorithms for an Environmental System

Environmental health indicators help plan and target interventions, assess human health susceptibility, and evaluate adaptation actions to climate change (Bishayee et al., 2021). Water usage has been shown to be extremely important in the agricultural and infrastructure project industries. The possibility of treating fluoridated potable water with zirconium-impregnated hybrid anion exchange resin has been thoroughly investigated (Mikulčić et al., 2021). Climate-specific EHIs strive to classify and investigate the overall health sound effects of climate fluctuation, alteration, which includes many possible influences regular and man-made mechanisms that could result in unintended additional or accidental health consequences (Ebi et al., 2017). Sensors will additionally offer preliminary information enabling assessing and tracking risks' chronological and geographical fluctuation, allowing projection scenarios of the current condition (Hambling et al., 2011).

Gómez et al. (2011) shared a study on the analysis of several approaches for atmospheric adjustment in CHRIS mode 2 and the MERIS pictures and the creation of algorithms to pinpoint the locations of blooms and the amounts of chlorophyll-a and phycocyanin. These entire algorithms enable the Instruments were utilized enabling keeping track of ecological water quality in the tiniest Coastal lakes and rivers of European Union water framework directive inter-calibration experiment. They also enable the generation of surveillance maps for toxic cyanobacteria (WFD). We continue to use it to track Spain's inland waters' Ecological Quality Ratio (EQR) (Gómez et al., 2011).

19.5 Outlining a Proposed Set of Indicators

Environmental health indicators (EHIs) analyze, monitor and quantify human health susceptibility, help develop and target treatments, and evaluate climate change adaptation and mitigation efforts (Hambling et al., 2011). We sought the best structure meant for building EHIs to assess and track the medical consequences of changes in the climate influence intervention development (Hambling et al., 2011). For such development of EHIs on climate change and health, researchers established a perhaps a particularly applicable paradigm within Driving force-pressure-state-exposure-effect action (DPSEEA). We suggest that using EHIs may be used to analyze, measure, or track consumer health vulnerability, plan, and target treatments, and analyze the efficacy of efforts to adapt as well as combat the effects of change within climate (Hambling et al., 2011).

Climate variability and alteration will exacerbate the economic and health impacts of vector-borne, food-borne, and water-borne infections, as well as breathing issues. To organize, authorities should quantify, forecast, and monitor human health sensitivity to climatic fluctuations and adaptation for or prevent these events (Hambling et al., 2011). The indicators are chosen in light of a case report, and usefulness of one among the indications examined provides an overview surface, in addition to the technical details that are necessary. To comprehend the impact of various inductors on this performance, a proposal is made. A methodology for creating a system of inductors is outlined in order to achieve this goal, and then a list of indicators is constructed in order to suggest a dashboard for the problem-solving process. The authors' goal is to gather enough data from tests to model the impact of inductors on performance. Numerous metrics are available to evaluate social, environmental, and economic elements of an eco-industrial park (EIP). The inquiry began with the management of sustainability evaluation over these parks must complete a crucial task: choosing indicators. The task is to list and organize a wide range of sustainability indicators to help with this effort. Using the four criteria, the 249 indicators have been sorted into three sustainability-related categories (social, environmental, and economic dimensions) (Hambling et al., 2011). The indicators are chosen in light of an example presented, and the usefulness was evaluated with an executive summary, together with all obligatory methodological adjustments (Peano et al., 2011). The SLAM observation process is significantly impacted when a significant quantity of outliers (nonstationary feature point pairs) are introduced. It immediately causes inaccuracies in estimating the pose of mobile robots and the positions of 3D elements, and it also causes drift in keyframe trajectories (Yu et al., 2020). 26 basic indicators were chosen after an analysis of their relevance to issues including food, energy, water, land, and pollution, and selection based on quantitative, stress-sensitive, predictable, anticipatory, controlled, responsive, and stable criteria. It is proposed that these form the basis for a new UA environmental ranking. This will be built using the Delphi technique and is intended to help municipalities create resilient and environmentally friendly towns (Alilou et al., 2018; Yu et al., 2020).

19.6 Metal–organic Framework Materials

The most current developments in fluoride removal using MOFs are covered in this section. The partition coefficient (PC), which is independent of functional circumstances, will be used to expand the real adsorption effectiveness of various MOFs toward fluoride (Ahmadijokani et al., 2021).

Electrochemical methods provide some advantages over other ways of making MOFs, such as milder synthesis conditions, faster synthesis times, and control over the reaction conditions in real time, which give the MOFs made the necessary properties. Due to their applications in the power sector, electrochemistry, electrocatalysis sensing, electrochromic, etc., MOFs have become a popular research topic for electrochemists all over the world (Manjunatha et al., 2021). The unique Mixed Matrix Membrane (MMM) was constructed through aluminum fumarate MOF, a superior fluoride adsorbent, where the base monomer is a form of cellulose sulfonate for the purpose of removing fluoride in polluted aquifers. Membrane fluoride adsorption capabilities ranged from 107 to 179 mg g1 with MOF attentions ranging from 2 to 10% wt. Fluoride rejection was greater than 99% in 10% AlFu (Karmakar et al., 2017). Repairing fluoride-polluted water effectively is an important undertaking. MOF materials are recognized as potential adsorption constituents among several strategies to eliminate fluoride from water (Haldar et al., 2020). Fluoride (F^-) detention within the aqueous system was investigated using hydrothermally synthesized Hossien Saghi et al. (2021) used MIL-53 (Fe), UIO-66, AP-UIO-66 and MOF-235 to construct a numerical framework to predict elimination under specific environmental conditions, and MIL-53 (Fe) to develop a calculus template forecasting elimination of fluoride in specific surroundings (Hossien Saghi et al., 2021).

Adil et al. (2022) includes the potential and difficulties that will occur within the field of MOF-nanofiber combinations formation of the MOF-nanofiber combinations in the coming years, as well as the synthesis, characteristics, and most current developments. Speedy adsorption kinetics, best adsorption ability, great discernment, and outstanding ability to reuse material are all characteristics of MOFs decorated on nanofibers. Additionally, relationships among MOF nanofiber, destination ions, along with multifunctional interaction sites blends with extremely organized pores construction are primarily responsible for the significant adsorption capabilities (Adil et al., 2022).

Tang et al. (2022) discussed their investigation of MOF formation and development within aquatic de-fluoridation in the present era, and accordingly core metal ions (Tang et al., 2022). Amongst the various methods to eliminate fluoride in aqueous materials, MOF constituents are considered suitable adsorption resources. This article examines current advances in the formation of MOFs and its use within the elimination of fluorine amounts from aquatic systems, classified on the basis of fundamental metal atoms. The mechanism of adsorption and an assessment of potential real-world uses are discussed. Various MOFs for fluoride removal. Ultimately, possibilities for fluorine eradication products made from MOFs have been developed suggested along with future research potential (Tang et al., 2022). Through batch adsorption experiments, it is explored to investigate the fluorine precipitation action of zirconium fumarate metal–organic framework-801 (MOF-801) within the water. At pH 2–10, MOF-801 has an excellent and consistent rate of adsorption. High ion strength alternative ions such as Cl^-, NO_3^-, and SO_4^{2-} did not show an impact on the fluoride adsorption capabilities. MOF-801 was de-fluoridated as a result of effective ion exchange that involved fluoride ions and the hydroxyl group (Zhu et al., 2018). The characteristics of MOF substances, such as those with high tensile strength porosity, high specific area of surface, and designable structure, create these superior traditional porous materials. Materials made with MOFs can be used directly or after SPE treatment. Their outstanding performance has already been proven in a number of SPE modes, including magnetic extraction process within solid phase, dispersive extraction procedure of solid phase, micro-solid phase removal, and solid phase microextraction are all types of extraction methods (Li et al., 2020). The MOF-derived electrode materials can be separated into three categories: The authors of this paper derived MOF-derived PC, MOF-derived PC, and MOF-supported composites. In general, porous nanostructure electrodes generated from MOFs have high surface area, excellent electrochemical stability, and strong electrical conductivity. MOF-derived materials continue to be hampered by their difficult production and unstable structure in an extremely corrosive electrolyte (Gurusamy et al., 2021). In many applications where MOFs are potential options, the special property of MOFs has been a significant problem. The discussion in this article includes multiphoton engagement, shock wave chemistry, MOF electrical and vibrational characteristics, novel PVA/MOF nanofibers, including amine-functionalized MOF synthesis (Karimi et al., 2021). In a range of chemical transformations, multifunctional components with MOF composites can interact with MOFs to boost effectiveness, selection, and longevity. They will be the best option to detect many things including proteins, compounds of organic matter, molecules of gases, ions, and electrons chemicals by creating composites based on MOF (Ramesh & Deepa, 2021). Powder XRD examination showed that the AlFu MOF's constitution had been altered as a result of fluorine adsorption. Thermogravimetric analysis was used to confirm the thermal stability up to 700 K. By X-ray fluorescence analysis, during adsorption, the fluoridated solution was demonstrated. As seen from FTIR analysis, fluoride replaced hydroxyl ions in the AlFu MOF (Karmakar et al., 2016).

Including its huge unique surface area, great stability, and its capacity to eliminate particular water contaminants, the metal–organic framework (MOF), a novel issue experienced, has a lot of promise for use in the water treatment sector, focuses on the best performing MOF-based composites and three-dimensional MOFs that have been investigated evaluated so far abolition of inorganic contaminants (IOCs) as well as organic contaminants (EOCs) throughout water, but also sums up the foundational adsorption but rather catalytic parameters of the system of these toxicants (Yan et al., 2022). Adsorption is a proven and efficient technique for removing fluoride and other water contaminants. Tang et al. (2023) shared information in their study, we used the solvothermal method to modify trimeric acid, organic ligands (BTC), 1,2,4,5-benzene tetra carboxylic three different types of cerium-based metal–organic frameworks (Ce-MOFs) were created, varying arrangements and possessions using terephthalic acid (BDC), phthalic acid (PMA), and phthalic acid. The Ce-MOFs' fluoride adsorption kinetics will conform to both the pseudo second order framework and Langmuir isothermal framework (Tang et al., 2023). Rhodamine B (RhB) was incorporated on site to produce radiometric fluorescence detection in a zirconium MOF, UiO-66-NH2. The nanocomposite UiO-66-NH2@RhB produced 450 and 585 nm emission peaks, exhibiting an octahedral building similar to that of UiO-66-NH2 through a significant amount of BET surface area. To improve the detection's precision, the reference RhB's red fluorescence was maintained unaltered. The blue intensity of fluorescence was measured. Increased when F was added, but the improved absorption spectrum decreased when Al^{3+} was added (Zhang et al., 2022). An essential objective is the creation of reliable, inexpensive, and equipment-free F identification techniques. With the use of a triazine-based planar ligand called 4,4′,4″-s-triazine-2,4,6-triyltribenzoate (TATB) and mixed lanthanide ions (Tb^{3+} and Eu^{3+}), Xiaoliang Zeng et al. created a number of combined metal–organic lanthanide frameworks inside. Zeng et al. (2020) in their study (Zeng et al., 2020).

A straightforward template-directed layer-by-layer (LBL) approach was used to create Ce-BTC MOFs decorated with hydroxyapatite (HAp), also known as HAp-Ce-BTC MOFs, which were successfully used to extract fluoride from freshwater (Jeyaseelan et al., 2022). Nanoscale DCPBA sheets with metal surface flaws allow for fluoride adsorption with substantial capacity over a broad acidity range. With a partition coefficient of up to 18.805 Lg^{-1}, Ce-DCPBA has great fluoride selectivity, even in composites with very high salt concentrations (8 mol/L of nitrate). This is due to the complexation of boron-fluoride at metallic faults (Su et al., 2023). For efficient fluoride cleanup, As HAp-Ce-BTC MOFs@Alg-CS beads (biohybrid beads), cerium-based metal–organic frameworks (Ce-BTC MOFs) with hydroxyapatite (HAp) and chitosan (CS) have been developed. The biohybrid particles had a considerable DC of 4865 mg/kg fluoride after 20 min of retention time at 25 °C with a dose of 0.1 g. Pseudo second-order kinetics and Langmuir models suit the results of adsorption kinetics and isotherms investigate fine, correspondingly (Jeyaseelan & Viswanathan, 2022). Using faulty CeMOFs, which were made using affordable quick water that was supplied from a reservoir to a stream in the drought technique in RT, a new technique is presented to effectively remove fluoride. Due to the specific molecular interactions here between CeMOFs exhibit ultrahigh halotolerant and Fluoride ion attraction over other anions across a wide pH range (5 mol L-1 of nitrate), OH^-/H_2O bound at the Ce defect sites and dissociative fluoride ions (Xie et al., 2022). Benzene-1,4-dicarboxylic acid (BDC) and 2-aminobenzene-1,4-dicarboxylic acid (ABDC), which are founded on MOFs for selective elimination of fluoride, were created by a trivalent metal atom, iron (Fe^{3+}). Fe@BDC and Fe@ABDC MOF composites are available that produce improved defluoridation capabilities of 4.90 and 4.92 mg/g (DC) for various adsorbents (Jeyaseelan et al., 2021). We created The solvothermal method was used to create Due to the particular identification of F by categories of boronic acid while the substantial connection among F with negative and Fe positive 3 ions, nanosized iron-based metal–organic frameworks (Fe-MOFs) using 3,5-dicarboxybenzeneboronic acid as organic linkers and Fe^{3+} as metal ions could be used to identify F with significant sensitivity or selective ability (Hou et al., 2021).

To avoid confusion of the recognition of fluoride ions, a luminous MOF-constructed chemiluminescence resonance energy transfer (CRET) stage was built. Strongly fluorescing 2′,7′-dichlorofluorescein (DCF) was encapsulated inside the structures of the NH2-MIL-101 (Al) MOFs creating a hybrid MOF, significantly suppressing the DCF fluorescence signal of DCF (Sun et al., 2020). The solvothermal process used to create a metal–organic framework (MOF-801) adsorbent based on zirconium and its effectiveness in removing fluoride ions from water. PXRD, FESEM, and XPS were used to describe the morphology of MOF-801, while BET was used to determine the pore structure and the target region. When MOF-801 was made, it was found to have a distinct octahedral form of a particle having a 0.304 nm lattice spacing. This demonstrates the existence of (011) ZrO_2 planes (Tan et al., 2020).

19.7 Efficient and Reliable Solutions to Remove Excess Fluoride from Water Environments

Early studies showed that water fluoridation reduced dental cavities; however, these findings have been disputed. Early support was predicated on fluoride's systemic decay-reduction role (Carstairs, 2015; Peckham & Awofeso, 2014). To increase the removal capability, a number of working factors were investigated; including fluoride concentration, pH, applied voltage, reaction duration, electrode separation, electrode reactive area, and the effect of coexisting ions are all important factors. The correct amount of sodium fluoride was added to tap water to create fluoride solutions in various amounts (1, 5 and 10 mg L^{-1}), the consequences of innovative approaches and aqueous sample with different pH (5, 7, and 9) on the effectiveness of removing fluoride were also examined (Solanki et al., 2021). Adsorption is a traditional approach that works with a variety of adsorbents, including clays, soils, and minerals with calcium content, synthetic substances, and carbon-based materials. Recently, DE metals have attracted interest as an adsorbent for eliminating fluoride (Meenakshi & Maheshwari, 2006). Tolkou et al. (2021) discussed the study on the use of innovative activated carbon; graphene oxide and carbon nanotubes are examples of carbon-based adsorbents (Fig. 19.2). Other examples include nanostructured compounds, minerals combined with their oxides or hydroxides, and natural materials to achieve increased fluoride removal. Lanthanum (La) has been used extensively to modify materials, including hybrid materials with activated carbon (such as La/

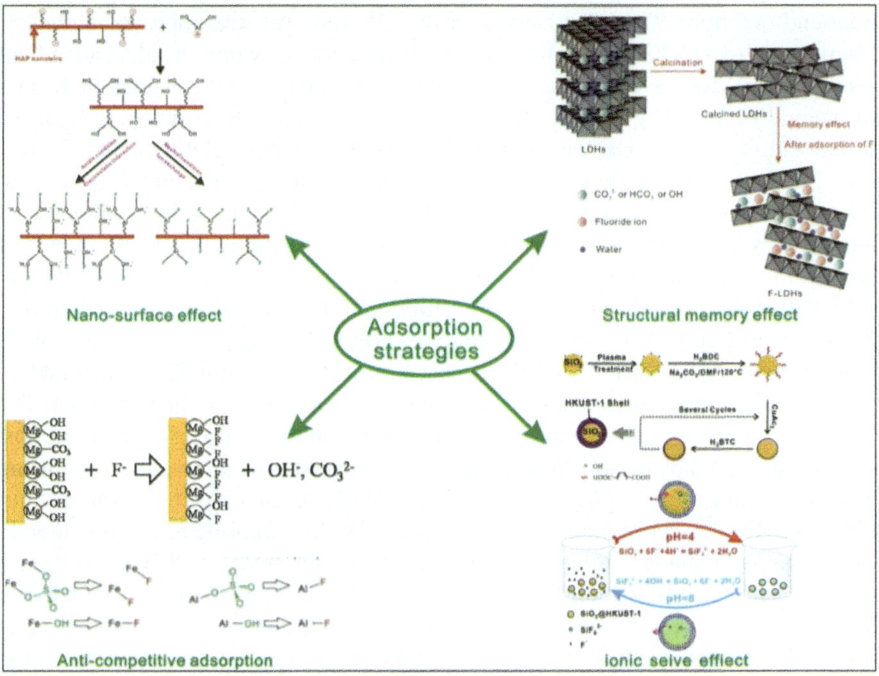

Fig. 19.2 Schematic diagram of adsorption strategies

Mg/Si-AC, La/MA and $LaFeO_3$ nanoparticles), as well as MgO nanostructures, to has been shown to have a maximum adsorption capability of 29,131 mg g^{-1} (Tolkou et al. 2021).

Mukherjee et al. (2019) has used a nanocomposite material made of three metals: Fe, Al, and Ti (Belhadi et al.), created via a flexible tunable chemical process, provide a novel method for treating water. For example, at pH 7, the soluble dye methylene blue (MB) is defluoridated and photodegraded. Their work discussed the production of trimetallic nanocomposite, FAT, using a tunable, rapid, and chemistry that saves energy as a pollutants or management process for waste within water (Mukherjee et al., 2019). Gai et al. have proposed aluminum hydroxide defluoridates. Hydrolysis of Al salts yields amorphous and crystalline AlOOH adsorbents, which are tested for defluoridation capacity, rate, pH sensitivity, and water quality. The defluorination efficacy of AlOOH is highly dependent on the hydrolysis pH, not the Al salt type. The adsorbent can remove > 95% fluoride in 2 min, reaches adsorption equilibrium in 2 h, and defluoridates 41.9 mg/g (Gai et al., 2021). The effects of coexisting anions, contact time, adsorbent dosage, heat pretreatment, neutralization, starting fluoride amount, solution pH, and overall mentioned variables are among the variables examined. According to the findings, fluoride adsorption starts quite quickly in the first five minutes before gradually increasing to achieve equilibrium in around one hour. The effectiveness of fluoride removal was improved by adsorbent dose. Almost 85% of the elimination effectiveness is within the ideal adsorbent dose of 16 g/L for the preliminary fluoride amount of 10 mg/L obtained in 10 mg/L obtained in 1 h (Nigussie et al., 2007). The removal of fluoride by manganese sulfate, chloride and MOE was best at a pH of 6.0, which is slightly acidic (Saikia et al., 2017). Using readily available and inexpensive ingredients, physicochemical techniques of adsorption and coagulation are employed to obtain the elimination of fluorine in water. A dose of 1000 mg/L results in a chloride removed amount of fluoride in a 5 mg L sample solution with removal percentages such as 92, 92, 94, and 91. Resolutely, with an equilibrium time of 3 h, rice husk at a concentration of 6 g/L was able to eliminate 83% fluoride through a solution containing 5 mg/L fluoride (Saikia et al., 2017). Fluoride adsorption starts rather quickly in the first five minutes, before gradually increasing to achieve equilibrium in around one hour. The efficacy of fluoride removal improved with adsorbent dose. The efficacy of almost 85% was achieved by establishing an optimal adsorbent dose of 16 g/L for a starting fluoride amount of 10 mg/L achieved in 1 h. With an increase in the initial fluoride concentration, more fluoride was adsorbed. Within, fluoride removal is largely constant as a percentage within the pH spectrum ranging from 3 to 8. The adsorption process in the Dubinin-Radushkevick (D-R) isotherm simulation corresponded to 332.5 mg/g in the adsorbent adequately. Ambient pH, discussed by Nigussie et al. (2007). Senewirathna et al. (2022) used to make cheap, because an adsorbent for the purpose, dehydrated palmyrah (*Borassus flabellifer*) carbonate with nut shells had been utilized, which are widely available, were used to prepare activated carbon. The effective result was chemically modified activated carbon. To remove fluoride, modified activated carbon was used as the absorbent. It works well to remove fluoride ions from drinking water. Consequently, the maximum fluoride removal efficacy of PAC was observed with 45 min of interaction, 0.2 g of adsorbent material, 1 mg/L of the initial amount of fluoride and the fluoride heat's amount, and heat of fluoride heat, and fluoride heat and 30 °C heat at adsorption efficiency (Senewirathna et al., 2022).

Recently, a compilation of several adsorbents was assembled throughout available research, including their ability to absorb abilities for the elimination of fluoride under different circumstances (pH, amount of starting fluoride, heat, duration of interaction, charges of the surface of the adsorbent, etc.). Furthermore, significant breakthroughs in the synthesis of new adsorbents that have been tried thus far for fluoride removal are highlighted and discussed (Bhatnagar et al., 2011). Fluoride overexposure can result in fluorosis of the disease, permanent demineralization of bone and dental tissues, along with chronic damage that affects brain activity, liver, thyroid, and even kidney. Fluoride must be removed from drinking water in order to be safe for human consumption. Numerous defluoridation procedures are discussed, including precipitation of coagulation, ion exchange, membrane separation and adsorption by Pillai et al. (2021). The best methods for dealing with fluoride-contaminated water have traditionally been nanofiltration and sorption techniques. Fluoride removal from aqueous solutions was researched using these methods. Valentukeviciene et al. (2019) in their study suggest defluoridation with nanofiltration method as a replacement for the reverse osmosis technique as a response to the unreasonable cost of RO-treated drinking water. To investigate the effectiveness of nanofiltration for removing fluoride, we used nanofiltration flat-sheet membranes in the laboratory (Valentukeviciene et al., 2019).

19.8 Public Health Strategies

Environmental health indicators (EHIs) analyse, monitor, and quantify human health susceptibility, help develop and target treatments, and evaluate climate change adaptation and mitigation efforts. Saliva, which contains 0.01–0.05 ppm fluoride, transports most of the fluorides ingested to the teeth. Acid-producing bacteria can erode calcium, magnesium, and phosphate hydroxyapatite in dental enamel. Fluorapatite, formed by fluoride with hydroxyapatite, resists acid erosion (Peckham & Awofeso, 2014). The Driving Force-Pressure-State-Effect-Affect-Action (DPSEEA) model was chosen and created by Hambling et al. (2011), is the best for dealing with climate change along with health security. Researchers propose that EHIs be used only as a piece of equipment for evaluating, measuring, and tracking consumer health vulnerability, planning and targeting actions, and assessing the efficacy of activities that react to threats effectively ameliorate climate change (Hambling et al., 2011).

To monitor systems with human health, it was critical to develop a method in the direction of identifying fluoride ions (F) within aqueous systems that is both rapid and precise. A new fluorescent probe called PCN-222, a zirconium porphyritic light metal organic framework (LMOF), was used to detect F in water in an ultrasensitive, quick, and selective manner (Chen et al., 2021).

Inside the Bengal rift valley, groundwater contamination of arsenic (As) and fluoride (F) is a concern for human health. West Bengal, India's south 24 Parganas region, Jaydhar et al. (2022), their study created a structure technique to evaluate groundwater susceptibility, including associated serious health risk concerns. The logistic regression technique (LR) and 15 hydrogeological factors were used to simulate groundwater potential zones. According to the assessment matrices used in the reported work, LR provides the best predictions for groundwater vulnerability (AUC-ROC is 0.891 and 0.861, and in training and validation, the kappa index is 0.810 and 0.720, respectively) (Chen et al., 2021).

Fluoride levels in water supplies that are higher than the suggested dose (> 1.5 ppm) is bad for the health of people. These effects include bone and teeth, Alzheimer's disease, DNA mutations, fertility problems, renal disorders, acute gastrointestinal difficulties, thyroid disease, brain damage, and most recently cancer (Ahmadijokani et al., 2021). WHO suggested that enamel be strengthened toward low level (1–1.5 mg/L). People who consume too much fluoride develop bone abnormalities, dental fluorosis, and skeletal fluorosis. In various regions of the world, groundwater can contain amounts of fluoride up to 30 mg/L (Jeyaseelan et al., 2021).

Fluoride poisoning that occurs suddenly can severely damage the stomach mucosa. Furthermore, people who inhale hydrogen fluoride have respiratory consequences that include haemorrhage, pulmonary oedema, tracheostomy, and shortness of breath. Researchers have shown that F^- may alter intracellular redox balance, lipid peroxidation, gene expression, and induction of apoptosis. It is also an oxidizing chemical and a well-known reversible enzyme inhibitor that inhibits the enzyme activity of at least 80 proteins. Increased fluoride ions in the water that is consumed can cause a variety of persistent illnesses; include fatigue, arthritis, cartilage, muscle deterioration, dentistry, and even spinal fluorosis, as well as osteoporosis. Heart, arteries, kidneys, liver, endocrine glands, and nervous systems that comprise a living thing being, and numerous sensitive portions, briefly described in this article, under severe conditions, we might all face un-favorable consequences. Ion-selective and colorimetric approaches with the use of water are used to examine the presence of fluoride molecules from many detection techniques. Additionally, various techniques for plasma exchange, nanofiltration, a process known as adsorption precipitation, coagulation, and reverse osmosis are all methods for eliminating fluorine residues through water mechanisms, and these have been reviewed by Solanki et al. (2022).

The review by Sankhla and Kumar (2018), which took into account over a century of fluoride poisoning in water, was the impact of contaminating components on India's water resources. To reduce the risk of developing fluorosis, it is recommended to drink water together with a smaller amount, such as 1.5 mg/L of fluoride ions. Supplementary research must be done regarding the numerous reports indicating elevated fluoride levels in water resources to determine whether there is a connection between fluoride and its effects on people's central nerve systems (Duggal & Sharma, 2022).

19.9 Conclusion and Future Suggestions

In this chapter, an effort is made to touch on the variability with adsorbents, which so far has induced the reduction of fluorine in wastewater and usable water. Fluoride reduces dental caries but causes cognitive decline, the thyroid gland does not produce enough hormones (called hypothyroidism), teeth, and even fluorosis in the skeleton system, enzymatic system, as well as imbalanced electrolytes, and even uterine cancer are all possible complications. Fluoride is abundant; therefore, humans absorb or inhale it from many sources. In the twenty-first century, A particularly inexpensive, productive, or cost-effective method of fluoride for teeth may not be artificial water fluoride. Alumina/aluminum-based materials, materials made from carbon, calcium-based substances, clays, soils, and chemical substances are all used because they absorb in the classical method of adsorption. Fluoride removal from solution form is being investigated using a variety of inverted ion-exchange resins consisting of cross-linked polystyrene, as well as multilayered double hydroxides, and redesigned zeolites comprising zeolites. Actual applications may deviate from the generalizations stated, since the effectiveness of fluoride removal varies depending on several site-specific chemicals and geographic and economic variables. Any given method that is appropriate in one region might not be able to satisfy the requirements in another.

Acknowledgements The authors thank our prestigious institute and other people who are involved in this study for their direct and indirect contributions.

References

Adil, H. I., Thalji, M. R., Yasin, S. A., Saeed, I. A., Assiri, M. A., Chong, K. F., & Ali, G. A. M. (2022). Metal–organic frameworks (mofs) based nanofiber architectures for the removal of heavy metal ions. *RSC Advances, 12*, 1433–1450. https://doi.org/10.1039/D1RA07034G

Ahmadijokani, F., Molavi, H., Rezakazemi, M., Aminabhavi, T. M., & Arjmand, M. (2021). Simultaneous detection and removal of fluoride from water using smart metal-organic framework-based adsorbents. *Coordination Chemistry Reviews, 445*, 214037. https://doi.org/10.1016/j.ccr.2021.214037

Alilou, H., Moghaddam Nia, A., Keshtkar, H., Han, D., & Bray, M. (2018). A cost-effective and efficient framework to determine water quality monitoring network locations. *Science of the Total Environment, 624*, 283–293. https://doi.org/10.1016/j.scitotenv.2017.12.121

Aoun, A., Darwiche, F., Al Hayek, S., & Doumit, J. (2018). The fluoride debate: The pros and cons of fluoridation. *Preventive Nutrition and Food Science, 23*, 171–180. https://doi.org/10.3746/pnf.2018.23.3.171

Armfield, J. M. (2010). Community effectiveness of public water fluoridation in reducing children's dental disease. *Public Health Reports, 125*, 655–664. https://doi.org/10.1177/003335491012500507

Belhadi, A., Kamble, S. S., Gunasekaran, A., Zkik, K., D. K. M., & Touriki, F. E. (2021). A big data analytics-driven lean six sigma framework for enhanced green performance: A case study of chemical company. *Production Planning & Control*, 1–24. https://doi.org/10.1080/09537287.2021.1964868

Benedict, S., Gowtham, N., Giri, D., & Sreelakshmi, N. (2018). Real time water quality analysis framework using monitoring and prediction mechanisms. In *2018 Conference on Information and Communication Technology (CICT)* (pp. 1–6).

Bhatnagar, A., Kumar, E., & Sillanpää, M. (2011). Fluoride removal from water by adsorption—A review. *Chemical Engineering Journal, 171*, 811–840. https://doi.org/10.1016/j.cej.2011.05.028

Bishayee, B., Ruj, B., Chakrabortty, S., & Nayak, J. (2021). Facile synthesis, characterization and application of heterogeneous al@si materials for adsorptive mitigation of fluoride: Optimization and cost analysis. *Environmental Nanotechnology, Monitoring & Management, 16*, 100490. https://doi.org/10.1016/j.enmm.2021.100490

Carstairs, C. (2015). Debating water fluoridation before Dr. Strangelove. *American Journal of Public Health, 105*, 1559–1569. https://doi.org/10.2105/ajph.2015.302660

Chen, B., Jiang, T., Fu, H., Qu, X., Xu, Z., & Zheng, S. (2021). Ultrasensitive, rapid and selective sensing of hazardous fluoride ion in aqueous solution using a zirconium porphyrinic luminescent metal-organic framework. *Analytica Chimica Acta, 1145*, 95–102. https://doi.org/10.1016/j.aca.2020.11.007

Chen, Q., Zhang, Y., & Hallikainen, M. (2007). Water quality monitoring using remote sensing in support of the eu water framework directive (wfd): A case study in the gulf of finland.

Environmental Monitoring and Assessment, 124, 157–166. https://doi.org/10.1007/s10661-006-9215-8

Chen, S., Li, B., Lin, S., Huang, Y., Zhao, X., Zhang, M., Xia, Y., Fang, X., Wang, J., Hwang, S.-A., & Yu, S. (2013). Change of urinary fluoride and bone metabolism indicators in the endemic fluorosis areas of southern china after supplying low fluoride public water. *BMC Public Health, 13*, 156. https://doi.org/10.1186/1471-2458-13-156

Duggal, V., & Sharma, S. (2022). Fluoride contamination in drinking water and associated health risk assessment in the Malwa belt of Punjab, India. *Environmental Advances, 8*, 100242. https://doi.org/10.1016/j.envadv.2022.100242

Ebi, K. L., Hess, J. J., & Watkiss, P. (2017). Health risks and costs of climate variability and change. In *Disease Control Priorities, Third Edition (Volume 7): Injury Prevention and Environmental Health* (pp. 153–169). The World Bank.

Eugenio Hernan, O., Manuela Leticia, K., & Mutsumi, K. (2022). Fluoride detection and quantification, an overview from traditional to innovative material-based methods. In W. Enos Wamalwa, J. L. Grace & K. Ayabei (Eds.), *Fluoride*, Ch. 3. IntechOpen.

Gai, W.-Z., Zhang, S.-H., Yang, Y., Zhang, X., & Deng, Z.-Y. (2021). Separation of excess fluoride from water using amorphous and crystalline alooh adsorbents. *ACS Omega, 6*, 16488–16497. https://doi.org/10.1021/acsomega.1c01620

Gómez, J. A. D., Alonso, C. A., & García, A. A. (2011). Remote sensing as a tool for monitoring water quality parameters for mediterranean lakes of european union water framework directive (wfd) and as a system of surveillance of cyanobacterial harmful algae blooms (scyanohabs). *Environmental Monitoring and Assessment, 181*, 317–334. https://doi.org/10.1007/s10661-010-1831-7

Gupta, P. K., & Maiti, S. (2022). Enhancing data-driven modeling of fluoride concentration using new data mining algorithms. *Environmental Earth Sciences, 81*, 89. https://doi.org/10.1007/s12665-022-10216-z

Gupta, S., Aga, D., Pruden, A., Zhang, L., & Vikesland, P. (2021). Data analytics for environmental science and engineering research. *Environmental Science & Technology, 55*, 10895–10907. https://doi.org/10.1021/acs.est.1c01026

Gurusamy, L., Anandan, S., & Wu, J. J. (2021). Nanomaterials derived from metal-organic frameworks for energy storage supercapacitor application. In A. Khan, F. Verpoort, A. M. Asiri, M. E. Hoque, A. L. Bilgrami, M. Azam & K. C. B. Naidu (Eds.), *Metal-organic frameworks for chemical reactions* (pp. 441–470). Elsevier.

Haldar, D., Duarah, P., & Purkait, M. K. (2020). Mofs for the treatment of arsenic, fluoride and iron contaminated drinking water: A review. *Chemosphere, 251*, 126388. https://doi.org/10.1016/j.chemosphere.2020.126388

Halliday, S. J., Skeffington, R. A., Wade, A. J., Bowes, M. J., Gozzard, E., Newman, J. R., Loewenthal, M., Palmer-Felgate, E. J., & Jarvie, H. P. (2015). *High-frequency water quality monitoring in an urban catchment: Hydrochemical dynamics, primary production and implications for the water framework directive* (Vol. 29, pp. 3388–3407). https://doi.org/10.1002/hyp.10453

Hambling, T., Weinstein, P., Slaney, D. (2011). A review of frameworks for developing environmental health indicators for climate change and health. *International Journal of Environmental Research and Public Health, 8*, 2854–2875.

Hassan, G., Shaheen, M. E., Taie, S. A. (2020). Prediction framework for water quality parameters monitoring via remote sensing. In *2020 First International Conference of Smart Systems and Emerging Technologies (SMARTTECH)* (pp. 59–64).

Hossien Saghi, M., Chabot, B., Rezania, S., Sillanpää, M., Akbar Mohammadi, A., Shams, M., & Alahabadi, A. (2021). Water-stable zirconium and iron-based metal-organic frameworks (mofs) as fluoride scavengers in aqueous medium. *Separation and Purification Technology, 270*, 118645. https://doi.org/10.1016/j.seppur.2021.118645

Hou, D., You, Y., Wu, X., Li, C., Wu, S., Zhang, C., & Xian, Y. (2021). A nanosized metal–organic framework for visual detection of fluoride ions with smartphone via colorimetric test kit. *Sensors and Actuators b: Chemical, 332*, 129508. https://doi.org/10.1016/j.snb.2021.129508

Hu, J., Stein, A., & Bühlmann, P. (2016). Rational design of all-solid-state ion-selective electrodes and reference electrodes. *TrAC Trends in Analytical Chemistry, 76*, 102–114. https://doi.org/10.1016/j.trac.2015.11.004

Iheozor-Ejiofor, Z., Worthington, H. V., Walsh, T., O'Malley, L., Clarkson, J. E., Macey, R., Alam, R., Tugwell, P., Welch, V., & Glenny, A. M. (2015). Water fluoridation for the prevention of dental caries. *Cochrane Database of Systematic Reviews*. https://doi.org/10.1002/14651858.CD010856.pub2

Jaydhar, A. K., Chandra Pal, S., Saha, A., Islam, A. R. M. T., Ruidas, D. (2022). Hydrogeochemical evaluation and corresponding health risk from elevated arsenic and fluoride contamination in recurrent coastal multi-aquifers of eastern India. *Journal of Cleaner Production, 369*, 133150. https://doi.org/10.1016/j.jclepro.2022.133150.

Jeyaseelan, A., Aswin Kumar, I., Naushad, M., & Viswanathan, N. (2022). Fabrication of hydroxyapatite embedded cerium-organic frameworks for fluoride capture from water. *Journal of Molecular Liquids, 354*, 118830. https://doi.org/10.1016/j.molliq.2022.118830

Jeyaseelan, A., Naushad, M., Ahamad, T., & Viswanathan, N. (2021). Design and development of amine functionalized iron based metal organic frameworks for selective fluoride removal from water environment. *Journal of Environmental Chemical Engineering, 9*, 104563. https://doi.org/10.1016/j.jece.2020.104563

Jeyaseelan, A., & Viswanathan, N. (2022). Investigation of hydroxyapatite-entrenched cerium organic frameworks incorporating biopolymeric beads for efficient fluoride removal. *Industrial & Engineering Chemistry Research, 61*, 7911–7925. https://doi.org/10.1021/acs.iecr.2c00487

Karimi, M., Mehrabadi, Z., Farsadrooh, M., Bafkary, R., Derikvandi, H., Hayati, P., & Mohammadi, P. (2021). Metal–organic framework. In M. Ghaedi (Ed.), *Interface science and technology* (pp. 279–387). Elsevier.

Karmakar, S., Bhattacharjee, S., & De, S. (2017). Experimental and modeling of fluoride removal using aluminum fumarate (alfu) metal organic framework incorporated cellulose acetate phthalate mixed matrix membrane. *Journal of Environmental Chemical Engineering, 5*, 6087–6097. https://doi.org/10.1016/j.jece.2017.11.035

Karmakar, S., Dechnik, J., Janiak, C., & De, S. (2016). Aluminium fumarate metal-organic framework: A super adsorbent for fluoride from water. *Journal of Hazardous Materials, 303*, 10–20. https://doi.org/10.1016/j.jhazmat.2015.10.030

Lakshmikantha, V., Hiriyannagowda, A., Manjunath, A., Patted, A., Basavaiah, J., & Anthony, A. A. (2021). Iot based smart water quality monitoring system. *Global Transitions Proceedings, 2*, 181–186. https://doi.org/10.1016/j.gltp.2021.08.062

Li, G., Xia, L., Dong, J., Chen, Y., Li, Y. (2020). Metal-organic frameworks. In C. F. Poole (Ed.), *Solid-phase extraction* (pp. 285–309). Elsevier.

Manjunatha, H., Janardan, S., Ratnamala, A., Ratnam, K. V., Rao, L. V., Ramesh, S., Naidu, K. C. B., Kumar, N. S., Khan, A., & Asiri, A. M. (2021). Electrochemical aspects of metal-organic frameworks. In A. Khan, F. Verpoort, A. M. Asiri, M. E. Hoque, A. L. Bilgrami, M. Azam & K. C. B. Naidu (Eds.), *Metal-organic frameworks for chemical reactions* (pp. 65–109). Elsevier.

Maroli, A. A., Narwane, V. S., Raut, R. D., & Narkhede, B. E. (2021). Framework for the implementation of an internet of things (iot)-based water distribution and management system. *Clean Technologies and Environmental Policy, 23*, 271–283. https://doi.org/10.1007/s10098-020-01975-z

Meenakshi & Maheshwari, R. C. (2006). Fluoride in drinking water and its removal. *Journal of Hazardous Materials, 137*, 456–463.https://doi.org/10.1016/j.jhazmat.2006.02.024

Mikulčić, H., Baleta, J., Wang, X., Duić, N., & Dewil, R. (2021). Green development challenges within the environmental management framework. *Journal of Environmental Management, 277*, 111477. https://doi.org/10.1016/j.jenvman.2020.111477.

Mukherjee, A., Adak, M. K., Upadhyay, S., Khatun, J., Dhak, P., Khawas, S., Ghorai, U. K., & Dhak, D. (2019). Efficient fluoride removal and dye degradation of contaminated water using fe/al/ti oxide nanocomposite. *ACS Omega, 4*, 9686–9696. https://doi.org/10.1021/acsomega.9b00252

Nayla Hassan, O. (2019). Water quality parameters. In Kevin, S., (Ed.), *Water quality*, Ch. 1. IntechOpen.

Nigussie, W., Zewge, F., & Chandravanshi, B. S. (2007). Removal of excess fluoride from water using waste residue from alum manufacturing process. *Journal of Hazardous Materials, 147*, 954–963. https://doi.org/10.1016/j.jhazmat.2007.01.126

Ozsvath, D. L. (2009). Fluoride and environmental health: A review. *Reviews in Environmental Science and Bio/technology, 8*, 59–79. https://doi.org/10.1007/s11157-008-9136-9

Palanichamy, Y., Kargar, M., & Zolfagharinia, H. (2021). Unearthing trends in environmental science and engineering research: Insights from a probabilistic topic modeling literature analysis. *Journal of Cleaner Production, 317*, 128322. https://doi.org/10.1016/j.jclepro.2021.128322

Peano, A., Bottero, M., & Cassatella, C. (2011). Proposal for a set of indicators. In C. Cassatella & A. Peano (Eds.), *Landscape indicators: Assessing and monitoring landscape quality* (pp. 193–215). Springer Netherlands.

Peckham, S., & Awofeso, N. (2014). Water fluoridation: A critical review of the physiological effects of ingested fluoride as a public health intervention. *The Scientific World Journal, 2014*, 293019. https://doi.org/10.1155/2014/293019

Pillai, P., Dharaskar, S., Pandian, S., & Panchal, H. (2021). Overview of fluoride removal from water using separation techniques. *Environmental Technology & Innovation, 21*, 101246. https://doi.org/10.1016/j.eti.2020.101246

Podgorski, J. E., Labhasetwar, P., Saha, D., & Berg, M. (2018). Prediction modeling and mapping of groundwater fluoride contamination throughout India. *Environmental Science & Technology, 52*, 9889–9898. https://doi.org/10.1021/acs.est.8b01679

Ramesh, M., & Deepa, C. (2021). Metal-organic frameworks and their composites. In A. Khan, F. Verpoort, A. M. Asiri, M. E. Hoque, A. L. Bilgrami, M. Azam & K. C. B. Naidu (Eds.), *Metal-organic frameworks for chemical reactions* (pp. 1–18). Elsevier.

Rathore, R. S., Sangwan, S., Kaiwartya, O., & Aggarwal, G. (2021). Green communication for next-generation wireless systems: Optimization strategies, challenges, solutions, and future aspects. *Wireless Communications and Mobile Computing, 2021*, 5528584. https://doi.org/10.1155/2021/5528584

Saikia, J., Sarmah, S., Ahmed, T. H., Kalita, P. J., & Goswamee, R. L. (2017). Removal of toxic fluoride ion from water using low cost ceramic nodules prepared from some locally available raw materials of Assam, India. *Journal of Environmental Chemical Engineering, 5*, 2488–2497. https://doi.org/10.1016/j.jece.2017.04.046

Sankhla, M. S., Kumar, R. (2018). Fluoride contamination of water in India and its impact on public health. *ARC Journal of Forensic Science, 3*, 10–15. https://www.arcjournals.org/journal-of-forensic-science/volume-3-issue-2/2.

Sarker, I. H. (2021). Machine learning: Algorithms, real-world applications and research directions. *SN Computer Science, 2*, 160. https://doi.org/10.1007/s42979-021-00592-x

Senewirathna, D. S. G. D., Thuraisingam, S., Prabagar, S., & Prabagar, J. (2022). Fluoride removal in drinking water using activated carbon prepared from palmyrah (*Borassus flabellifer*) nut shells. *Current Research in Green and Sustainable Chemistry, 5*, 100304. https://doi.org/10.1016/j.crgsc.2022.100304

Shrestha, S., Kazama, F., & Newham, L. T. H. (2008). A framework for estimating pollutant export coefficients from long-term in-stream water quality monitoring data. *Environmental Modelling & Software, 23*, 182–194. https://doi.org/10.1016/j.envsoft.2007.05.006

Sivasankar, V. (2016). Surface-modified carbons as scavengers for fluoride from water. *Springer Nature* I-XIII and 1–223.

Solanki, Y. S., Agarwal, M., Gupta, A. B., Gupta, S., & Shukla, P. (2022). Fluoride occurrences, health problems, detection, and remediation methods for drinking water: A comprehensive review. *Science of the Total Environment, 807*, 150601. https://doi.org/10.1016/j.scitotenv.2021.150601

Solanki, Y. S., Agarwal, M., Maheshwari, K., Gupta, S., Shukla, P., & Gupta, A. B. (2021). Removal of fluoride from water by using a coagulant (inorganic polymeric coagulant). *Environmental Science and Pollution Research, 28*, 3897–3905. https://doi.org/10.1007/s11356-020-09579-2

Su, J., Yuan, M., Han, L., Deng, H., Chang, J., Zhuang, Y., Wang, J., & Zhang, Y. (2023). Ultrathin metal organic framework nanosheets with rich defects for enhanced fluoride removal. *Chemical Engineering Journal, 451*, 138989. https://doi.org/10.1016/j.cej.2022.138989

Sun, Y., Xu, X., Zhao, Y., Tan, H., Li, Y., & Du, J. (2020). Luminescent metal organic frameworks–based chemiluminescence resonance energy transfer platform for turn–on detection of fluoride ion. *Talanta, 209*, 120582. https://doi.org/10.1016/j.talanta.2019.120582

Tan, T. L., Krusnamurthy, P. A. P., Nakajima, H., & Rashid, S. A. (2020). Adsorptive, kinetics and regeneration studies of fluoride removal from water using zirconium-based metal organic frameworks. *RSC Advances, 10*, 18740–18752. https://doi.org/10.1039/D0RA01268H

Tang, X., Xia, W., Qu, X., Wang, C., Wang, W., Liang, Y., Zeng, Y., Xiong, W., Cheng, M., Song, B., Zhou, C., & Zhao, X. (2023). Structure–performance correlation guided cerium-based metal–organic frameworks: Superior adsorbents for fluoride removal in water. *Chemosphere, 312*, 137335. https://doi.org/10.1016/j.chemosphere.2022.137335

Tang, X., Zhou, C., Xia, W., Liang, Y., Zeng, Y., Zhao, X., Xiong, W., Cheng, M., & Wang, Z. (2022). Recent advances in metal–organic framework-based materials for removal of fluoride in water: Performance, mechanism, and potential practical application. *Chemical Engineering Journal, 446*, 137299. https://doi.org/10.1016/j.cej.2022.137299

Timmerman, J. G., Ottens, J. J., & Ward, R. C. (2000). The information cycle as a framework for defining information goals for water-quality monitoring. *Environmental Management, 25*, 229–239. https://doi.org/10.1007/s002679910018

Tolkou, A. K., Manousi, N., Zachariadis, G. A., Katsoyiannis, I. A., & Deliyanni, E. A. (2021). Recently developed adsorbing materials for fluoride removal from water and fluoride analytical determination techniques: A review. *Sustainability, 13*, 7061.

USPHS. (2015). U.S. Public health service recommendation for fluoride concentration in drinking water for the prevention of dental caries. *Public Health Reports, 130*, 318–331. https://doi.org/10.1177/003335491513000408

Valentukeviciene, M., Zurauskiene, R., & Boussouga, Y. A. (2019). Fluoride removal from groundwater by technological process optimization. *Ecological Chemistry and Engineering, 26*, 133–147. https://doi.org/10.1515/eces-2019-0010

Xie, K., Wang, S., Yuan, M., Zhang, H., Deng, H., Zhang, Y., Wang, J., & Zhuang, Y. (2022). Tailored defect-rich cerium metal organic frameworks for efficient fluoride removal from wastewater. *Separation and Purification Technology, 302*, 122152. https://doi.org/10.1016/j.seppur.2022.122152

Yabushita, M., Imanishi, Y., Xiao, T., Osuga, R., Nishitoba, T., Maki, S., Kanie, K., Cao, W., Yokoi, T., & Muramatsu, A. (2021). Transcription-induced formation of paired al sites in high-silica cha-type zeolite framework using al-rich amorphous aluminosilicate. *Chemical Communications, 57*, 13301–13304. https://doi.org/10.1039/D1CC05401E

Yan, C., Jin, J., Wang, J., Zhang, F., Tian, Y., Liu, C., Zhang, F., Cao, L., Zhou, Y., & Han, Q. (2022). Metal–organic frameworks (mofs) for the efficient removal of contaminants from water: Underlying mechanisms, recent advances, challenges, and future prospects. *Coordination Chemistry Reviews, 468*, 214595. https://doi.org/10.1016/j.ccr.2022.214595

Yu, Z., Zhu, L., & Lu, G. (2020). An improved phase correlation method for stop detection of autonomous driving. *IEEE Access, 8*, 77972–77986. https://doi.org/10.1109/ACCESS.2020.2990227

Zeng, X., Hu, J., Zhang, M., Wang, F., Wu, L., & Hou, X. (2020). Visual detection of fluoride anions using mixed lanthanide metal–organic frameworks with a smartphone. *Analytical Chemistry, 92*, 2097–2102. https://doi.org/10.1021/acs.analchem.9b04598

Zhang, M., Liang, R., Li, K., Chen, T., Li, S., Zhang, Y., Zhang, D., & Chen, X. (2022). Dual-emitting metal–organic frameworks for ratiometric fluorescence detection of fluoride and al^{3+} in sequence. *Spectrochimica Acta Part a: Molecular and Biomolecular Spectroscopy, 271*, 120896. https://doi.org/10.1016/j.saa.2022.120896

Zhu, X.-H., Yang, C.-X., & Yan, X.-P. (2018). Metal-organic framework-801 for efficient removal of fluoride from water. *Microporous and Mesoporous Materials, 259*, 163–170. https://doi.org/10.1016/j.micromeso.2017.10.001

The manufacturer's authorised representative in the EU is Springer Nature Customer Service Centre GmbH, Europaplatz 3, 69115 Heidelberg, Germany. If you have any concerns regarding our products, please contact ProductSafety@springernature.com

Printed and bound by CPI Group (UK) Ltd, Croydon, CR0 4YY
30/09/2026
02225322-0014